PLANT EMBRYOLOGY

EDUARD STRASBURGER (1844–1912)
Founder of Gymnosperm embryology

below: — title page of STRASBURGER's classic account of the embryology of angiosperms and gymnosperms (Jena, 1879)

FLORA.

57. Jahrgang.

№ 24. Regensburg, 21. August 1874.

Inhalt. E. Fleischer: Beiträge zur Embryologie der Monokotylen und Dikotylen. (Mit Tafel VI, VII, VIII.) – F. Arnold: Lichenologische Fragmente. XVII. — Bruchia vogesiaca.
Beilage Tafel VI, VII, VIII.

Beiträge zur Embryologie der *Monokotylen* und *Dikotylen*.

Von

E. Fleischer.

(Mit Tafel VI, VII, VIII)

Die Embryologie der Pflanzen ist für die Lösung vieler Fragen aus dem Gebiete der Systematik sowohl, als auch der Morphologie und Physiologie von solcher Wichtigkeit, dass sie schon seit einer Reihe von Decennien das Interesse vieler Forscher gefesselt hat. Gleichwohl ist bis in die neueste Zeit ein nicht unwesentlicher Theil derselben fast vollständig brach liegen geblieben, nämlich derjenige, welcher sich mit der Feststellung und Vergleichung der einzelnen Entwickelungsschritte der Pflanze von dem Punkte an beschäftigt, wo letztere als Individuum zu existiren beginnt, und diese Entwickelung bis zu dem Ruhezustande im reifen Samen verfolgt.

Frühere Arbeiten haben in Bezug auf diese Periode der Entwickelung bisweilen ein, meist nur nebensächliches Interesse auf deren erste Stadien verwendet und betreffs der weiteren sich höchstens um die Ausbildung der äusseren Gestalt bekümmert,

Flora 1874 24

From the first publication by EMIL FLEISCHER, the founder of comparative embryogeny, who was then only twenty-five years old

DIE ANGIOSPERMEN

UND

DIE GYMNOSPERMEN.

VON

Dr. EDUARD STRASBURGER
PROFESSOR AN DER UNIVERSITÄT JENA.

MIT XXII TAFELN.

JENA.
VERLAG VON GUSTAV FISCHER.
(VORMALS FRIEDRICH MAUKE).
1879.

JOHANNES HANSTEIN (1822–1880)
Founder of embryotectonics and embryogenergy

PLANT EMBRYOLOGY

EMBRYOGENY OF THE SPERMATOPHYTA

BY

DONALD ALEXANDER JOHANSEN

ISBN 81-211-0043-7

BISHEN SINGH MAHENDRA PAL SINGH
23-A, Connaught Place
Dehra Dun-248 001
1990

PLANT EMBRYOLOGY

EMBRYOGENY OF THE SPERMATOPHYTA

BY

DONALD ALEXANDER JOHANSEN

WALTHAM, MASS., U.S.A.

Published by the Chronica Botanica Company

PREFACE

A variety of circumstances indicated that the time was ripe for the preparation of a comprehensive discussion and summarization of our knowledge of the embryology of gymnospermous and angiospermous plants. The available information is scattered through an amazing array of publications going back over a period of more than a century and printed in many languages. For neither group of plants has this information ever been brought together within the covers of a single volume, nor has it ever been correlated as a whole.

For many years the need for such a volume has been acute. The vast literature appears to be useless to most investigators without integration and classification, while to others it seems to be totally unknown. As an example, a recent textbook on plant anatomy includes the statement that there is particular need for intensive study of the origin of the root during embryogeny.

Again, there has recently arisen an attempt on the part of certain persons, displaying a lack of acquaintance with the entire history of plant morphology, first to belittle the work of the more prominent embryologists, and second to establish an analogy between plant embryology and animal embryology. They disregard the fact that primate embryology, which they apparently have in mind, is essentially ontogeny; no recognized plant embryologist has ever entertained a corresponding conception with regard to the higher plants. If the detractors of plant embryology would only take the trouble to inform themselves from readily available sources, they would find that both the question of analogies between plant and animal embryology and that of the relationship of plant embryology to the other fields of plant morphology were settled decades ago.

The treatment of embryological topics is restricted to a consideration of the following divisions: *embryogenesis*, *embryotectonics*, *embryogenergy* and particularly *embryonomy*. All other fields are excluded, although, for the sake of clearer comprehension, brief discussions of such topics as megagametogeny and fertilization are occasionally presented. Plant embryology is of considerably broader extent than is indicated by the four fields first mentioned. For instance, it includes such subjects as embryonic morphology, which is more conveniently considered from the standpoint of the related disciplines of anatomy and cytology, and embryonic physiology, which most embryologists are content to leave to the tender mercies of plant physiologists. On the other hand, plant embryology emphatically does not include such subjects as the structure and development of the ovule, megasporogeny and megagametogeny, microsporogeny and seedling anatomy. These come under the heading of plant morphology, precisely as embryology is itself a division of that science.

It should be noted that the description of the embryogeny of a given species may, for convenience, be terminated at any stage preceding the final one of maturation, provided that the more important and significant phases are discussed. This, therefore, explains the apparent inconsistency in the comprehensiveness of discussion of many species.

The Gymnosperms, as has been maintained by HAGERUP and others, do not constitute a natural group. They have, therefore, been separated into distinct phyla or divisions on much the same basis that the Algae have been divided into phyla. These phyla embrace the *Cycadophyta*, *Ginkgophyta*, *Coniferophyta* and *Ephedrophyta*. The Anthophyta are treated as a unit, without segregation into the classical taxonomic groups, Dicotyledons and Monocotyledons. As will be explained in the text, there is no rigid embryological distinction between the two. Those who argue for the separation of the two ignore the fact that there are numerous species in both groups which are acotyledonous.

Since emphasis has been placed on the rigid definition of terms in the present text, a glossary is appended at the end.

In every instance cited the original publication was consulted. No secondhand information is included; sources such as abstracts have been ignored since they are too often erroneous or do not present the requisite data. The sources that were consulted were confined to such as were available in the Stanford University Libraries, in the Biology Library of the University of California at Berkeley, and in the reprint collections of Drs. D. H. CAMPBELL, L. L. BURLINGAME, G. M. SMITH and the author. These sources were ordinarily adequate, but certain obscure or discontinued serial publications were unavailable. A number of references which might be deemed to be pertinent were therefore necessarily omitted from consideration; this refers particularly to theses from European universities. In addition, volumes or other publications were loaned by the following: Drs. FRANS VERDOORN, A. S. FOSTER, E. W. SINNOTT, H. B. TUKEY and J. T. BUCHHOLZ. Drs. RENÉ SOUÈGES, and PIERRE CRÉTÉ generously provided complete series of reprints of their numerous publications.

Permission to utilize material which appeared in the standard botanical journals was granted by the editors or publishers. Since the manuscript was prepared during the war years, contact was made with the Office of the Alien Property Custodian regarding abstracts from foreign publications under its jurisdiction.

It should be mentioned that the original manuscript was completed in November, 1944. Advantage was taken of the intermission between completion of the original manuscript and publication to make certain additions and to bring the literature up to the end of 1946.

A prominent morphologist who had read the page proofs made the suggestion that it would be very desirable to have record kept at a central source of newly investigated species and particularly of proposed Variations under the angiospermous Types. The author is undertaking this task but, because of very poor library facilities, will include only publications of which reprints are forwarded to him. Such publications should be embryonomic in nature, not in other fields since such are useless for comparative purposes. Whenever sufficient material accumulates, summaries will be presented at intervals in one of the regular botanical journals.

Dr. L. L. Burlingame kindly turned over all of his remaining research material of *Araucaria angustifolia* (originally described as *A. braziliana*); in the main, the account and illustrations for this species are based on this material. Dr. J. T. Buchholz contributed material of various species of *Pinus*. Dr. H. F. Copeland loaned his slides of *Styrax californica*. Dr. C. L. Huskins and Mr. Auray Blain permitted incorporation of the latter's investigations on *Trillium erectum*. Dr. Ladley Husted contributed a summary of his studies on *Crepis capillaris*.

To Dr. G. M. Smith the author will always be deeply grateful for the use of his laboratory over a period of many years and also for the privilege of discussing numerous debatable points. Mr. Richard Soper assisted with the arduous task of compiling the plant-name indexes. Finally, gratitude must be expressed to the writer's wife, Elizabeth Burgess Johansen, who assisted with many details but above all cheerfully managed to put up with the inconvenience of being a "laboratory widow" while the manuscript was being prepared.

Donald A. Johansen

861 East Columbia Avenue
Pomona, California, U.S.A.
February, 1950

To

My Undergraduate Major Professor

Douglas Houghton Campbell

Pioneer American Plant Embryologist

CONTENTS

INTRODUCTION

Chapter I — CYCADOPHYTA: CYCADACEAE

Chapter II — GINKGOPHYTA: GINKGOACEAE

Chapter III — CONIFEROPHYTA: PINACEAE

Chapter IV — CONIFEROPHYTA: ARAUCARIACEAE

Chapter V — CONIFEROPHYTA: SCIADOPITACEAE

Chapter VI — CONIFEROPHYTA: TAXODIACEAE

Chapter VII — CONIFEROPHYTA: CUPRESSACEAE

Chapter VIII — CONIFEROPHYTA: SAXEGOTHAEACEAE

Chapter IX — CONIFEROPHYTA: PODOCARPACEAE

Chapter X — CONIFEROPHYTA: PHEROSPHAERACEAE

Chapter XI — CONIFEROPHYTA: CEPHALOTAXACEAE

Chapter XII — CONIFEROPHYTA: TAXACEAE

Chapter XIII — EPHEDROPHYTA: EPHEDRACEAE

Chapter XIV — EPHEDROPHYTA: GNETACEAE

Chapter XV — EPHEDROPHYTA: WELWITSCHIACEAE

Chapter XVI — ANTHOPHYTA: GENERAL CONSIDERATIONS

Chapter XVII — ANTHOPHYTA: TYPES AND VARIATIONS

Chapter XVIII — ANTHOPHYTA: SPECIAL AND COMPARATIVE EMBRYOLOGY

Chapter XIX — ANTHOPHYTA: APOMICTIC EMBRYOGENY

Chapter XX — ANTHOPHYTA: ADVENTITIOUS EMBRYOGENY

Chapter XXI — ANTHOPHYTA: POLYEMBRYONY

LIST OF ILLUSTRATIONS

MELCHIOR TREUB (1851–1910)
Pioneer in Orchid embryogeny

DOUGLAS HOUGHTON CAMPBELL (*b.* 1859)
Pioneer in the embryogeny of primitive Angiosperms

JOHN THEODORE BUCHHOLZ (*b.* 1888)
Author of noteworthy contributions to Gymnosperm embryology

ETIENNE CHARLES RENÉ SOUÈGES (*b.* 1876)
Founder of Angiosperm embryonomy

INTRODUCTION

Prefatory Remarks:- It has been said that curiosity is the beginning of science, that curiosity leads to observation and the recording of facts concerning a particular subject, and that the next step in the succession of events is classification of the facts according to their similarities or dissimilarities. When the observations are made by a great many different investigators, scattered throughout the world and possessed of more or less individualistic viewpoints, the time eventually arrives when all the facts that have been accumulated must be integrated as a whole in order that every aspect may be viewed in the correct perspective. So long as such a correlation and evaluation is not made, real progress is at a standstill. When a logical classification of the facts has been made available to other investigators in the field, they are provided with the means whereby more rapid and more precisely scientific progress is rendered possible.

Each new fact that is made known, without reference to any scheme of classification, may be compared to the delivery of a load of bricks, girders or lumber to a building site, each being provided by a different source. The mere fact of delivery does not contribute anything to the actual construction of the building. Everything awaits the advent of an architect who can design the building on the basis of the materials available and supervise the erection. The utility, the artistic appearance and the general acceptability of the completed structure all depend greatly on the integrative ability, resourcefulness and experience of the architect. He can not, however, hope to please everyone who views or wishes to make use of the completed building; there will always be, precisely as there have always been, critics who will exercise their freedom of criticizing one aspect or another, on valid or imaginary grounds, or who might wish to make alterations to suit their own convenience. So long as he conforms to the engineering principles and civil codes involved in the erection of the building and does not violate accepted artistic conceptions, that architect succeeds best who strives to please no one but himself, provided he acquires during the process a feeling that he is making a distinct contribution to the well-being of his fellows. Not until the building is fully completed does it acquire an individuality of its own, nor can it be classified as to the purpose for which it is to serve.

So it has been, and is, with plant embryology. The stages of curiosity and recording of facts have been going on for centuries; the time to assemble the facts and to discover to what conclusions they lead when they have been integrated and compared has arrived. In other words, the period of classification is at hand, if not somewhat overdue.

The final step in the process is that of verification of both the facts and the conclusions drawn from a consideration of the former; that is to say, the classification proposed needs to be confirmed or disproved and corrections made where necessary. Since innumerable conclusions are based upon facts or observations provided by others, there is always the possibility that errors may have been made by the latter, which might or might not invalidate the bases upon which the classification was erected. Verification of every recorded fact and conclusion can only come with the future.

History of Angiospermic Embryology:- The history of plant embryology, in general, may be divided into three epochs, each of which in turn is clearly separable into periods marked by the publication of certain fundamental contributions to the advancement of botanical knowledge. The first epoch extends from the earliest beginnings up to the time when HANSTEIN (1870) observed the progress of segmentation in the embryo and STRASBURGER (1877) discovered the phenomenon of fertilization. The second epoch commenced with the same dates and extends to the time of commencement of exact descriptive plant embryogeny (1910). The third epoch starts at that period and continues into the present time. The first epoch witnessed the hazy origin of the science, the second epoch encompassed its growth and development, and the third epoch is one of maturation.

I. First Epoch

The first epoch, according to DUCHARTRE, may be divided into three periods. The first period has a nebulous beginning in Greek antiquity and ends in the last years of the Seventeenth Century when CAMERARIUS (1694) sent his famous letter on sexuality in plants to VALENTINI. The second period covers the time from the beginning of the Eighteenth Century to the end of the third decade of the Nineteenth Century, when AMICI (1822, 1830) and BRONGNIART (1827, 1828) determined the true function of pollen, studied its germination, and followed the path of the microgametophyte from the stigma to the micropyle of the ovule. The third period is a prolongation of the second and extends to 1870–77. Investigations on the ovule, megagametogenesis and microsporogenesis multiplied and became more precise. At the same time the exact origin of the embryo was determined; earlier and earlier stages were studied; the sequence of successive stages was described; the proembryo was defined; the suspensor was distinguished; and the forms and dimensions of the cotyledons were noted in both monocotyledons and dicotyledons. That division of Embryology now designated as Embryography attained its greatest development, while Hanstein laid the foundations of Embryotectonics and Embryogenergy.

First period. — The first period of the first epoch is best described as a long period of extremely slow incubation. In fact, some twenty centuries were required for the most elementary and haziest notions concerning the multiplication of plant species to attain a reasonable degree of certitude.

The conception of sexuality in plants reaches back into the remote past to the horticulturalists and agriculturalists of the ancient Asiatic civilizations, who had definite ideas on the subject. Their conceptions were actually more concrete than those of the later Greeks, which were decidedly confused. To EMPEDOCLES OF ARIGENTUM (circa 495–435 B.C.), who left fragments of a philosophical poem which was also a sort of treatise on nature, fruits and seeds were the eggs of plants; the plant necessarily possessed male and female reproductive organs; and the sexes were united in the same individual. ARISTOTLE (384–322 B.C.) did not believe that sexuality existed in plants, while his famed disciple, THEOPHRASTOS OF ERESOS (373–286 B.C.), the acknowledged "Father of Botany," appeared to be somewhat confused by all the notions regarding sexuality although he admitted the existence of fertilization. THEOPHRASTOS first distinguished cotyledons from ordinary leaves.

During the long period of the Middle Ages (476–1453) embryological investigation was, to use a modern expression, "blacked out." The only noteworthy contribution to appear during this time was made by ALBERT OF BOLLSTADT (1206–1280), Bishop of Ratisbonne, perhaps better known as ALBERTUS

MAGNUS, who recognized the fundamental distinctions between plant and animal life.

At the beginning of the Sixteenth Century, observations began anew and multiplied rapidly. These were signaled by the appearance of a Latin poem by GIOVANNI PONTANO (1426–1503), perceptor of ALPHONSE, King of Naples, which was the first modern publication to mention sexuality in plants. However, the true savant of the epoch was ANDREAS CAESALPINIUS (1519–1603), born at Arezzo in Tuscany, the greatest botanist of the Renaissance and who undoubtedly merits recognition as the founder of descriptive botany. CAESALPINIUS studied all plant organs minutely, but unfortunately he was not entirely freed from Aristotelian dogmatism and attributed a soul to plants, locating it in the depths of the pith.

CLUSIUS (or DE L'ECLUSE) (1526–1609) distinguished two sorts of flowers in *Carica papaya*, calling them respectively staminate and pistillate flowers since they represented two different sexes. ADAM ZALUZIANSKI A ZALUZIAN in 1592 published his "Methodi herbariae libri tres," in which he recognized that the majority of plant species are androgynous and that in others the sexes are separated. He also clearly distinguished hermaphroditic and dioecious plants.

The invention of the microscope at the beginning of the Seventeenth Century gave a great impetus to the advancement of plant morphology.

If CAESALPINIUS is to be considered as the originator of systematic botany and organography, MARCELLO MALPIGHI (1628–1694) is incontestibly the founder of plant anatomy. There has long raged a controversy as to whether MALPIGHI or his contemporary, NEHEMIAH GREW (1641–1711) merits the honor: the weight of opinion, the evidence as to when each commenced his work, and a comparison of their publications, in the opinion of SACCARDO, all favor MALPIGHI. On the other hand, there is general agreement that the two illustrious men share the honor of laying the solid base on which the present science of embryology rests.

MALPIGHI discovered the ovule and the megagametophyte (which he called the "amniotic sac"), figured and described a large number of seeds, microspores, fruits and grains, and demonstrated the foliar character of the cotyledons. He presented the first rough outline of the development of the embryo, and observed rather precisely the principal stages in the germination of the seeds of *Zea mays*, *Pisum sativum* and *Vicia faba*. Unfortunately, however, he believed that the ovule and megagametophyte were useless secretions and did not admit the sexuality of flowers. GREW introduced the term "albumen" (now known as the endosperm), discovered the micropyle, recognized the existence of the two sexes, and noted the necessity for the pollen to be applied to the stigma before the seed could develop.

JOHN RAY (1628–1705) in 1686 attempted to establish a natural system of classification in which he compared and united various species by means of diverse characters. Among these he included the number of seeds and the number of cotyledons. He distinguished the "albumen" of mature seeds, called the embryo the "medulla," and, most significantly, discerned between embryos with one and two cotyledons.

Despite all the observations outlined above, as well as numerous others not mentioned, it is apparent that little real knowledge was adduced concerning the exact structure and functions of all parts of the flower representing the respective sexes. It still remained necessary that the organs of generation be identified with certainty and their influence on one another be determined. To RUDOLPH JACOB CAMERARIUS (1665–1721) fell the honor of making both the identification and the determination. He was the real discoverer of sexuality in plants. In a letter to VALENTINI, dated 1694, he dealt at length, in categorical terms, on the subject, and left no doubt concerning the rigor of his observations and the exactitude of his conclusions, as SOUÈGES justly remarks.

Second period. — The second period of the first epoch encompassed some 130 years and is noteworthy in that it witnessed the rise of systematic botany. Most of the botanists of the period were more interested in the development of that division of the botanical sciences, but there was some progress on the embryological side.

CAMERARIUS devoted careful attention to the various parts of the stamen and pistil and all the circumstances surrounding pollination. He distinguished between staminate and pistillate flowers, as well as between monoecious and dioecious plants. He established that the anthers and pistil were the fundamental parts of the flower.

For more than a century the remarkable observations of CAMERARIUS were either totally ignored or stirred up controversies. The first resulted because people for a long time put faith in the incredible conceptions of MALEBRANCHE (1638–1715), who contended that all seeds, successively smaller down to infinity, were preformed in the mother plant. These seeds, inserted one in the other at creation, were separated successively in proportion to the birth of the individuals. These conceptions are still known as the "theory of the nesting of germs" (*i.e.*, germ-plasms). The controversies need not be discussed, since they added nothing to embryological knowledge.

In 1703 SAMUEL MORLAND advanced the idea that the pollen traversed the style, passed into the ovary and penetrated the ovule. He believed that the pollen contained the rudiments of the embryo of the future plant. A year later a Parisian apothecary, ETIENNE FRANÇOIS GEOFFROY, proposed essentially the same ideas. MORLAND and GEOFFROY, because of the role they attributed to the pollen, are the precursors of those who, a century later, were known as the "pollinists."

SEBASTIAN VAILLANT (1669–1722) repulsed the Morland-Geoffroy hypotheses. Among other things, in 1717 he described the explosive opening of the anthers and demonstrated the necessity for the intervention of pollen for the fertilization of the ovules. Unfortunately, however, he entertained a weird conception of the latter phenomenon. He called the style a pump and claimed that it "did not transmit the powder [pollen] to the eggs, but only the vapor or the volatile spirit which was liberated from these grains of powder and which fecundated the eggs."

JOSEPH KOELREUTER (1733–1806) first determined the relation between the various floral organs and their reproductive functions. He noted the crests and pores present in microspores and the exudation of an oily matter from the pores, which he believed constituted the fertilizing substance. Absorbed by the stigma and mixed with the stigmatic secretions, the substance penetrated the style and then the ovules; true fertilization, therefore, occurred on the stigma. KOELREUTER carried on many interesting experiments involving pollen. The most important of these was his placing pollen of *Nicotiana paniculata* on the stigma of *N. rustica*, whereby he obtained the first hybrid with viable seeds.

MICHEL ADANSON (1727–1806) contended that the embryo existed before fertilization, that it received preliminary excitation from a vapor ("aura") produced by the microspores, transmitted by the stylar vessels and which entered the ovule by way of the "umbiblical cord" (funiculus). FRIEDRICH WILHELM GLEICHEN, called RUSWORM (1717–1783), an exponent of the sexual theory, in 1764 expressed the opinion, based on microscopic examination of microspores, that they corresponded to animal spermatozoids and penetrated the ovules where they were transformed into embryos. He figured microgametophytes of *Asclepias*, but failed to realize the significance of these structures.

That some seeds contained endosperm and others lacked it was first established by JOSEPH GAERTNER (1732–1791) in 1788. In 1811–18 LOUIS RICHARD described microgametophytes without determining their function; he also established the notion of "macropodial" and "brachypodial" embryos and of "exorhizoidal" (dicotyledonous) and "endorhizoidal" (monocotyledonous) embryos.

An attempt to elucidate the more important problems concerned with the structure and development of the ovule and seed was made by LUDWIG CHRISTIAN TREVIRANUS (1779–1864) in 1815 in a remarkable work on the development of the embryo. He dealt with a large number of both monocotyledonous and dicotyledonous species; among other observations, he discerned the constitution of the ovule and illuminated the role of the cotyledons in the nutrition and growth of the seedling. HENRI JOAQUIM DUTROCHET (1776–1847) in 1824 confirmed the observations of TREVIRANUS regarding the structure of the ovule, but, unlike the latter, he failed to discern the inner integument. In 1821 JOHAN AUGUST TITTMAN presented a series of well chosen figures and many details of various modes of seed germination.

Third period. — The third period of the first epoch commences in 1822 with the preliminary discoveries of GIAMBATTISTA AMICI (1786–1864) and extends to the time of publication, in 1870, of HANSTEIN's memoir on the angiospermic embryo. It thus extends over some 50 years and witnessed the partial solution of the following problems: (*1*) the course of fertilization; (*2*) the nature of the contents of the microspore and the microgametophyte, and (*3*) the origin and organization of the megagametophyte. On the other hand, at the end of the period the development of the embryo subsequent to fertilization still remained to be determined.

Progress during this period, despite the abundance of publications, was not considerable as many sterile polemics appeared and a number of workers (*e.g.*, SCHACHT, TULASNE and UNGER) were several times led astray by inadequate observations. However, the ground was prepared for the future and the fundamental questions of sexuality and fertilization, as we now understand them, were definitely settled and universally accepted. Again, two important developments in the general scientific field occurred, which immeasurably aided in morphological investigation. One was the perfection of the microscope; the other was the founding of both general scientific periodicals and circumscribed botanical journals.

AMICI's first observations concerned the microgametophyte of *Portulaca oleracea*; in it he noted a circulation of granules in the tube as it developed on the stigma and that the granules disappeared after the lapse of about three hours. These studies attracted the attention of ADOLPHE BRONGNIART (1801–1876), who worked out the intimate structure and development of the microspore, the relation between pollen and the stigma, the connections between the stigma and ovule, the structure of the latter, the introduction of a fertilizing substance into its interior, the formation and development of the embryo, and the relation of the latter to its surroundings. BRONGNIART recognized and named the "sac embryonnaire," substituting this term for the "amniotic sac" of MALPIGHI. However, he did not recognize the pre-existence of what is now known as the egg; he believed that it came into existence as a direct consequence of fertilization, resulting from an invagination of the wall of the sac under the action of the fecundating substance.

What is probably the first authentic demonstration of parthenogenesis was made in 1839 by JOHANN JAKOB BERNHARDI, who noted that seed could be developed in *Cannabis* without pollination.

That the ovule generally possessed two integuments was first recognized by ROBERT BROWN, who also first enunciated the fact that the microgametophyte penetrates the ovule through the micropyle.

The famous theory of JOHANN HORKEL (1769–1846) and his nephew MATHIAS JAKOB SCHLEIDEN (1804–1881) on the fertilization process and the mode of origin of the embryo was propounded in 1836. This theory admitted that the microgametophyte penetrated more or less deeply into the interior of the megagametophyte, there to be transformed into an embryo; and that, in order to penetrate into the cavity of the sac, sometimes it pierced the latter, at other times it merely pressed against the membrane and remained enveloped. This theory started a controversy which divided investigators into two camps. On one side were those who contended that the microgametophyte penetrated the sac and were called the "Horkelians" or "pollinists" (WYDLER, GRIFFITH, GELEZNOFF, DEECKE) and those who believed that it merely pressed against the sac (SCHLEIDEN, SCHACHT, KNORZ); on the other side were the adversaries, called the "preformists," who supported the contention of AMICI that the microgametophyte did not quite approach the megagametophyte and that the embryo developed at the expense of an embryonal vesicle which pre-existed in the sac (HARTIG, HUGO VON MOHL, KARL MÜLLER, particularly HOFMEISTER and finally RADLKOFER). SCHACHT, TULASNE and UNGER, at first "pollinists," soon ranged themselves on the side of the "preformists."

SCHLEIDEN, incidentally, presented excellent figures of the embryos of *Echium*, *Oenothera*, *Phormium* and *Potamogeton*; in fact, the initial stages of development in *Oenothera* are accurately represented. At the same time, ADRIEN DE JUSSIEU (1797–1853) developed with more success a knowledge of the external morphology of the monocotyledonous embryo. F. G. F. MEYEN in 1841 took the suspensor to be a continuation of the microgametophyte. He also presented several figures of the embryo of *Capsella bursa-pastoris*, *Alsine* and *Helianthemum*, in which the quadrant stages are correctly represented. In his other figures the cells are schematically outlined and arbitrarily disposed.

GELEZNOFF defended and developed the theory of HORKEL and SCHLEIDEN. WILLIAM GRIFFITH, like GELEZNOFF, SCHLEIDEN and WYDLER, believed that the embryonic vesicle was borne at the extremity of the microgametophyte. TH. DEECKE pretended to demonstrate that there was no breach of continuity between the microgametophyte and the tube developed inside the embryo sac, and consequently that they were not one and the same thing. In another publication, he presented an excellent figure of the quadrant proembryo of *Stachys sylvatica*, but his other stages are inexact.

H. SCHACHT was the most outspoken exponent of the Horkel-Schleiden theory. He published many figures of embryos at various stages of development, but was too preoccupied with his controversies to devote close attention to what he presented, with the result that many of the figures are from very oblique sections and are difficult to interpret.

The egg of the megagametophyte was first discovered and named by THEODOR HARTIG in 1838, who believed that the microgametophyte transported some substance which served to fertilize this egg. GERBERT GIRAUD in 1844 observed the presence of an embryonal vesicle in *Tropaeolum majus* before fecundation occurred and before the microgametophyte had reached the megagametophyte. Unfortunately, the contradictory observations of GUGLIELMO GASPARRINI on *Citrus* in 1846 placed him on the side of the "pollinists." KNORZ, a privatdocent at Marburg, in 1848 recognized that a microgametophyte arrived at the megagametophyte, but he contested the pre-existence of an embryonal vesicle in the sac. FRANZ UNGER, at first numbered among the "pollinists," later changed his ideas when he wrote on the embryogeny of *Hippurus vulgaris* and admitted the presence of a vesicle at the upper end of the megagametophyte which elongated into an "embryonic tube" under external influence. He clearly indicated the formation of the quadrant and octant stages in that species and also the delimitation of the epidermal initials. His figures are more precise than those of HOFMEISTER or TULASNE and most other authors preceding HANSTEIN.

The morphological publications of WILHELM HOFMEISTER covered a period of about 15 years. His first paper, on *Oenothera*, which appeared in 1847, placed him amongst the "preformists." He observed the penetration of the megagametophyte by the microgametophyte in *Canna*. He described various types of ovules, determined the position, form and dimensions of the megagametophyte. In the latter he differentiated between the sister cells of the egg (the synergids), the polar nucleus and the antipodals, fixing the number of the latter. His complete figure of the megagametophyte of *Bulbocodium vernum* was reproduced in all the classical works. He recognized and named the proembryo ("Vorkeim"). He noted that the zygote was soon divided into two cells by a transverse wall; that one cell multiplied but little, while the other developed into a spherical embryo. The development of the endosperm was also studied and the fact that the antipodals remained quiescent was noted. It was with regard to the monocotyledonous embryo that HOFMEISTER provided the most information.

LOUIS RENÉ TULASNE, at first a "pollinist," later changed his viewpoint because of the weight of evidence presented by HOFMEISTER. TULASNE'S observations on the megagametophyte are inferior to those of HOFMEISTER, but those on embryogeny are superior to those of the latter. In fact, TULASNE'S illustrations are better than his descriptions thereof.

It is difficult to assign the proper places to TULASNE and HOFMEISTER in the history of plant embryology, as SOUÈGES observes. HOFMEISTER deserves the credit for casting the most light on the constitution of the megagametophyte and the origin of the embryo, while TULASNE called attention to the great diversity in the form of the embryo and therefore laid the foundations of the modality underlying embryogeny.

Several interesting modifications in the structure of the megagametophyte of *Mirabilis jalapa* were noted by PIERRE ETIENNE DUCHARTRE in 1848, as well as the precocious separation of the suspensor from the embryo proper. GEORGE DICKIE, in the same year, first demonstrated that the microgametophyte and the embryo were two separate formations. LUDWIG RADLKOFER in 1856 proved that the suspensor was an embryonal structure and thus could no longer be considered as a prolongation of the microgametophyte.

ROBERT CASPARY, in 1854, and THELO IRMISCH, in 1858, occupied themselves with germination of the seed and the constitution of the seedling. The latter defined the cotyledonary node and indicated the limits of the hypocotyledonary axis.

SERGIUS ROSANOFF, a student of HOFMEISTER, in 1866 described precisely the first segmentations in *Tiaridium* and *Heliotropium* and compared the embryogeny of these genera with those in other genera in related families. This was probably the first study in comparative embryogeny.

II. Second Epoch

The publication in 1870 of the famous memoir by J. HANSTEIN on the development of monocotyledonous and dicotyledonous embryos marks the commencement of the second epoch in the history of plant embryology. The epoch may be divided into two periods, the first extending over some twenty-five years (from 1870 to about 1895), the second over some fifteen years (from about 1895 to 1910).

First period. — The HANSTEIN memoir is remarkable in that it is the first which was concerned solely with the ovule, the megagametophyte, fertilization, the endosperm and the sequence of segmentation within the proembryo and embryo. Beginning at about this time, investigators came to realize the significance and importance of segmentation in the life history of all living organisms since it forms the basis of their classification. This was essentially an amplification of the cell theory of SCHLEIDEN and SCHWANN, enunciated in 1838: that there was a division of work among cells which were united to one another to constitute an organism, and it was further realized that these cells were derived from other cells ("omnia cellula e cellula"). Segmentation, in essence, is cell division in cells which, subsequent to their division, remain united; moreover, if these cells remain united, it is according to an evident object of the association, which necessarily implies a division of work.

In evaluating the work of the investigators of this period, it is necessary to consider the microtechnique methods which they employed. These were crude in the extreme, compared to the procedures of the present day, and involved a high degree of manual dexterity, together with the use of acids and alkalis to render the tissues sufficiently transparent. A worker of the present time, dependent as he is on paraffin embedding, the precision rotary microtome and coal-tar dyes of known dye content, can have nothing but admiration for his predecessors and wonder how they ever accomplished their painstaking labors.

HANSTEIN was the first who defined precisely the position of the walls of segmentation in the embryo and the form of the cells thus separated. He introduced into the new science of embryology a terminology which, for the most part, is still followed. He gave a well defined embryogenic significance to such terms as quadrants and octants, to the upper hemisphere or cotyledonary and stem regions, and to the lower hemisphere or hypocotyledonary and root regions, all of which are parts of the embryo proper, which, by reason of its general form, may be compared to a globe or sphere. He designated the hypophysis, determined its position in the lower portion of the sphere, and indicated its functions in the construction of the radicular region. To him we also owe our conception of the epidermal initials, periblem and plerome, three fundamental embryonic tissues which give rise to the epidermis, cortex and stele, respectively, of the adult plant. He is the creator of embryotectonics, in that he devised the notion of tiers in the distribution of the embryonic elements. He also laid the foundations of embryogenergy by demonstrating the destinations of these tiers and also those of the primordial embryonic cells.

Many of HANSTEIN'S conclusions or generalizations were premature or erroneous, owing mainly to the insufficient number of species which he studied. The only species which he studied in adequate detail were *Capsella bursa-pastoris* and *Alisma plantago*. Several other species were investigated less carefully. He generalized that the first wall in the dicotyledonous proembryo marked the separation of the two future cotyledons; that the upper hemisphere always engendered the cotyledons and epicotyl, while the lower hemisphere gave rise to the hypocotyl and root; and that, in *Alisma plantago*, the first three of the four superposed cells of the monocotyledonous proembryo corresponded respectively to the upper hemisphere of the dicotyledonous embryo, the lower hemisphere and the hypophysis. However, his observations on the origin of the initials of the cortex at the apex of the radicle in *Nicotiana tabacum* are erroneous. Finally, his conclusions that the rules of embryogeny are identical among all dicotyledons and equally analogous throughout the monocotyledons, but fundamentally different for the two groups, are not at present tenable.

Most of the investigations which subsequently appeared were patterned after those of HANSTEIN. For instance, in 1873, G. HIERONYMUS duplicated for *Centrolepis tenuior* the earlier work on *Alisma plantago*. CARL HAGEN, in the same year, presented several stages in *Mesembryanthemum* which appeared to resemble the corresponding stages in *Capsella*, but HAGEN repeated certain errors of interpretation originally made by HANSTEIN and which were also later made by KNY and WESTERMAIER. IGNATZ URBAN, again in the same year, was the first to observe (in *Medicago*) that the first wall to be erected in the terminal cell could be oriented diagonally.

The first comparative embryologist was E. FLEISCHER, who presented an extensive paper in 1874 covering several orchids, *Juncus glaucus*, *Luzula multiflora*, *Iris gueldenstaedtiana*, *Ornithogalum nutans* and *Leucojum aestivum* among monocotyledonous species, and *Oxalis valdiviensis*, *Helianthus annuus* and *Asclepias cornuti* among the dicotyledons. He demonstrated that there was not a sharp demarcation between the monocotyledonous and dicotyledonous types established by HANSTEIN, but a series of analogies. In *Ornithogalum nutans*, particularly, he presented a series of embryogenic forms clearly comparable to those of dicotyledons.

All comparative embryologists who have carefully considered the entire problem of presumable differences in the fundamental organization of the proembryo (which, after all, is of greater significance than the organization of the mature embryo) have followed FLEISCHER, although there have been attempts as late as 1946 on the part of uninformed persons to resurrect the mistaken conceptions of HANSTEIN.

L. KNY, also in 1874, examined a number of the *Brassicaceae*, particularly *Brassica napus*, and *Alisma plantago*, following HANSTEIN'S procedures and repeating many of his errors.

In the same year Count SOLMS-LAUBACH investigated representatives of the *Hydnoraceae* and *Rafflesiaceae*, noting that the embryos in these families were excessively reduced in development. Five years later he published rather complete and precise observations on the *Dioscoreaceae* and *Commelinaceae*.

F. HEGELMAIER is generally considered to be one of the outstanding embryologists of the period. In 1874 there appeared his detailed paper on the embryogeny of several monocotyledonous species, including *Pistia stratiotes*, *Canna indica*, *Sparganium ramosum* and *Triticum vulgare*. While he failed to establish the precise origin of the middle cell of the row of three proembryonic cells, he did note the differences in the regions of the mature embryo to which each cell gave rise. He further established the origin of the stem apex as well as the morphological value and the organization of the radicle in these species. Later he investigated four dicotyledonous families — the *Ranunculaceae*, *Umbelliferae*, *Geraniaceae* and *Papaveraceae*. Considering the inadequacy of his technique methods, his observations were rather accurate, though there were gaps and some errors in interpretation. In brief, it is to HEGELMAIER that we owe the current conception that there can exist a close embryological connection between very diverse plant families at the same time that there occur sharp divergencies within the same family or even genus.

The peculiarly interesting embryogeny of the *Papilionaceae* first came to light through an unfortunately incomplete study of HEGELMAIER'S on *Lupinus*. This subject was further pursued by STRASBURGER in 1880 and later presented in greater detail by GUIGNARD.

Study of the embryogeny of parasitic plants was initiated by L. KOCH's paper on the Cuscutas in 1877 and another on the *Orobanchaceae* the following year. These constituted the first observations of embryos which are acotyledonous at maturity.

The great majority of embryological papers which appeared before 1875 emanated from the European continent, particularly from France and Germany. ALEXANDER DICKSON appears to be the first to present a modern embryological paper in English. He was concerned with the origin of the root in three species of *Tropaeolum.*

To MAX WESTERMAIER is due the credit for correcting the erroneous conclusions of HANSTEIN regarding the embryogeny of *Capsella bursa-pastoris.* He established this species as a classical type to serve as a basis for comparison in embryogenic researches. Later investigators, unfortunately, did not compare their species any too carefully and, as a consequence, made all sorts of ludicrous mistakes. Some even went so far as to consider practically all dicotyledonous species as developing in a manner identical with that of *Capsella.* The *Capsella* mode of embryogeny, in fact, is now known to occur in only a few genera all belonging to the *Brassicaceae.*

The remarkable work of MELCHIOR TREUB on the embryogeny of the *Orchidaceae* appeared in 1879. He, unfortunately, oriented his investigations from the physiological standpoint, and was particularly interested in the role played by the suspensor in that respect. In 1882 he discovered that the zygote may be segmented vertically (in certain species of the *Loranthaceae*).

C. NORNER, in 1881, attempted to distinguish three types of embryonic development in certain species of the *Poaceae.* This was one of the pioneer attempts to delimit embryogenic types.

One of the most extensive and detailed embryological investigations was that of LEON GUIGNARD in 1882 on almost a hundred species — some thirty in the *Mimosaceae*, several from the *Caesalpiniaceae*, but with most from the *Papilionaceae.* The researches of GUIGNARD on these dicotyledonous families, together with those of TREUB on the *Orchidaceae*, are major landmarks in the history of plant embryogeny.

Second period. — The second period of the second epoch, extending from about 1895 to 1910, was mainly one during which species not hitherto investigated were studied; little that was particularly worthy of note was produced. It was a period of tedious addition of mostly mediocre papers to the rapidly expanding literature in the field.

III. Third Epoch

The Third Epoch comprises the years from 1910 onward. It may be divided into two periods, one extending from 1910 to 1937, when RENÉ SOUÈGES promulgated the laws of embryogeny; the other from 1939, when the same renowned embryologist presented a system of classification of embryogenic types, to the present time.

First period. — This was, for most practical purposes, a continuation of the second period of the preceding epoch. SOUÈGES' first detailed embryonomic investigation appeared in 1910; it was rapidly followed by others until the total number of titles reached approximately 150 by the end of 1946. The great tragedy of plant embryology, a serious indictment of most other workers, has been, and to a great extent still is, that so little attention was paid to the fundamental contributions of SOUÈGES. During all of the third epoch, only three embryologists — BHADURI, CRÉTÉ and NOLL — attained the sublime and precise mode of embryonomic procedure established by SOUÈGES.

Under the influence of STRASBURGER in Germany, of GUIGNARD in France, and of TREUB in Java, numerous papers appeared in the European literature. The impulse generated by these great investigators spread to other continental countries and presently to the United States. However, in none of these other countries did embryologists of the first rank appear; they were, for the most part, imitators who contributed little of fundamental importance. In England, for instance, one did not appear on the scene until 1936 when METCALFE published his important paper on *Ranunculus ficaria*.

Among the German investigators, despite their self-bestowed reputation for detailed accuracy, a disposition toward the misuse of morphological terminology soon appeared, then spread to their Swedish colleagues and later to those from other countries who blindly admired German methods. This tendency has had most unfortunate effects on the advancement of plant embryology. At the present time, embryological terms are so mishandled that it is extremely difficult to understand the average German, Swedish and Indian paper, which is more often purely morphological and not at all embryological.

The most prominent German morphologist of the period was K. SCHNARF, whose "Embryologie der Angiospermen" appeared in 1927, and his "Vergleichende Embryologie der Angiospermen" in 1931. Both are partially mistitled; they are morphological rather than purely embryological works. However, the first is important in that it includes a preliminary classificatory scheme which was adopted and expanded by JOHANSEN in 1945.

In the United States embryological studies arose in three distinct centers. The first and most important investigator was D. H. CAMPBELL at Stanford University; his studies on the *Araceae*, *Zannichellia* and *Naias* were fundamental. The second center comprised the group at the University of Chicago headed by J. M. COULTER and C. J. CHAMBERLAIN, and the third centered around J. H. SCHAFFNER at the Ohio State University. There were also a number of more or less independent investigators, including, among others, M. T. COOK, W. A. CANNON, J. E. GOW, R. G. LEAVITT, F. E. LLOYD, F. H. BILLINGS and J. E. KIRKWOOD.

Russian and Italian workers who published in their native languages suffered from the fact that their confreres in other countries had little or no acquaintance with their media of communication and thus tended to ignore their contributions. Prominent Italian embryologists of the period included A. CHIARUGI, E. CARANO and E. FRANCINI. Other conspicuous embryologists who must be mentioned were F. A. F. C. WENT in The Netherlands and S. V. MURBECK in Sweden.

In 1934 SOUÈGES presented a resume of his work up to that time, followed by another three years later. Finally, in 1937, there appeared his "Les lois du développement," a work whose supreme significance in plant embryology has been sadly overlooked.

Second period. — This period is too uncomfortably close to the present time for one to evaluate it without incurring accusation of bias.

In 1939 SOUÈGES published his "Embryogénie et Classification," in which is presented a detailed scheme for the delimitation of the various embryonomic types. This pamphlet is equally as significant as the earlier one on the laws of development, but, like the latter, it has not received the serious recognition that it deserves.

Investigations continued apace, principally on species not hitherto studied. Most such studies are characterized by five features: (*1*) ignorance of the laws of embryogeny, (*2*) lack of thorough acquaintance with the publications of other investigators, (*3*) utter confusion in terminology, (*4*) failure to work out the proembryonic development in sufficient detail, and (5) the solemn pronouncement of weirdly absurd conceptions of embryonomic relationships. The present text constitutes an attempt to remedy all these deficiencies in the current output of embryological investigations.

The period is noteworthy for the development of plant morphology and embryology in India. Such studies were first initiated by an American, W. Dudgeon, at Allahabad University, then after his death were carried forward by his pupil, P. Maheshwari, and by A. C. Joshi, as well as by students of these two. This phenomenal growth is clearly indicated in Chapter XVIII by the number of papers cited from Indian scientific journals.

At the present time, Souèges continues to be the foremost plant embryologist, although now retired from professional duties. His work is being continued by his pupil and successor, Pierre Crété.

History of Gymnosperm Embryology:- Many of the phases of the rise and development of Gymnosperm embryology are bound up with the corresponding growth on the Angiosperm side, and need not be repeated here.

The first comprehensive treatise on the subject was E. Strasburger's "Die Coniferen und die Gnetaceen," which appeared in 1872, and was followed in 1879 by his "Die Angiospermen und die Gymnospermen." These two monumental works mark the real commencement of investigation on gymnospermous plants.

In 1883 Goroschankin demonstrated the rupture of the microgametophyte in *Pinus pumilis*, but erroneously supposed that both microgametes fused with the archegonial nucleus. This mistake was corrected by Strasburger the following year; in *Picea* he proved that only one microgamete fuses with the archegonial nucleus. Interest in the microgametes that was thus engendered led to the studies of Belajeff in 1891 on *Taxus*, and these were followed by the discovery of ciliated male cells in *Ginkgo* by Hirase in 1895, in *Cycas* by Ikeno in 1896 and in *Zamia* by Webber in 1897.

The development of the archegonium was first elucidated by Strasburger in 1879; there was some difference of opinion concerning the nature of the ventral nucleus or cell. Blackman (1891) considered that of *Pinus sylvestris* to be an arrested gamete and Chamberlain (1895) came to the same conclusion; that this conception is correct has been demonstrated by the discovery of the fusion of the ventral nucleus with one of the microgametes in other species of *Pinus*.

Earlier accounts of the fertilization process are fragmentary and somewhat unsatisfactory from the cytological standpoint. The entire question was ably resolved by Haupt in 1941 in two species of *Pinus*, and it is now probable that essentially the same process prevails in nearly all other Gymnosperms.

Although Strasburger presented most of the significant details of embryogeny and some gaps were filled by intervening investigators, it remained for Buchholz to place the entire field on a sound basis. Buchholz is to gymnospermic embryogeny what Souèges is to angiospermic embryogeny. Other great names in the history of gymnospermous embryogeny include those of A. A. Lawson, M. Ferguson, C. J. Chamberlain, W. J. G. Land, W. T. Saxton, H. H. W. Pearson, A. H. Hutchinson, S. Hirase, E. M. Berridge, W. C. Coker, L. L. Burlingame and M. Tahara. The contributions made by each of these persons are best ascertained from the bibliographies contained in the following chapters.

At the present time there are two main centers of research on Gymnosperm embryology: one at University College, Dublin, under J. Doyle and W. J. Looby, and the other at the University of Illinois under J. T. Buchholz.

CYCADOPHYTA

Chapter I

CYCADACEAE

The *Cycadaceae* comprise nine living genera: *Zamia*, *Macrozamia*, *Ceratozamia*, *Encephalartos*, *Dioon*,* *Cycas*, *Microcycas*, *Bowenia* and *Stangeria*.

Fertilization:- Accounts of the fertilization process are still rather fragmentary. *Bowenia serrulata* is the most thoroughly investigated species.

The number of archegonia is very variable. There are usually a large number of archegonial initials, with many hundreds in *Microcycas*, but ordinarily only three or four mature.

So many microgametophytes usually reach the archegonial chamber that all mature archegonia are fertilized. The archegonial cytoplasm becomes extremely turgid as the fertilization period approaches and would spurt into the archegonial chamber if not prevented from doing so by the enlarged and turgid neck cells. A curious phenomenon occurs in *Bowenia* at this time: each archegonium secretes a drop of water through the neck cells; this water, together with liquid from the microgametophytes, furnishes the medium in which the microgametes swim actively. In this same genus the outer wall of each neck cell is very thin and halfway up on the inner side of each neck cell the wall is quite thick; with a narrow cell supporting the two neck cells, a hinge-like arrangement is thus erected, permitting the cells to swing open momentarily for the entrance of the microgametes. The neck cells are in intimate contact at the apex only and the active movements of the microgametes constitutes sufficient pressure to cause the cells to open. The neck cells are not demolished as happens at the corresponding period in most Gymnosperms, but they do collapse soon after the free-nuclear divisions commence. The number of microgametes which enter a given archegonium may vary from one to six or even more, but only one actually penetrates the cytoplasm to fuse with the archegonial nucleus. The microgamete enters with considerable force, strong enough to strip off the cytoplasm and ciliated band in the upper part of the archegonium, and leaving the naked nucleus to travel the remaining short distance to contact the archegonial nucleus.

The most satisfactory accounts of the fusion of the male and female nuclei indicate that the process is essentially similar to that more fully known for *Pinus* and other Coniferophyta.

In *Bowenia*, for instance, the microgamete nucleus enlarges considerably after becoming freed. After contacting the archegonial nucleus, it presses into the nuclear membrane of the latter, without losing its spherical shape, and produces a concave indentation (FIG. 1A). The archegonial nucleus is approximately three times the diameter of the microgamete nucleus. The membrane of the archegonial nucleus apparently becomes dissolved where the two are in

---

* Certain authors (*e.g.*, BUCHHOLZ in the article on Gymnosperms in the Encyclopedia Britannica) have recently attempted to resurrect the name originally proposed by LINDLEY, namely, *Dion*. This spelling is clearly an orthographic error; under Sect. 13, Art. 70 of the International Rules of Botanical Nomenclature, the corrected form *Dioon* (of MIQUEL) must be employed.

contact, but that of the microgamete nucleus persists until the next stage is nearly completed. In *Zamia* it has been stated that the microgamete nucleus becomes entirely surrounded by the archegonial nucleus, but in view of what happens next this is undoubtedly a misinterpretation. In any case, a true

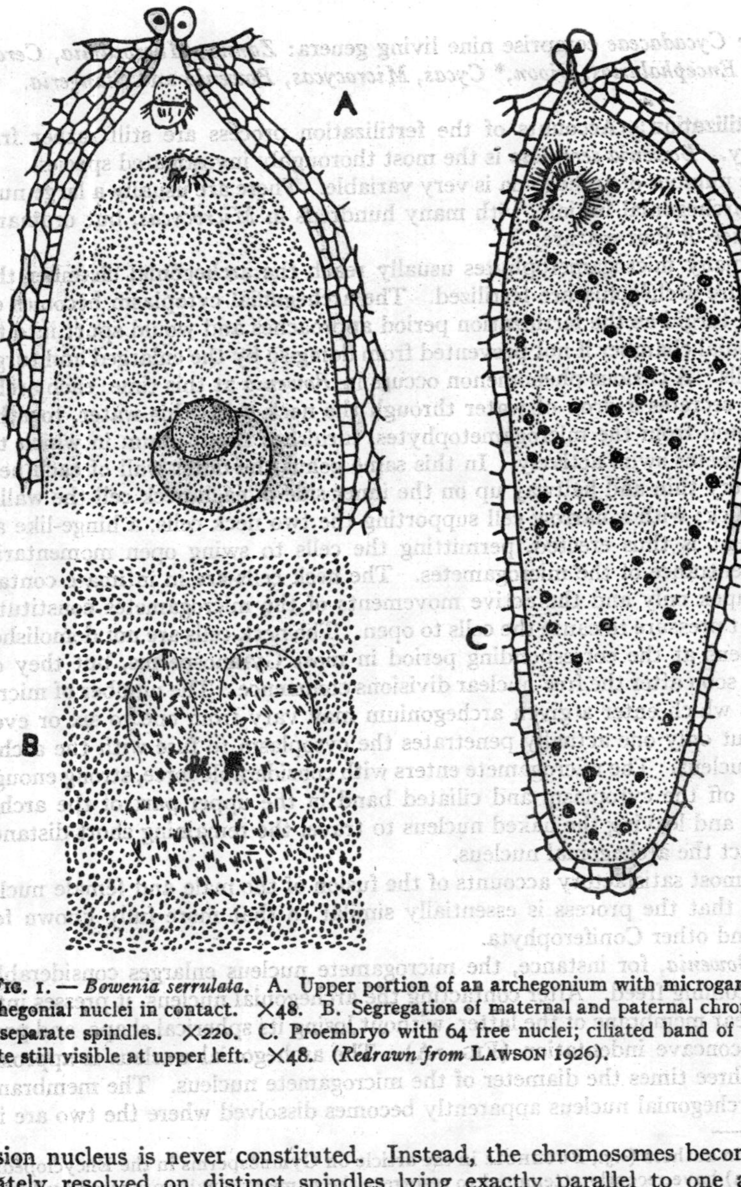

FIG. 1. — *Bowenia serrulata*. A. Upper portion of an archegonium with microgamete and archegonial nuclei in contact. ×48. B. Segregation of maternal and paternal chromosomes on separate spindles. ×220. C. Proembryo with 64 free nuclei; ciliated band of microgamete still visible at upper left. ×48. (*Redrawn from* LAWSON 1926).

fusion nucleus is never constituted. Instead, the chromosomes become separately resolved on distinct spindles lying exactly parallel to one another (FIG. 1B). The two sets of chromosomes next divide and pass to the respective poles, but what subsequently happens has been too vaguely described to be acceptable.

Proembryo:- A prolonged period of simultaneous divisions next ensues. The most striking feature of these mitoses is their remarkably small size in comparison with the resting nuclei. In nearly all investigated species the mitoses are not followed by the formation of any kind of cell walls. In *Dioon edule* complete but evanescent segmentation of the entire archegonial cytoplasm may occur as early as the 64-nucleate stage; these walls disappear completely before the next in the series of mitoses commences. The number of free-nuclear divisions apparently varies slightly according to the species, but is most frequently eight or nine, with 256 or 512 nuclei respectively. After about the seventh series of divisions, the mitoses become somewhat irregular; consequently, the exact number of nuclei may be greater or lesser than the presumable arithmetical number.

The free nuclei are usually evenly dispersed throughout the archegonial cytoplasm (FIG. 1C). Most of the earlier accounts describing the formation of a central vacuole or a central disorganization of the cytoplasm and nuclei now appear to be inaccurate. What happens in the archegonial cytoplasm just previous to wall formation varies somewhat according to either genus or species. In *Stangeria* many of the nuclei in the central and upper portions of the proembryo migrate to the basal portion, where the cytoplasm becomes more and more dense, with a corresponding vacuolization of the cytoplasm in the upper portion. A second period of two or three free-nuclear mitoses now occurs, followed by wall formation. Like *Stangeria*, *Cycas revoluta* also has a few simultaneous free mitoses at the base of the proembryo, while there are amitotic divisions in the upper portion in *Cycas* and resting nuclei in *Stangeria*. Walls appear simultaneously throughout the entire proembryo in *Macrozamia moorei*, *Encephalartos*, *Dioon edule* and *Stangeria paradoxa*, but in *Macrozamia reidlii* a small central region remains noncellular with free nuclei. Wall formation occurs only at the base of the proembryo in *Bowenia*, *Zamia* and *Ceratozamia*. The walls in many species are often weak and evanescent save at the base of the proembryo. A vacuole appears in the central region sooner or later and the former archegonium finally becomes empty but its outline is preserved by its thick membrane. The permanent cell walls eventually are developed at the base of the proembryo and the cells thus formed become differentiated into three regions.

Embryo:- Differentiation in the cellular region of *Stangeria* (FIG. 2D) commences almost immediately after delimitation but in most of the other genera many hundreds of cells are produced before recognizable differentiation becomes apparent. At first there are two general regions: an upper one, in contact with the free-nuclear region and with the largest cells; and a lower one composed of smaller, denser and more numerous cells. In *Bowenia* a simultaneous mitosis occurs as soon as the free nuclei have migrated to the base of the proembryo and permanent cell walls result, but the central region remains free-nucleate (FIG. 2A, B).

Suspensor development commences quite early. The large cells furthest removed from the base of the proembryo become very much elongated (FIG. 2B, C). During the earlier formative stages the suspensors are separate but all those from the several embryos soon become united and curve, coil and twist as they slowly grow deeper and deeper into the endosperm. The suspensors, in fact, are so intricately associated that it is impossible to examine them separately. Growth, however, is rather uneven, as the suspensors from some embryos elongate more than do those from other embryos. Finally only the suspensor system belonging to the one embryo that is to reach maturity attains full development. The suspensors of the cycads are remarkably long. Those of *Ceratozamia*, when fully stretched out, are 7 to 8 cm. long, and those of *Bowenia*, *Cycas* and *Macrozamia* may exceed 10 cm.

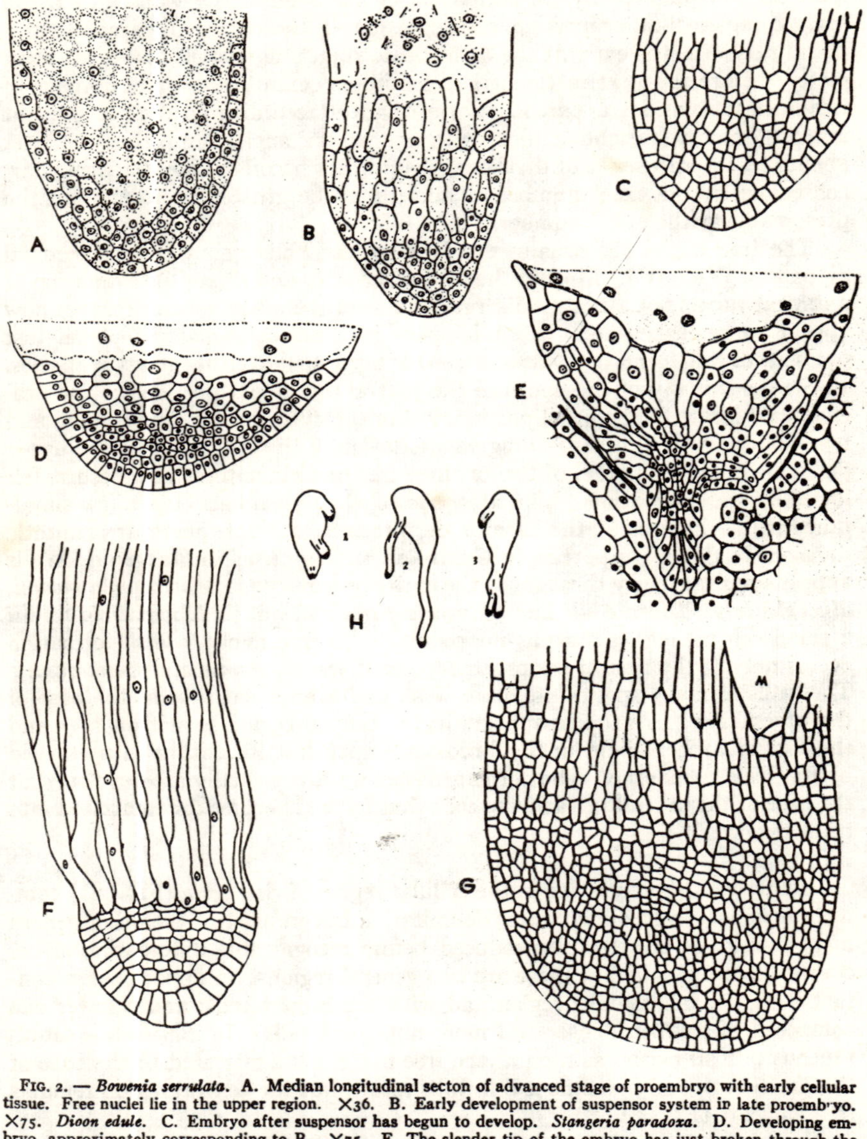

FIG. 2. — ***Bowenia serrulata.*** **A. Median longitudinal secton of advanced stage of proembryo with early cellular tissue. Free nuclei lie in the upper region. ×36. B. Early development of suspensor system in late proembryo. ×75. *Dioon edule.* C. Embryo after suspensor has begun to develop. *Stangeria paradoxa.* D. Developing embryo, approximately corresponding to B. ×75. E. The slender tip of the embryo has just broken through the archegonial membrane and is composed largely of haustorial cells. The upper cells are also distinctly haustorial. ×75. *Bowenia serrulata.* F. Developing embryo with suspensor. The epidermal cells are considerably larger than the central cells. ×75. *Dioon edule.* G. Developing embryo, with meristems that are to produce coleorhiza and later root cap differentiated; *m*, mucilage cavity. *Macrozamia reidlei.* H. Three different examples of several embryos developing from one zygote, *3* shows simple polyembryony resulting from splitting of one proembryo. (A, B, F *redrawn from* LAWSON 1926; C, G *from* CHAMBERLAIN 1910; D, E *from* CHAMBERLAIN 1916; H *from* BAIRD 1938–39).**

The smaller, denser cells at the base of the proembryo become differentiated into two regions. The outermost consists of a single layer of large cells which doubtless is equivalent to an epidermal layer, and above it are several tiers of small, dense cells (FIG. 2F). The outer cells, particularly in the upper portion of the proembryo, may become haustorial (FIG. 2E). The region of small, dense cells contributes most of the suspensor tissue in *Stangeria*, and all of the periblem and plerome in all genera (FIG. 2G).

Cotyledon development, as far as described, is essentially similar in all genera except *Ceratozamia*. The latter genus normally has only a single cotyledon; the others have two, or occasionally three. The seeds are shed in *Cycas*, *Macrozamia* and *Ceratozamia* before the cotyledons are developed. Since the seeds lie in a horizontal position after being shed, gravity may have something to do with the development of the one cotyledon in *Ceratozamia*, but certainly not in the other two genera. When seeds of *Ceratozamia* were placed on a clinostat for germination, the suppressed cotyledon underwent full development. In many species the cotyledons have frequently been reported as being more or less unequal in size. The cotyledons close over the stem apex, and in *Zamia* and *Bowenia* they fuse together at an early stage along the upper half of their length, while in *Encephalartos* they are fused throughout their entire length, by both margins at the base and by either one or both margins near the apex.

At approximately the central region of the embryo a conspicuous and characteristic structure, called the coleorhiza, is developed. Its function is mainly protective; after the embryo attains full growth, the coleorhiza becomes very hard. During germination the coleorhiza is the first portion to become extruded from the seed, then the root tip digests its way through the base of the coleorhiza, followed by the cotyledons.

Simple polyembryony prevails among the cycads, but the extra embryos are commonly overlooked since they are still within the archegonial membrane at the time they have just begun to elongate. In *Macrozamia reidlei* from one to three lateral embryos have been observed in addition to the normal basal embryo (FIG. 2H1–3). These develop in exactly the same way as the basal embryo — from a small group of actively dividing cells in one particular peripheral region of the proembryo. The primary suspensor may also fork and give rise to two equally developed basal embryos. Branching of the suspensor is also known in *Encephalartos*.

BAIRD, A. M. 1938–39. A contribution to the life-history of *Macrozamia reidlei*. Jour. Roy. Soc. West Australia 25: 153–175.

CHAMBERLAIN, C. J. 1910. Fertilization and embryogenesis in *Dioon edule*. Bot. Gaz. 50: 415–429.

CHAMBERLAIN, C. J. 1916. *Stangeria paradoxa*. Bot. Gaz. 61: 353–372.

DORETY, H. A. 1908. The embryo of *Ceratozamia*. Bot. Gaz. 45: 412–416.

LAWSON, A. A. 1926. A contribution to the life-history of *Bowenia*. Trans. Roy. Soc. Edinburgh 54: 357–394.

SAXTON, W. T. 1910. The development of the embryo of *Encephalartos*. Bot. Gaz. 49: 13–18.

SEDGWICK, P. J. 1924. Life history of *Encephalartos*. Bot. Gaz. 77: 300–310.

WEBBER, H. J. 1901. Spermatogenesis and fecundation of *Zamia*. U. S. Dept. Agr., Bureau Pl. Ind. Bull. No. 2, 92 p.

GINKGOPHYTA

Chapter II

GINKGOACEAE

There is only one living species, *Ginkgo biloba*. Nothing has been described concerning the embryogeny of the several known fossil genera.

Fertilization:- The microspore at maturity is four-celled, consisting of a generative cell, two prothallial cells (one of which aborts) and a tube cell. After the microgametophyte has reached the pollen chamber of the ovule, the generative cell divides to produce a body cell and a stalk cell. The body cell divides into two microgametes or sperms; blepharoplasts later appear and produce a multitude of cilia. A single sperm enters each archegonium and unites with the nucleus of the latter. The sperm nucleus, when it comes into contact with the archegonial nucleus, is about one-sixth the volume of the latter (FIG. 3A). The sperm nucleus apparently sinks into the other nucleus, but precisely what occurs between the time the two nuclei come into contact and the resolution of the chromosomes is obscure in the published accounts, which are all rather old. The chromosomes from each nucleus, in any event, are segregated on separate intranuclear spindles (FIG. 3B) and presently move to common opposite poles. The fertilization process in *Ginkgo*, therefore, is essentially similar to that occurring in most Gymnosperms.

Fusion of the ventral canal nucleus with the nucleus of the archegonium has been described, but whether this can result in the development of a normal embryo is unknown although not beyond the realm of possibility.

Proembryo:- Two free nuclei are formed as a result of the first mitosis; subsequently, about eight free nuclear divisions occur, with a total of around 256 nuclei (FIG. 4A). During the successive nuclear divisions in the proembryo the size of the nuclei decreases. Thus, while the fusion nucleus measures 85 to 100μ in diameter, the nuclei after the seventh division measure 20 to 25μ only. This decrease of nuclear size is evidently closely related to the chromatin-elimination which occurs up to the 32-nucleate stage. The manner of chromatin-elimination is as follows: during nuclear division deeply staining basophilic granules are found scattered among the spireme; when the chromosomes form the equatorial plate, the granules are scattered mostly toward both poles; they then form larger masses of less basophilic nature and finally disappear. During the later mitoses some of the nuclei may fail to divide. The nuclei are evenly distributed through the archegonial cytoplasm. After either the eighth or ninth series of divisions, wall formation occurs simultaneously throughout the entire archegonium in such a manner as to enclose a single nucleus within each cell. The resulting cells at first do not exhibit any pronounced dissimilarities in size, shape or contents (FIG. 4B). Later the cells in the upper two-thirds or more of the spherical proembryo divide only a few times or not at all, and take no part in the development of the embryo proper. Their contents become thin and watery and the cell walls become wavy and even break down. The cells in the lower portion of the proembryo divide repeatedly, with activity greatest toward the base, and a small-celled tissue becomes organized (FIG. 4C). There are thus three imperfectly delimited regions in the developing proembryo. An organized, recognizable suspensor is never developed.

Embryo :- At first the tissues of the young embryo consist entirely of small isodiametric meristematic cells; but presently two regions of differentiation may be distinguished, one located immediately behind the other in the longitudinal axis of the embryo, the two being separated by only a few undifferentiated cells. The region at the apex gives rise to the cotyledons, primary leaves the stem apex (FIG. 4D). The inner region becomes the growing point of the primary root. This region first becomes noticeable by the delimitation of irregular rows of rectangular cells which resemble the characteristic parallel rows of cells of the central root-cap region.

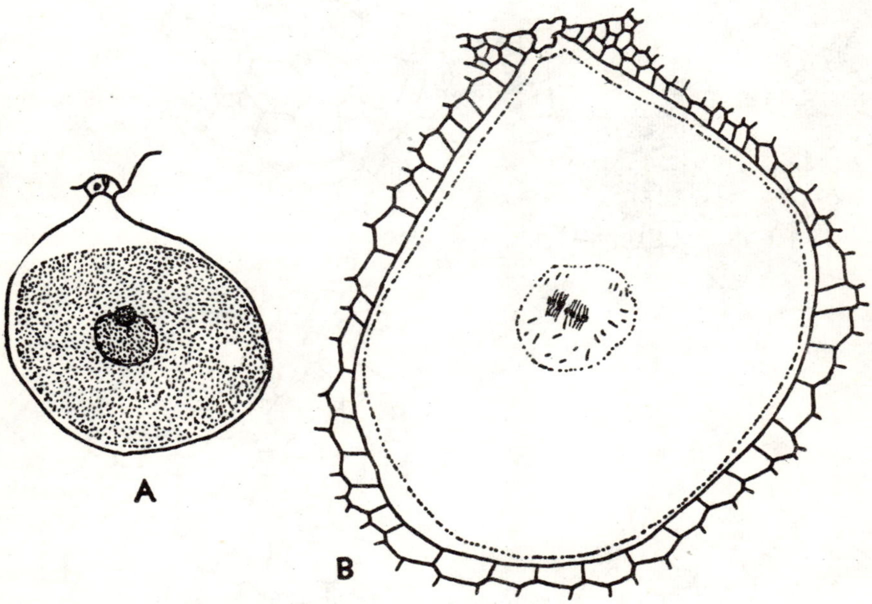

FIG. 3. — *Ginkgo biloba*. A. Sperm and archegonial nuclei united. ×77. B. First mitosis after contact of male and female nuclei. (A *redrawn from* HIRASE 1918; B *from* HIRASE 1895).

The cotyledon primordia originate in the marginal region of the broad apical end and are first distinguishable as crescentic mounds of tissue which rapidly grow ahead of the stem apex. Each of the cotyledons has its own apical meristem. Normally two cotyledons are produced, but the number is not constant as three cotyledons are commonly found. Freakish and deformed embryos have also been described. During their earlier development one of the two cotyledons is usually longer than the other and is notched at the apex while the shorter one is deeply cleft. At maturity the cotyledons are of about equal length.

The primary leaves originate shortly after the cotyledons have become well differentiated and arise in the same manner. The periblem and plerome are not sharply delimited and can be recognized as general regions only. The first procambial tissue is differentiated in the cotyledons and passes from these back into the body of the embryo. The next set of procambial strands arise in con-

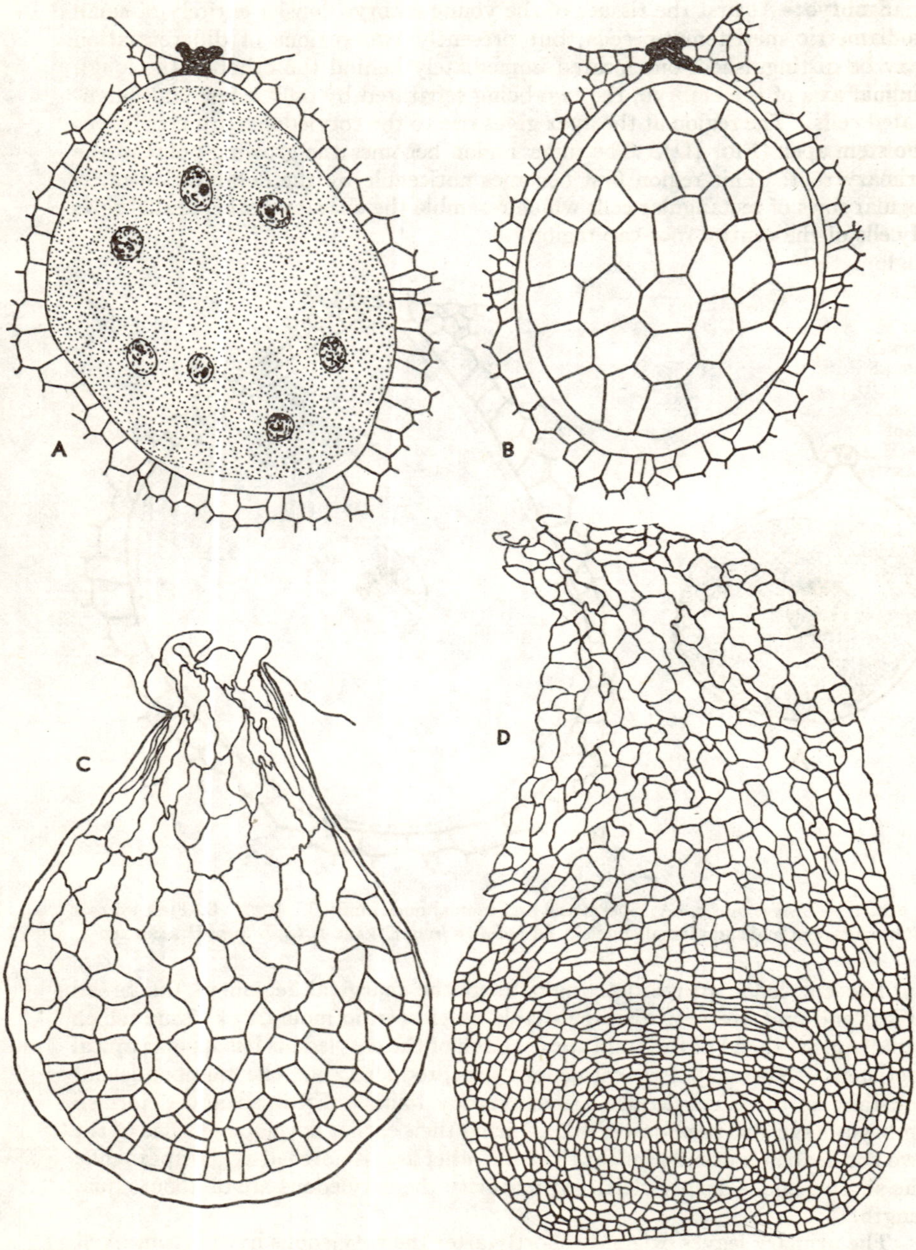

FIG. 4. — *Ginkgo biloba*. A. Proembryo with sixteen free nuclei (seven shown). ×105. B. Cell formation completed in proembryo. ×105. C. Developing embryo with meristematic cells being organized in basal portion. ×120. D. Median longitudinal section of older embryo with origin of cotyledon primordia. ×98. (C, D *redrawn from* LYON 1904).

nection with the primary leaves and are closer together than the set originating from the cotyledons. The shape of the embryonal stele is determined by these two sets of procambial strands: if there are two cotyledons, the stele is ellipsoidal as seen in transverse section, but if there are three cotyledons the stele is triangularly prismoidal. At the root meristem an ellipsoidal plerome narrows down into a broad wedge, whereas a prismoidal plerome narrows down into a blunt pyramid. In the mature embryo five leaf primordia are usually distinguishable.

Polyembryony:- Generally only one of the two archegonia normally present is fertilized to develop into an embryo. The other archegonium is often also fertilized, but one of the two embryos presently aborts. Three archegonia may rarely occur and all three may produce embryos. In several recorded instances a single proembryo may produce two (twin) embryos, of which one usually aborts before maturity is attained.

COOK, M. T. 1902. Polyembryony in *Ginkgo*. Bot. Gaz. 34: 64–65.
HERZFELD, S. 1927. Beiträge zur Kenntnis von *Ginkgo*. Jahrb. Wiss. Bot. 66: 814–862.
HIRASE, S. 1895. Études sur la fécondation et l'embryogénie du *Ginkgo biloba*. Jour. Coll. Sci., Imp. Univ. Tokyo 8: 1–16.
HIRASE, S. 1918. Nouvelles recherches sur la fécondation et l'embryogénie du *Ginkgo biloba*. Bot. Mag. Tokyo 32: 139–143.
LYON, H. G. 1904. The embryogeny of *Ginkgo*. Minn. Bot. Studies 3: 275–290.
SHIMAMURA, T. 1928. On the formation of proembryo of *Ginkgo biloba* L. Bot. Mag. Tokyo 42: 71–76.
SHIMAMURA, T. 1931. A note on the mitotic division in the proembryo of *Ginkgo*, with special reference to a chromatin-elimination. (In Japanese, with English résumé.) Bot. Mag. Tokyo 45: 525–530.

CONIFEROPHYTA

Chapter III

PINACEAE

Essential Features of the Embryogeny of the Pinaceae:- All genera included in the family are polyembryonic. This results from the fertilization of several archegonia, and the condition is described as simple polyembryony. In many conifers, moreover, each zygote splits into several to many embryos, giving rise to the condition known as cleavage polyembryony. There is ample evidence in the organization of the young embryo to indicate that in those

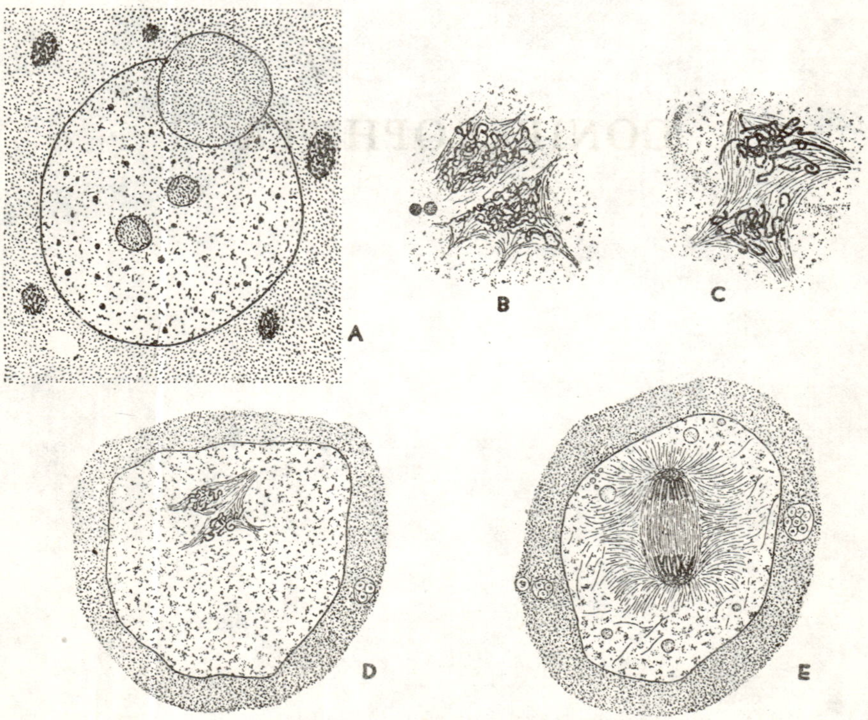

FIG. 5. — *Pinus lambertiana.* A. Microgamete sinking into archegonial nucleus. ×353. B. Segregation of male and female chromosomes on separate spindles. ×450. *Pinus monophylla.* C. Later stage than preceding, with two groups of chromosomes on common polyarch spindle. ×450. D. Archegonial nucleus, at a stage about intermediate between B and C. ×270. E. Anaphase of first embryonal mitosis. ×270. (B–E *from* HAUPT 1941).

species in which cleavage of the zygote does not occur they were derived from forms in which cleavage polyembryony prevailed.

Most species in the *Pinaceae* possess several archegonia. In *Pinus* there are from two to six in most of the species, but in *P. radiata* and perhaps one or two other species there is usually only a single archegonium, which probably indicates an advanced and highly specialized condition. All the archegonia in a given ovule are ordinarily fertilized; even if fertilization does not take place, the ovulate strobilus may nevertheless continue apparently normal growth but viable seeds naturally are not produced. Examples of this condition are common in *Pinus*, *Cedrus* and *Sequoia*. In cultivated specimens of *Cedrus* the staminate strobili do not appear until the ovulate strobili are almost fully developed so that fertilization is not effected at all.

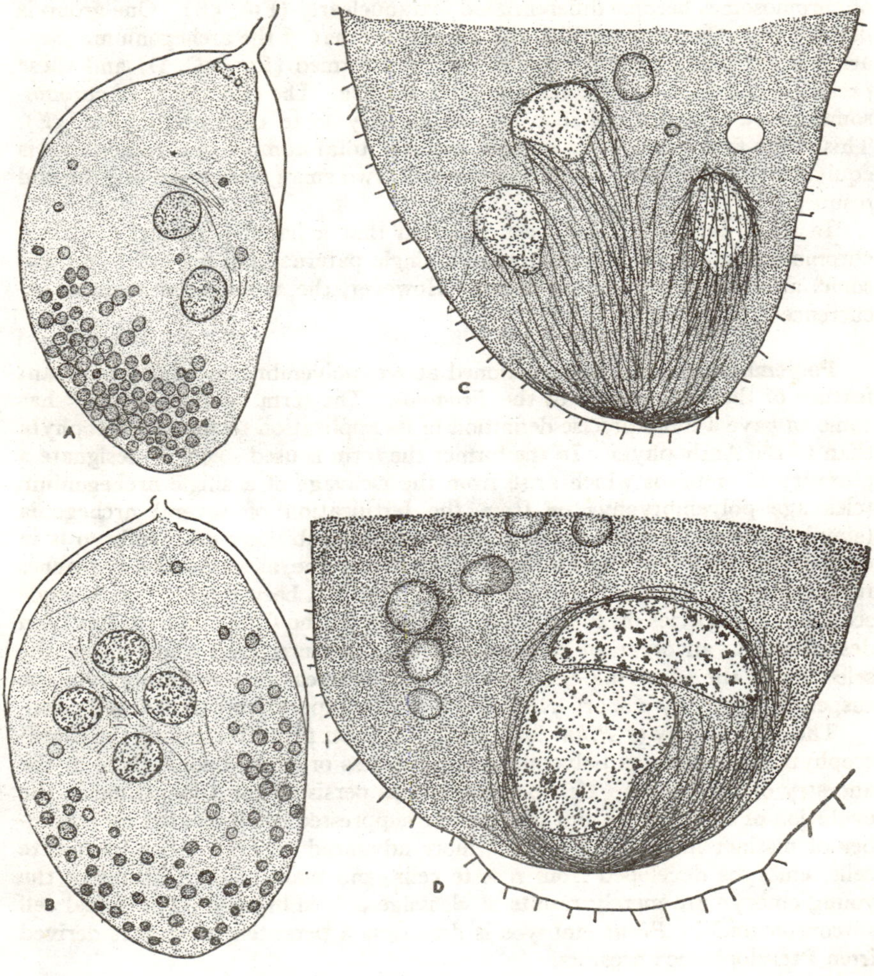

FIG. 6. — *Pinus lambertiana*. A. Two free nuclei in center of archegonium. (The small rounded bodies are presumably protein granules.) ×70. B. Four free nuclei in center. ×70. C. Free nuclei (three shown) being pulled to base of archegonium by fibrils. ×235. D. Free nuclei (two shown) at base. ×235. (*Original*).

Fertilization:- The process of syngamy has been accurately described in its entirety for only a few species. The most comprehensive account is that for two species of *Pinus*, *P. lambertiana* and *P. monophylla*. Descriptions for the other genera are mostly fragmentary or so old as to be unreliable.

The microgametophyte of *Pinus* pours its contents — dense cytoplasm gorged with starch, small tube nucleus, stalk cell and two large male nuclei of unequal size — into the archegonium. The larger male nucleus ordinarily is the functional one. When the nucleus reaches the archegonial nucleus, the two become flattened along the plane of contact. The nucleoplasm of the male nucleus is much denser than that of the archegonial nucleus (FIG. 5A). As the former sinks into the latter the nucleoplasm becomes less dense until the density of the two is finally equal. The membranes separating the fusing nuclei gradually disappear where they are in contact and simultaneously two groups of chromosomes become differentiated intranuclearly (FIG. 5B). One group is derived from the male nucleus, the other from that of the archegonium. Separate multipolar polyarch spindles are then formed (FIG. 5C, D) and these presently converge into a common diarch spindle. The two groups of chromosomes move together and soon lose all identity as to their origin (FIG. 5E). This is the first sporophytic mitosis since the total number of chromosomes is equivalent to the normal diploid number. Two small free nuclei are the end result.

In *Abies balsamea* it has been reported that a haploid number of paired chromosomes, each pair consisting of a single paternal and maternal chromosome, arise following nuclear fusion. However, the actuality of such an occurrence is questionable.

Polyembryony:- As was mentioned above, polyembryony is the dominant feature of the embryogeny of the *Pinaceae*. The term "polyembryony" has come to have a more precise definition in its application to the Coniferophyta than to the Anthophyta. In the former the term is used solely to designate a plurality of embryos which arise from the cleavage of a single archegonium (cleavage polyembryony) or from the fertilization of several archegonia (simple polyembryony). In both types of polyembryony each embryo is in competition with the others for survival and only one, as a rule, finally reaches full maturity although more than one seedling has been raised from a single seed in a few species of *Pinus*. The suspensor in the Coniferophyta has been described as "an organ of competition, the structure upon whose merit the selection of the surviving embryo depends," whereas in the Anthophyta the suspensor is usually a short-lived structure of comparatively unchanged cells.

The embryogeny of *Pinus* is probably the most primitive among the Coniferophyta, and cleavage polyembryony doubtless originated somewhere in the ancestry of this genus. This peculiarity has persisted for a long time in the evolution of the conifers, but has become suppressed or eliminated by a number of distinct methods in various more advanced genera. Aborted rosette cells, embryos developed from rosette cells, and many other features of the young embryo are merely results of cleavage polyembryony. The apical cell so conspicuous in *Pinus* embryos is doubtless a persistent character derived from Pteridophytic ancestors.

Larix, *Picea* and *Pseudotsuga* have only simple polyembryony; *Pinus*, *Cedrus* and *Tsuga* have cleavage polyembryony. *Abies* (as based on *A. balsamea*) normally shows only simple polyembryony, but an occasional embryo has been found with cleavage polyembryony. The remaining genus, *Keteleeria*, remains uninvestigated.

Proembryo:- The earlier stages in the embryogeny of the *Pinaceae* are best understood from the description of that of *Pinus*, which may therefore be taken as representative of the other genera.

The daughter nuclei resulting from the first zygotic mitosis are rather small at first but rapidly increase in size and presently divide simultaneously (FIG. 6A). The four free daughter nuclei (FIG. 6B) again are small at first, then increase about twice in bulk. Fine fibrils appear in the archegonial cytoplasm around each nucleus, then increase in diameter and length and become extended to the base of the archegonium. There some of the fibrils become trans-

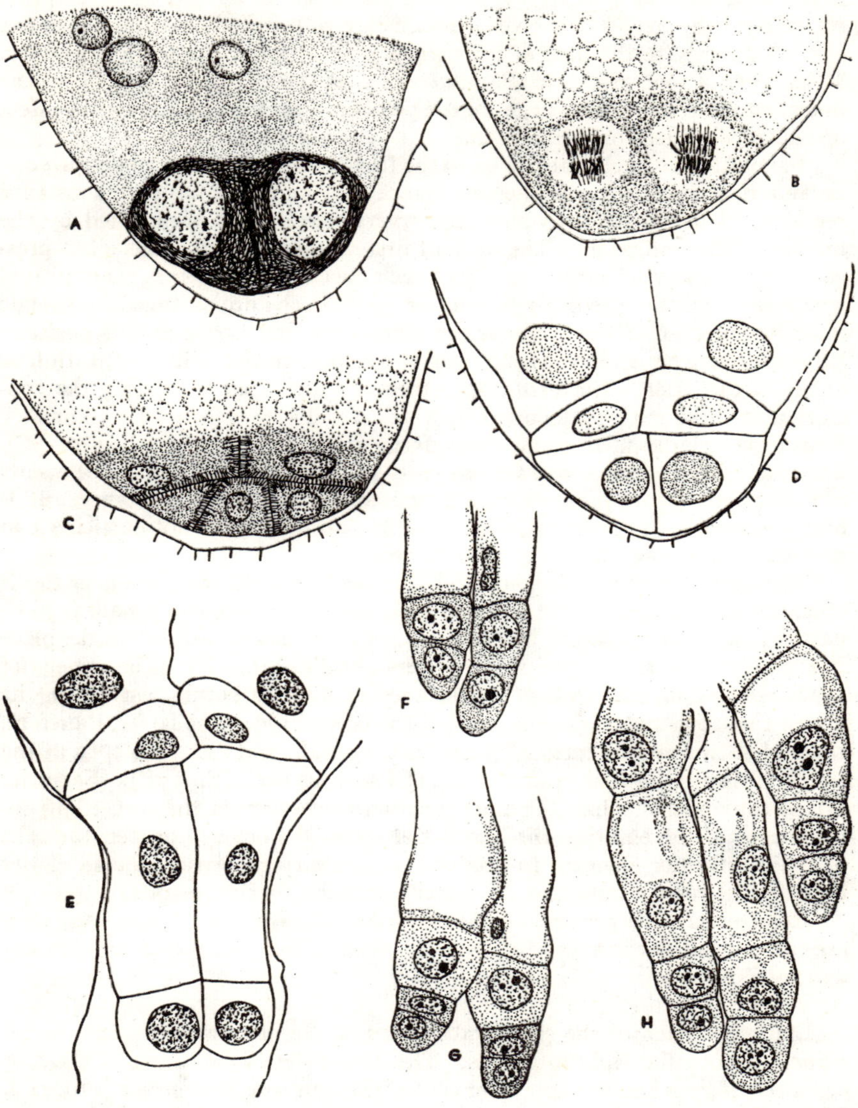

FIG. 7.—*Pinus lambertiana*. A. Four free nuclei (two shown) at base of archegonium, surrounded by fibrils. ×185. *Pinus radiata*. B. First division of free nuclei. ×235. C. Proembryo with two tiers of four cells each. Cell formation in process of completion. ×235. *Pinus laricio*. D. Three-tiered proembryo. ×235. E. Four-tiered proembryo. Early elongation of suspensor cells. ×235. *Pinus radiata*. F. Two of the four primary embryos have delimited the primary suspensor initials. ×235. G. Apical cells of primary embryos (two shown) have delimited first embryonal tubes; primary suspensor initials commencing to elongate. ×235. H. Third segments delimited. ×235. (*Original*).

versely appressed to the archegonial membrane and soon the nuclei become pulled to the base of the archegonium by means of the tractive force exerted by the fibrils (FIG. 6C, D). This traction mechanism is always present, though demonstrable only by special staining methods, whether the ovulate strobilus is pendant, erect or oriented in any which direction in nature, since the great majority of the ovules in a given strobilus are so oriented that the nuclei can not fall directly to the base of the archegonium through the force of gravity. It has been demonstrated in *P. thunbergii* that the change in position of the ovulate strobili from the vertical to the pendant after fertilization has no effect on the development of the embryo.

Soon after the nuclei are pulled to the base the fibrils appear to become disorganized, or at least to lose the appearance of being fibrils, but their substance remains. This substance, which stains very deeply, has been noted by the majority of investigators. The mass at first surrounds all the nuclei but presently moves away from the vicinity of each nucleus and becomes concentrated where cell walls are presently developed. The nuclei are arranged in a single plane at the base of the archegonium when they first arrive at this position. All next divide simultaneously to form two tiers of nuclei (FIG. 7B), with four nuclei in each tier. The first transverse walls are produced during the conclusion of these mitoses, followed by vertical walls in each tier by secondary transverse spindle fiber connections formed between the nuclei (FIG. 7C). The exact mechanism of wall formation is still not perfectly clear since events are obscured by the presence of the remnants of the fiber substance. It is highly probable, in any event, that the fibril substance does not participate in any way in the construction of the cell walls.

The lower tier of cells is completely enclosed by walls but the upper tier is open toward the apex of the archegonium, hence is incompletely walled. The next division usually occurs in the upper tier, but may occasionally take place in the lower tier, and results in three tiers of cells (FIG. 7D). The upper tier again remains unwalled above. Still another division occurs, usually in the lowest tier, to complete development of the proembryo. The latter, therefore, consists of twelve cells completely enclosed by walls and four cells open at the top, all arranged in four uniform tiers of four cells each (FIG. 7E). Each tier except the upper one has a definite function to perform in the succeeding development of the embryo: the lowest tier forms the embryo proper, the cells of the second tier elongate to produce the primary suspensor, the third tier becomes the rosette cells, and the upper tier presently disintegrates.

Rosette cells are present in *Pinus*, *Cedrus*, *Tsuga* and *Pseudolarix;* they are present but become aborted early in *Abies*, *Picea* and *Larix*, and are absent in *Pseudotsuga*.

Embryo:- Each of the eight walled cells of the first two tiers may be regarded as potential embryo initials. The lowest tier is the source of the four primary embryos and the upper tier that of the four rosette embryos (FIG. 7D). The four primary embryos are developed promptly, while those of the rosette tier are somewhat delayed in growth. Occasionally one or more of the eight cells — most frequently those of the rosette tier — may either fail to grow further or disintegrate completely. From the beginning each of the eight cells which grows further develops by apical cell growth. The four cells of the lowest tier divide once and interpose another tier of cells (the primary suspensor initials) which represent the first segments delimited by apical cells each with a single cutting face (FIG. 7F). A sixteen-celled late proembryo stage is the result. Simultaneously the cells of the primary suspensor tier commence to elongate and thrust the terminal cells forward into the corrosion cavity which has just been formed through enzymatic action. Shortly after the suspensor cells have begun to elongate, the apical cells delimit the second segments (the first embryonal tubes) of their respective embryos (FIG. 7G); then the third and subsequent segments are cut off (FIG. 7H). In certain species of *Pinus*, the four apical cells of the lower embryos have become separated and distinctly rounded off at the tip by this time, and the older segments elongate successively to produce suspensor-like secondary additions which are more commonly called embryonal tubes (FIG. 8A, B). In other species the four embryos still adhere closely for some time by their primary suspensor cells, in that actual separation occurs later.

Those embryonal tubes first formed may elongate as a single unit of cells but sooner or later the successive additions divide by vertical walls before elongation commences and then elongate as two collateral embryonal tubes. There is probably no actual difference between the primary suspensor cells and the embryonal tubes, except in the order of their origin. The elongation of both types of suspensor cells is so rapid that it presently exceeds the rate of enlargement of the corrosion cavity and as a consequence the suspensors and embryonal tubes become coiled and convoluted. The suspensor cells first formed soon collapse.

The embryo initials of the rosette tier remain quiescent during the earlier stages of suspensor elongation, but soon commence to divide. Each initial of this tier divides independently of its fellows and forms a group composed of several cells which assume a polarity in one direction or another. Some of the rosette initials may disintegrate and disappear while still unicellular or the process may not begin until several cell divisions have been undergone. Occasionally the initials may elongate to resemble suspensor cells. After the rosette embryos have become well developed they sometimes manifest a characteristic apical cell growth and short suspensor cells or embryonal tubes may be produced. Nevertheless, rosette embryos with extensively developed and elongated suspensors are rarely encountered. As a rule, all of them become aborted when they have grown to include about twelve cells, but well developed embryos have been observed in *P. montana* (*P. mugo*) which were presumably derived from rosette cells since they occupied the space above the primary embryo and were laterally oriented.

A struggle for existence and supremacy now sets in between both the four lower embryos derived from each archegonium and those derived from other archegonia (the latter condition exists only when two or more archegonia are present in the same ovule). A variation in the degree of growth and development of the different embryos soon becomes manifest. Ordinarily those embryos most deeply seated at the apex of the corrosion cavity grow fastest; at least it is apparent that they occupy their positions in the forefront because their cells had multiplied more rapidly and their suspensors had elongated more quickly. It is rare that more than one embryo attains maturity in a given ovule, but the existence of two fully developed embryos has been reported for *Pinus monophylla*, *P. edulis* and *P. torreyana*.

In most species of *Pinus* the activity of the apical cell does not continue throughout the entire period of differentiation of the embryo. At first it is a hemispherical cell with a single cutting face and resembles that of *Chara* and of filamentous fern gametophytes (FIG. 8A, B, D). After from three to five segments have been delimited it becomes transformed into an apical cell with three cutting faces and is then like that at the stem or rhizome apex of *Equisetum* (FIG. 8C, F). The segments which it cuts off proliferate rapidly in various planes. In *P. banksiana* the apical cell was still recognizable in embryos with about 500 cells and in *P. montana* it is clearly distinct in fully mature embryos. The latter is an extreme instance; generally the apical cell becomes indistinguishable before the cotyledons make their appearance. There is some evidence in *P. radiata* that the apical cell ends its career by producing the torus and corpus of the stem apex, probably by a final periclinal division, but proof of the actual occurrence of such a division has not been adduced.

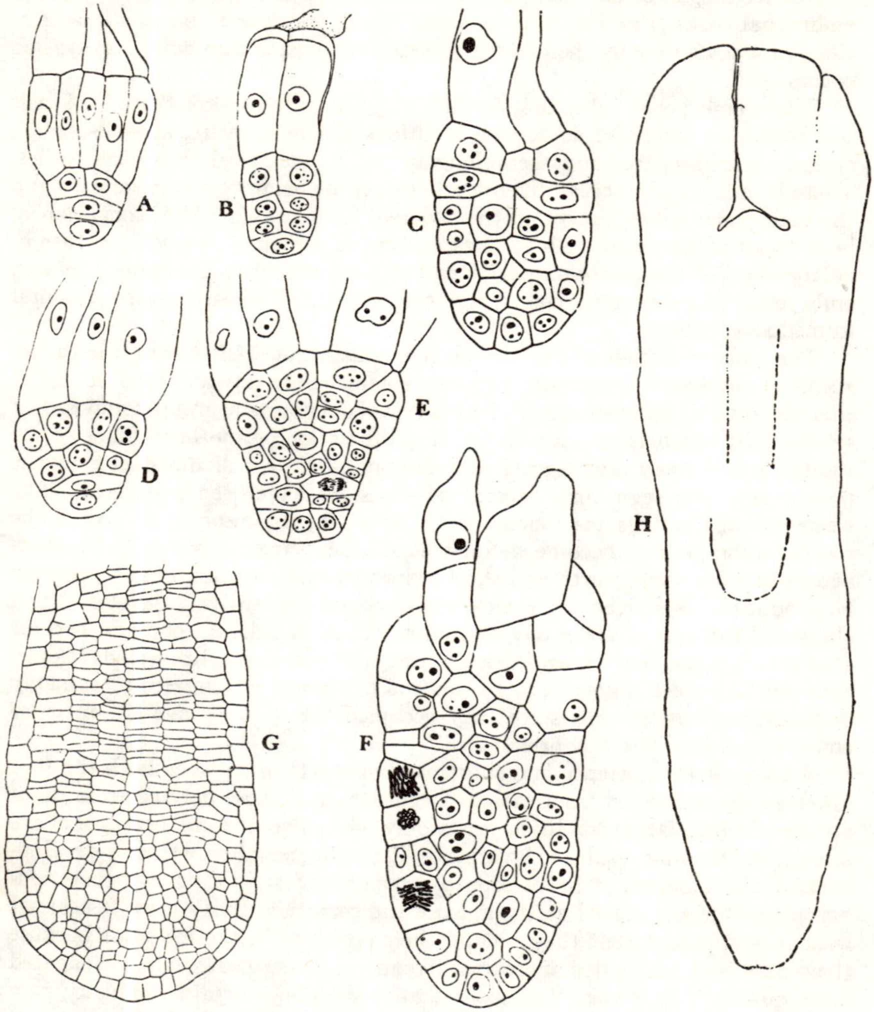

FIG. 8. — *Pinus ponderosa.* A–B, D–E. Four successive stages in development of young embryo (drawn from whole mount dissected preparations). ×175. *Pinus nigra.* C, F. Two successive stages of the developing embryo; apical cell present. ×141. *Pinus banksiana.* G. Embryo shortly before cotyledon formation begins; apical cell has disappeared. ×175. *Pinus radiata.* H. Median longitudinal section of mature embryo. The cotyledons are proportionately shorter than in many other species. (*Original*).

The mean number of cotyledons in *Pinus* is eight. The low species are *P. banksiana* with from three to six and *P. contorta* with from two to eight. The high species are *P. lambertiana* with from twelve to eighteen and *P. sabiniana* with from seven to eighteen.

Comparative Embryogeny:- The following key indicates to a certain degree the relationships of the various genera from the embryological standpoint. The one exception is that *Larix* is more closely related to *Pinus* and *Cedrus* than is indicated by its position in the key.

Each fertilized archegonium (with rare exceptions in *Abies*) gives rise to only one embryo (Simple Polyembryony); apical cell absent.
  Tier of rosette cells present.
    Primary suspensor tier is followed by a succession of three or four similar additions .... **Picea**
    Primary suspensor becomes very long and is succeeded directly by secondary suspensors .... **Abies**
  Tier of rosette cells is not formed .... **Pseudotsuga**
Each fertilized archegonium gives rise to more than one embryo (Cleavage Polyembryony); apical cell usually present.
  Rosette embryos present.
    Polarity units (capable of developing into embryos) eight in number.
      Cleavage of four primary embryo initials occurs early .... **Pinus**
      Cleavage of four primary embryo initials occurs late .... **Cedrus**
    Polarity units indefinite in number, increasing during early embryo growth .... **Keteleeria**
  Rosette embryos absent.
    Apical cell present; suspensor cells not embryo-forming.
      Rosette cells very ephemeral .... **Tsuga**
      Rosette cells persistent .... **Larix**
    Apical cell not definitely known; several successive suspensor cells are embryo-forming .... **Pseudolarix**

The series of diagrams presented in FIGS. 9 and 10 will serve to elucidate the above features more clearly.

The various genera, with the exception of *Pinus*, will now be discussed in slightly greater detail.

Picea.— As in *Pinus*, the late proembryo of *P. smithiana* consists of four tiers of four cells each. The uppermost tier of incompletely walled cells soon disintegrates. The next tier consists of rosette cells; the third is composed of suspensor initials which elongate as a unit. The lowest tier of embryonic cells divides shortly after elongation of the suspensor has begun.

The first division of the embryonic cells is soon followed by another series of divisions in the lower cells. The result is that each of the original four cells of the lowest tier of the proembryo is supplanted by three cells, making a total of twelve. The first two tiers are embryonal tubes and subsequent to further development constitute the secondary suspensor. The tier adjacent to the suspensor begins to elongate and is followed by the next tier and later by those of succeeding tiers to be added by further divisions of the apical tier. It is probable that the third tier of embryonal tubes is the final set to be delimited in most instances, but one or two more may be produced. Later the embryonal tubes lose their symmetry to become interlaced and irregular in their rate of elongation as they erect the massive secondary suspensor.

The precocious division of the embryonic cells to form an additional tier is a feature which characterizes *Picea*. In the related *Abies* the embryonic tier usually does not divide until after the suspensor has become very much elongated.

The rosette tier remains extant for some time but the nuclei disintegrate without undergoing division, at about the time the nuclei of the primary suspensor cells disappear.

The successive suspensor cells collapse in the order of their origin. The cells of the successive segments become from twenty to thirty times as long as their original diameter before the nuclei disappear and the individual elements collapse.

The median number of cotyledons in the genus is six; *P. mariana* and *P. abies* are low with three and *P. orientalis* is high with a report of fourteen cotyledons.

Abies.— The species studied include *A. balsamea*, *A. grandis*, *A. pinsapo*, *A. venusta* and *A. amabilis*.

There is no account of the development of the proembryo which is wholly acceptable. An older description of proembryonic development in *A. balsamea*

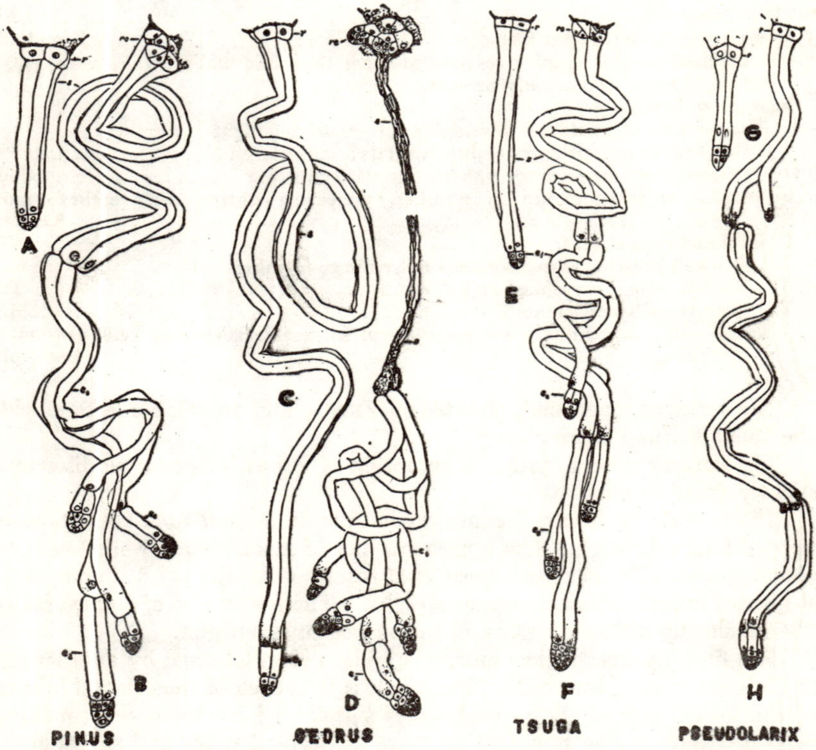

FIG. 9. — Comparative series of similar stages in the development of the embryos of *Pinus laricio* (*P. nigra*) (A, B), *Cedrus libani* (C, D), *Tsuga mertensiana* (E, F) and *Pseudolarix amabilis* (G, H) Symbols: *s*, primary suspensor; e_1, e_2, embryonal tubes; *r*, rosette cells which may give rise to rosette embryos, *re*. (*From* BUCHHOLZ 1931) (*See also* FIG. 10).

has been thrown open to question as a result of a later study of the growth of the embryo. Proembryo development probably is very similar to that in *Pinus* and agrees with that of *Picea*.

Abies differs from the related *Picea* in that the three tiers — suspensor, rosette and embryonic — of cells do not divide until after the suspensor tier has elongated to a considerable extent. All trace of the tier of relict nuclei vanishes at an early stage. Presently the embryonic tier successively divides to produce three tiers of four cells each.

The suspensor tier elongates to a remarkable extent in *A. venusta* and *A. pinsapo*, but less so in *A. balsamea*. Each cell has elongated to approximately sixty times its original diameter by the time the first division occurs in the embryonic tier and finally becomes 150 to 200 times this length before collapse of the secondary suspensor commences. The nuclei are no longer recognizable when elongation ceases.

Elongation of the embryonal tubes produces the secondary suspensor. The developing embryo becomes wider and the number of elongating cells increases in proportion so that the secondary suspensor becomes progressively more massive. The latest additions to the mass elongate most rapidly; the system becomes considerably coiled and convoluted.

Rosette cells are usually present in *A. venusta* and the majority may produce embryos. Such cells have been described as being absent in other species, but this is a matter of some doubt. The rosette embryos are not developed until after the primary suspensor has fully elongated.

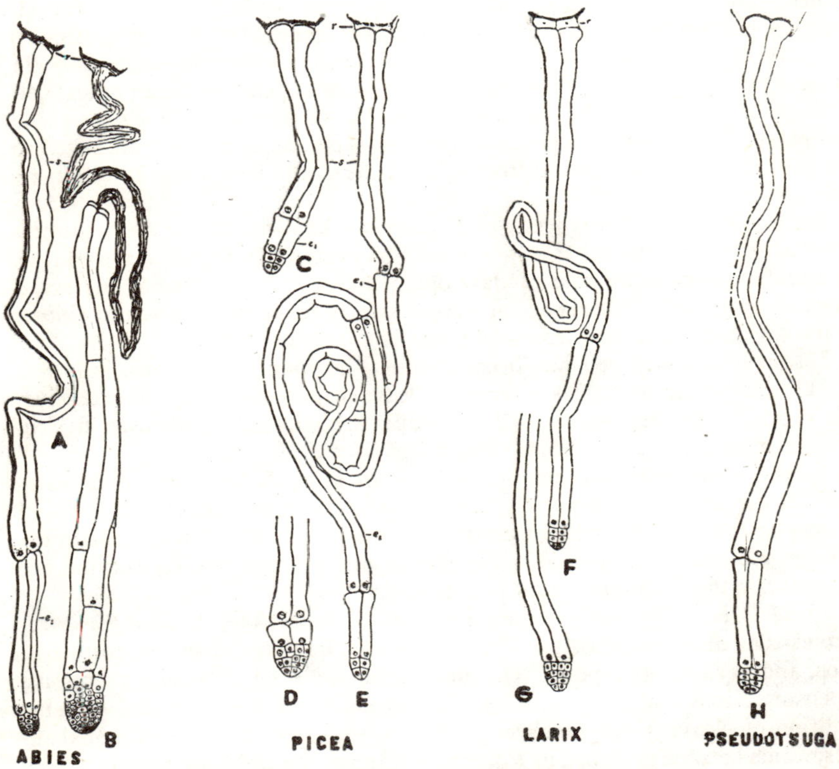

FIG. 10. — Stages similar to those in Fig. 9, for *Abies balsamea* (A, B), *Picea excelsa* (C, D, E), *Larix kaempferi* (F, G) and *Pseudotsuga taxifolia* (H). Symbols are the same. Rosette cells are usually aborted or are entirely absent, and the embryo initials in the zygote combine to form a single embryo. (*From* BUCHHOLZ 1931).

The number of cotyledons on the mature embryo varies between five and eight. A conspicuous calyptroperiblem, which is the conifer equivalent of a root cap, is present. It merges with the secondary suspensor and surrounds the plerome of the root tip.

Cleavage polyembryony occurs in from 10% to 13% of all observed instances and as many as four embryos may attain the eight- to sixteen-celled stage before elimination commences. Whether cleavage is to occur or not depends entirely on the completeness of the fiber system developed during the eight-celled stage. When it is complete, there results a variable degree of separation between the adjoining cells and the latter may thereupon develop independently into embryos. If fiber formation is incomplete, a single embryo is the result. Generally speaking, simple polyembryony characterizes the genus *Abies*.

Pseudotsuga.— The proembryonic stages are essentially the same as in *Pinus*. While the suspensors are elongating a mucilaginous plug, similar to the cellulose plug in *Podocarpus*, is formed at the base of the archegonium above where the rosette tier might be located if such were formed. An apical cell has not been observed. The cotyledons on the mature embryo vary in number from five to nine.

Cedrus.— The embryogeny of *C. deodara* and *C. atlantica* is precisely like that of *Pinus*. The primary suspensors elongate tremendously and the embryonal tubes may elongate before the lower embryos become separated. Cleavage polyembryony is characteristic of the genus, but the separation is relatively delayed in comparison to *Pinus*. Rosette embryos occur; they do not develop much beyond those of *Pinus*. There appears to be an apical cell in the younger embryos, but it is not as regularly observed as in *Pinus*. The mature embryo generally has nine cotyledons; the number ranges from seven to eleven in *C. atlantica*, nine to fifteen in *C. deodara* and from eight to ten in *C. libani*.

Keteleeria.— The embryogeny of *Keteleeria* is known only for the anatomy of the mature embryo, but from scattered observations it might be stated that the number of cells capable of developing into embryos becomes increased during early growth. This genus otherwise greatly resembles *Pinus* and *Cedrus* embryologically. The cotyledons number from two to four.

Tsuga.—An apical cell has been observed in younger embryos, in which respect this genus resembles *Cedrus*. Cleavage polyembryony is extensive. The rosette cells degenerate early. While the cotyledons average five in number, most species have from three to four.

Larix.—This genus, as exemplified by *L. decidua*, *L. laricina* and *L. leptolepis* (*L. kaempferi*), is essentially like *Pinus* in its embryogeny. The cells of the rosette tier collapse without producing embryos, but they may remain extant for a considerable period. The cotyledons average five on the mature embryo, ranging in number from four to seven.

Pseudolarix.—The early embryogeny of *P. amabilis* (*P. kaempferi*) resembles that of *Pinus* closely. The rosette tier, however, does not form embryos, and perishes undivided. On the other hand, each of several successive suspensor cells is embryo-producing. This constitutes a unique and extreme condition of cleavage polyembryony. The lowest tier of four embryonal initials, which become detached in *Pinus* to form separate embryos, remain united in *Pseudolarix* and become combined to produce a single embryo. The presence of an apical cell has not been demonstrated. In *P. amabilis* the cotyledons number from four to five.

ALLEN, G. S. 1945. Embryogeny and the development of the apical meristems of *Pseudotsuga taxifolia* (Lamb.) Britt. Diss., Univ. of California, Berkeley.

BUCHHOLZ, J. T. 1920. Embryo development and polyembryony in relation to the phylogeny of conifers. Amer. Jour. Bot. 7: 125–145.

BUCHHOLZ, J. T. 1920. The embryogeny of the conifers. Proc. Int. Congr. Pl. Sciences 1: 359–392.

BUCHHOLZ, J. T. 1942. A comparison of the embryogeny of *Picea* and *Abies*. Madroño 6: 156–167.

BUCHHOLZ, J. T., and E. M. OLD. 1933. The anatomy of the embryo of *Cedrus* in the dormant stage. Amer. Jour. Bot. 20: 35–44.

BUTTS, D., and J. T. BUCHHOLZ. 1940. Cotyledon numbers in conifers. Trans. Ill. Acad. Sci. 33: 53–62.

FERGUSON, M. C. 1904. Contributions to the life history of *Pinus*, with special reference to sporogenesis, the development of the gametophyte and fertilization. Proc. Wash. Acad Sci. 6: 1–202.

HAUPT, A. W. 1941. Oögenesis and fertilization in *Pinus lambertiana* and *P. monophylla*. Bot. Gaz. 102: 482–498.

HUTCHINSON, A. H. 1917. Morphology of *Keteleeria fortunei*. Bot. Gaz. 63: 124–134.

HUTCHINSON, A. H. 1920. Fertilization in *Abies balsamea*. Bot. Gaz. 69: 153–163.

HUTCHINSON, A. H. 1924. Embryogeny of *Abies*. Bot. Gaz. 77: 280–289.

KILDAHL, N. J. 1907. Development of the walls in the proembryo of *Pinus laricio*. Bot. Gaz. 44: 102–107.

LAWSON, A. A. 1909. The gametophyte and embryo of *Pseudotsuga douglasii*. Ann. Bot. 23: 163–180.

MATHEWS, A. C. 1932. The seed development of *Pinus palustris*. Jour. Elisha Mitchell Sci. Soc. 48: 101–118.

MIYAKE, K. 1903. On the development of the sexual organs and fertilization in *Picea excelsa*. Ann. Bot. 17: 351–372.

MIYAKE, K., and K. YASUI. 1911. On the gametophytes and embryo of *Pseudolariz*. Ann. Bot. 25: 639–647.

SAXTON, W. T. 1922. Preliminary note on the life history of *Cedrus deodara* Loud. Indian Jour. Bot. 3: 90.

SCHOPF, J. M. 1943. The embryology of *Larix*. Illinois Biol. Monogr 19: 5–97.

SMITH, R. W. 1923. Life history of *Cedrus atlantica*. Bot. Gaz. 75: 203–208.

SMOLSKA, A. 1929. Die Entwicklung des Archegoniums und der Befruchtungsprozess bei *Larix europaea* DC. Bull. Acad. Polonaise des Sci. et des Lett., Cl. Sci., Math. et Nat., B, 1927: 993–1038.

STRASBURGER, E. 1879. Die Angiospermen und die Gymnospermen. Jena.

WAKAYAMA K. 1929. On the influence of gravity upon the development of embryo of *Pinus thunbergii* Parl. Cytologia 1: 68–75.

Many of the observations recorded for *Pinus* are original.

Chapter IV

ARAUCARIACEAE

Fertilization:- The microgametophyte contains two large male cells or sperm nuclei and numerous smaller nuclei, consisting of prothallial, stalk and tube nuclei, which look much alike.

The actual process of syngamy has been described, probably only partially, for *Araucaria angustifolia* (*A. braziliana*), in which a male cell comes into contact with the nucleus of the archegonium and the cytoplasm of the former envelops both nuclei. The latter phenomenon is a feature characteristic of the family. The two nuclei remain in contact for some time, then a perforation at the point of contact occurs and the gap stretches until the contents of the two nuclei are contained in a common cavity. The fusion nucleus divides shortly thereafter; this mitosis has been described as being intranuclear in *Agathis australis*, hence it is probable that fertilization in the *Araucariaceae* is essentially similar to the more fully described process occurring in *Pinus*.

Proembryo:- The zygote nucleus, surrounded by the dense sheath derived from the cytoplasm of the male cell, is generally located slightly below the center of the archegonium. The daughter nuclei resulting from the first zygotic mitosis assume various positions (FIG. 11A), then divide simultaneously to produce a proembryo with four free nuclei (FIG. 11B). The *Araucariaceae* differ sharply from all other Coniferophyta in that the nuclei do not migrate to the basal end of the archegonium and there continue the development of the proembryo, but retain their central positions. Subsequent divisions of the free nuclei appear to follow no definite order and are not simultaneous. It is a most remarkable fact that mitoses in the proembryo of any araucarian have never been observed. In *Araucaria angustifolia* the total number of free nuclei, before wall formation sets in, is not less than 32 nor more than 45 (FIG. 11C–G); in *Agathis australis* the number may be anywhere from 32 to 64. It is therefore apparent that the number of nuclei present when wall formation commences is very variable. At this juncture the nuclei are evenly distributed in a concentric fashion throughout the proembryonic cytoplasm, which in the meantime has increased in volume but slightly.

Elongation commences simultaneously in both the upper and lower groups of cells immediately upon completion of wall formation (FIG. 11H). The lower group of cells is destined to produce the cap; generally no further divisions take place at any subsequent period in these cells. The upper group of cells produces the suspensors; these cells divide longitudinally to increase their number to about twice what it originally was (FIG. 11I). The upper ends of the suspensor cells are forced backward and upward until they are stopped by the intact apex of the archegonium (*Araucaria*) (FIG. 12A), or may be forced through the ruptured top (*Agathis*). The central group of cells constitute the embryonic initials; they remain inactive until after the primary suspensor cells complete development (FIG. 12B, C).

The growth of the suspensor system is rapid; in *Araucaria angustifolia* the total number of cells is close to twenty, and in *Agathis australis* the number ranges from eleven to thirty-eight. Usually there are three or four proembryos developing in a given ovule; their suspensors become tightly coiled together. One proembryo finally emerges below the others, the victor in the struggle for survival, and the latter slowly disintegrate. No evidence of any splitting which could result in polyembryony has been observed, although two subjoined mature embryos have been found in *Araucaria angustifolia*. The

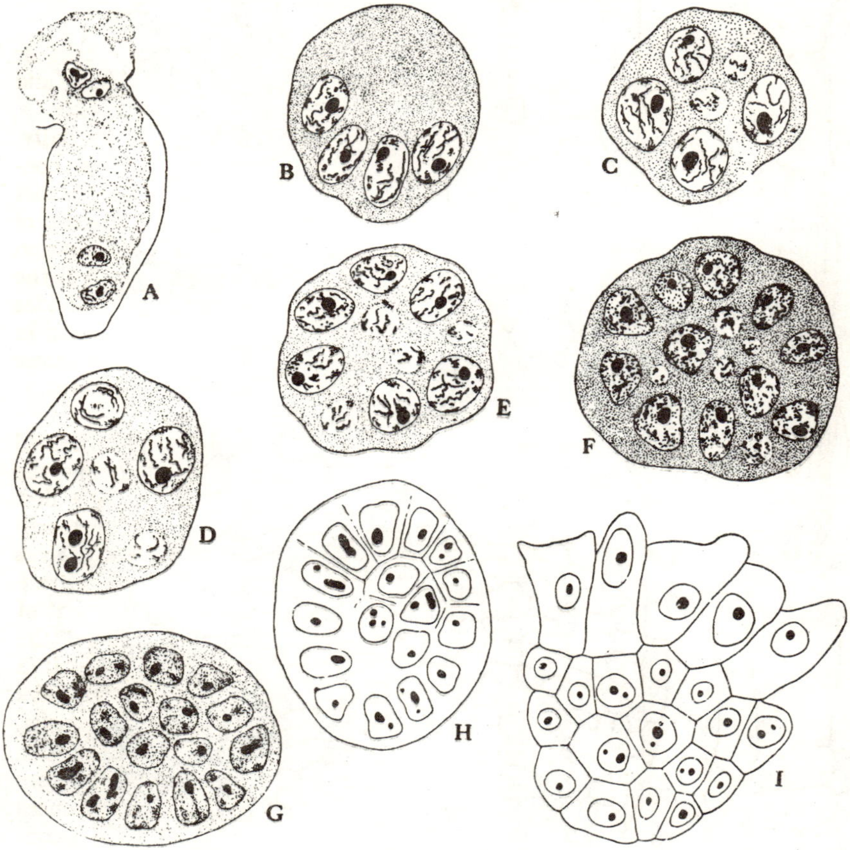

FIG. 11. — *Araucaria angustifolia* (*A. braziliana*). A. Archegonium with 2-nucleate proembryo near base. ×247. B. Four-nucleate proembryo, with nuclei disposed near base. ×727. C. Six-nucleate proembryo, with nuclei evenly distributed. ×727. D. Nine-nucleate proembryo (six nuclei shown). ×727. E. Ca. 21-nucleate proembryo (ten nuclei shown). ×727. F. Ca. 32-nucleate proembryo. ×727. G. Ca. 45-nucleate proembryo. The nuclei are arranged in position for wall formation; the apical nuclei (at base) have become slightly elongated. ×727. H. Beginning of wall formation in proembryo. ×727. I. Late proembryo; early elongation of suspensor initials. ×727. (*Original*).

primary suspensor cells begin to disintegrate shortly after the embryonic initials commence growth and presently disappear, their place being taken by a massive secondary suspensor system (FIG. 12D).

The nature of the proembryonic cap of the araucarians is a puzzling feature. It remains unchanged from the time of its maturity until completion of primary suspensor development. The cap has the appearance of a highly specialized structure, but its actual functions remain conjectural. It may serve to protect the embryonal initials, to elaborate digestive enzymes, or to prevent the occurrence of polyembryony, according to various authors. None of these conjectures is by itself wholly convincing. When development of the embryonal

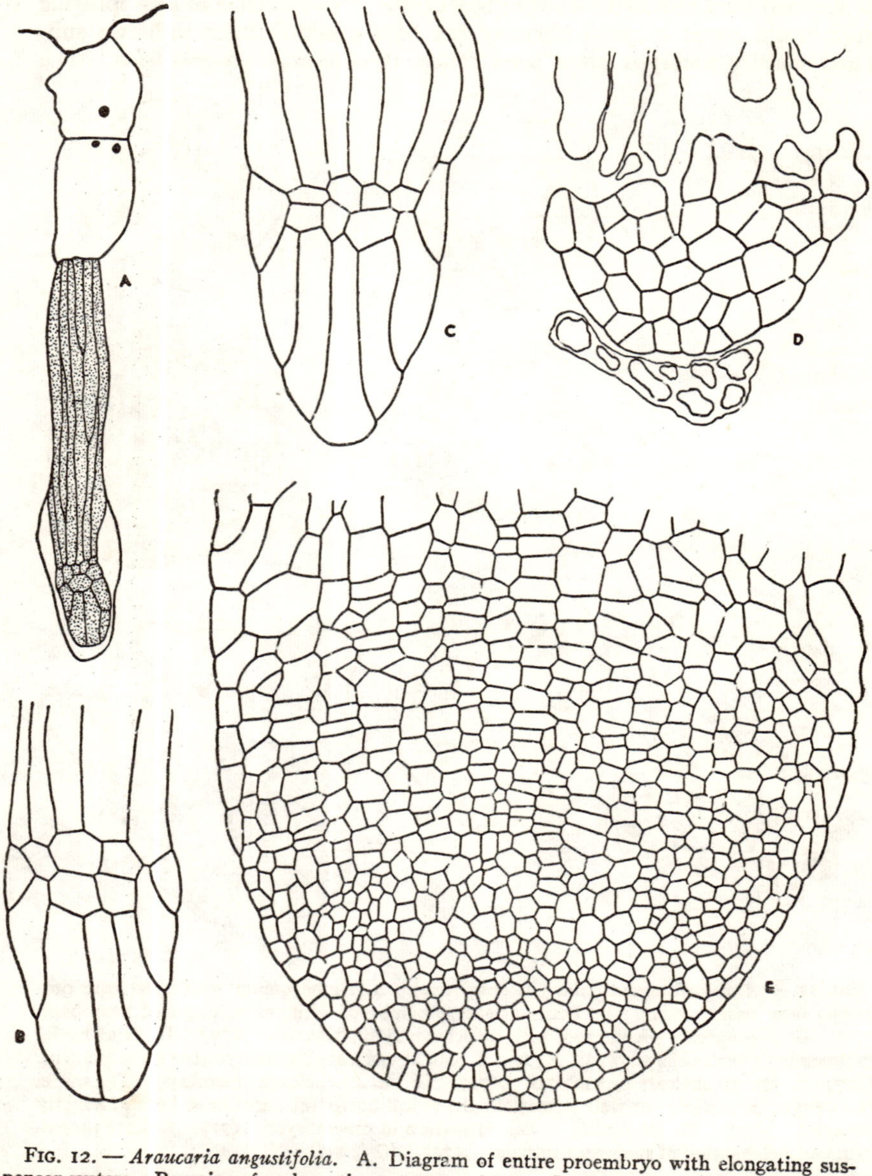

FIG. 12. — *Araucaria angustifolia.* A. Diagram of entire proembryo with elongating suspensor system. Remains of archegonium at apex. ×105. B. Proembryo with cap and portion of suspensor. ×352. C. Slightly older stage than preceding. ×352. D. Early embryo, with suspensor and cap cells shed and in process of disorganization. ×352. E. Developing embryo, before origin of cotyledons, in median longitudinal section. ×278. (*Original*).

initials commences, the cap cells begin to disintegrate, then become crushed by the growth of the embryo and shortly disappear (FIG. 12D).

The embryonic initials of the central part of the proembryo which have remained inactive until the development of the primary suspensor was concluded now begin rapid growth. At first all cells grow and divide about equally, then the upper ones begin to enlarge and elongate, while divisions continue interruptedly in the lower cells. The upper cells gradually cease dividing and elongate to produce a massive secondary suspensor which pushes the embryo still further down into the endosperm. The lower cells produce a large cylindrical body of meristematic tissues after which the activity of the apical meristem practically ceases (FIG. 12E).

Embryo:- Development of the embryo now commences with the delimitation of three new meristems. One of these originates in the region where the secondary suspensor joins the main body and constitutes the growing point of the hypocotyl and root. The other two develop at opposite sides of the original apical meristem and quickly produce the two cotyledons. The remains of the original apical meristem continue dormant and later form the meristems of the growing point when the seed germinates. Growth extends over a period of about two months and is confined largely to the cotyledons, which are exceptionally large in comparison to the remainder of the embryo. The vascular tissues are early delimited. Seven vascular bundles traverse the cotyledons and each is traceable back to its separate union with the vascular cylinder of the hypocotyl. The procambial strands form a hollow cylinder in the latter.

Resin canals are abundant only in the cortex of the embryo and in the cotyledons and do not occur elsewhere. Tissues of the stele and cortex are all alike at first and no definite boundary (a pericycle) between the two tissues can be perceived.

Summarizing, it might be said that the features which distinguish the araucarian proembryo from those of other Coniferophyta include: (*1*) the retention of the free nuclei in the approximate center of the archegonium; (*2*) concentric arrangement of the free nuclei at the time of wall formation; (*3*) complete encirclement of the embryonal initials by the cap and suspensor cells; and (*4*) mode of formation of the secondary suspensor system.

BURLINGAME, L. L. 1915. The morphology of *Araucaria braziliensis*. III. Fertilization, the embryo and the seed. Bot. Gaz. 59: 1–39.
EAMES, A. J. 1913. The morphology of *Agathis australis*. Ann. Bot. 27: 1–38.

To
My Undergraduate Major Professor
Douglas Houghton Campbell
Pioneer American Plant Embryologist

Chapter V

SCIADOPITACEAE

The family includes only the genus *Sciadopitys*.

Fertilization:- All nuclei contained in a microgametophyte (tube nucleus, stalk nucleus and two microgametes) are discharged into an archegonium. The larger sperm nucleus alone unites with the archegonial nucleus. The position of the latter may be in the center of the archegonium or a little higher up. The chromosomes, as in other carefully investigated Coniferophyta, are resolved into two separate groups whose spindles presently fuse into one.

Proembryo:- The polarity of the spindle of the first mitosis appears to be variable. It is sometimes parallel to the longitudinal axis of the archegonium, sometimes transverse to this axis, and occasionally oblique. The first two nuclei are free and remain near the center of the archegonium. The second mitosis follows soon after the first. Four free nuclei result and all then pass to the base of the archegonium and there enlarge. Earlier accounts described the formation of cell walls succeeding the third division, but it now appears to be more probable that there are five free nuclear divisions which result in 32 nuclei. These 32 nuclei then become arranged in three tiers, each containing about ten nuclei. Cell wall formation commences at this stage, and is completed after the sixth simultaneous series of nuclear divisions. In the uppermost tier, some of the cells remain open toward the apex of the archegonium.

Embryo:- There is some question as to which of the tiers of cells participates in the formation of the prosuspensor and rosette cells: the earlier accounts of the first stage of embryo development are undoubtedly based upon misinterpretations and no subsequent investigation of these stages has been made. The lowest tier comprises the embryo initials. Not all embryo systems, as observed at a later stage of growth, have a rosette group; when the rosette cells are absent, it is probably the uppermost tier of cells which elongates as a group of prosuspensor cells. When the rosette cells are present, the prosuspensor cells are undoubtedly derived from the second tier of cells and the rosette cells from the third tier.

The prosuspensor grows rapidly and becomes exceedingly long, coiled and twisted as it pushes the terminal group of embryonic initials deep into the gametophytic tissues (FIG. 13A). The mechanical pressure is so great that embryonic cells are occasionally crushed. The prosuspensor usually consists of from seven to nine cells. When the prosuspensor has become fully elongated the nuclei disappear and the cells soon begin to collapse. Later the primary suspensors force the remnants of the prosuspensor back into the archegonial region.

The embryonic initials remain one-celled during the elongation of the prosuspensor, then become two-celled. From these two-celled initials primary suspensors may commence elongation before the prosuspensors have attained their maximum development. These primary suspensors elongate rapidly,

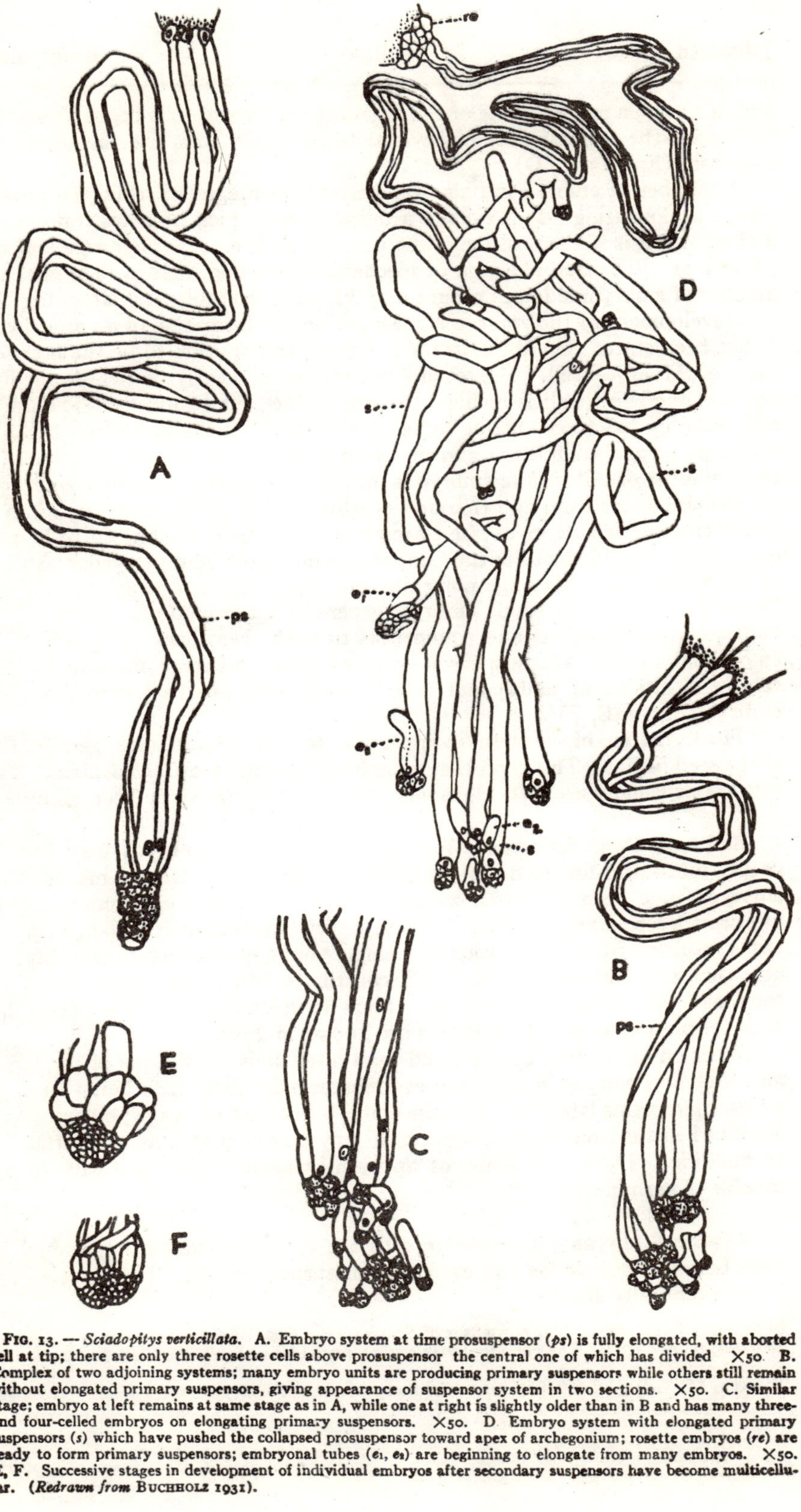

Fig. 13. — *Sciadopitys verticillata.* **A.** Embryo system at time prosuspensor (*ps*) is fully elongated, with aborted cell at tip; there are only three rosette cells above prosuspensor the central one of which has divided ×50. **B.** Complex of two adjoining systems; many embryo units are producing primary suspensors while others still remain without elongated primary suspensors, giving appearance of suspensor system in two sections. ×50. **C.** Similar stage; embryo at left remains at same stage as in A, while one at right is slightly older than in B and has many three- and four-celled embryos on elongating primary suspensors. ×50. **D** Embryo system with elongated primary suspensors (*s*) which have pushed the collapsed prosuspensor toward apex of archegonium; rosette embryos (*re*) are ready to form primary suspensors; embryonal tubes (e_1, e_2) are beginning to elongate from many embryos. ×50. **E, F.** Successive stages in development of individual embryos after secondary suspensors have become multicellular. (*Redrawn from* Buchholz 1931).

and during this process some of the adjacent embryonal cells begin elongation to initiate the formation of embryonal tubes which comprise the secondary suspensor (FIG. 13, C, D).

A number of embryo initials may become disintegrated while the prosuspensor is elongating. It is usually a terminal cell which becomes transparent and empty; it is in the position of the cap cell of *Podocarpus spicatus* and *Cephalotaxus*. Through abortion or mechanical crushing, it is evident that the number of embryonic initial groups may vary considerably at a later stage.

Development of the embryo is initiated by the formation of an apical cell. It has been stated that the primary suspensor cell appears to be the first segment of this apical cell. The second wall is oriented nearly vertical to the first one and the third wall is inclined to both of the first two walls. The apical cell probably has three cutting faces. Precisely when the apical cell becomes no longer recognizable is not known, but it appears to cease activity when the total number of cells in the embryo is much less than that in *Pinus* embryos.

Despite losses from abortion and crushing, each fertilized archegonium in *Sciadopitys verticillata* gives rise to more embryos than are known to be produced by any other genus in the Coniferophyta. Each embryo system exhibits an extreme type of cleavage polyembryony, splitting into many distinct embryos (FIG. 13D). A single embryo, generally a terminal one, gains an advantage over the others and commences to push them back by means of its embryonal tubes. The suspensor becomes more and more massive by successive additions of embryonal tubes, which elongate backward from the embryo (FIG. 13E, F).

The formation of the stem tip appears to be delayed until after germination of the seed begins. This condition resembles that found in *Cephalotaxus*. Two cotyledons usually are formed, though three have occasionally been observed.

Bud and Twin Embryos:- Well developed young embryos may become deeply lobed and finally divide to form twin embryos. Nearly mature twin embryos have been observed, which were on the end of a common, closely interwoven secondary suspensor. These embryos are considered to be identical twins resulting from the division of a single young embryo, since it is improbable that two embryo systems could contribute two mature embryos to the combined polyembryonic complex. Such twin embryos could conceivably both germinate, as similar instances are known in *Pinus*.

Other embryos may be produced by a later division of some of the larger multicellular embryos that have grown more rapidly than the others. In order to distinguish this late division of the embryo from the cleavage polyembryony described in the preceding paragraph, it may be designated as bud formation or budding. These bud embryos apparently never attain maturity or are capable of germinating.

Rosette Embryos:- The rosette cells are plainly groups of embryonal initials because they divide and even put out suspensor cells. The rosette complexes eventually abort.

BUCHHOLZ, J. T. 1931. The suspensor of *Sciadopitys*. Bot. Gaz. 92: 243–262.
LAWSON, A. A. 1910. The gametophytes and embryo of *Sciadopitys verticillata*. Ann. Bot. 24: 403–421.
TAHARA, M. 1937. Contributions to the morphology of *Sciadopitys verticillata*. Cytologia: Fujii Jubilee Vol., 14–19.

Chapter VI

TAXODIACEAE

The division of the *Taxodiaceae* into two subfamilies, in each of which there is more or less agreement in developmental details, simplifies the discussion of the embryology of the family:

Sequoidae: including *Sequoia*, *Sequoiadendron* * and *Athrotaxis*.

Taxodioideae: *Cryptomeria*, *Taxodium*, *Cunninghamia*, and probably the still uninvestigated genera *Glyptostrobus* and *Taiwania*.

However, the suggestion that *Sequoia* and *Sequoiadendron* be combined into one family and that all the other genera be constituted into monogeneric families warrants serious consideration at some future time when knowledge of the morphology of all species is completely known.

Sequoidae

Fertilization:- The older descriptions of the fertilization process are decidedly questionable in their cytological aspects. The more recent accounts, though accurate, unfortunately omit essential cytological details. It appears, however, that the two microgametes from a single microgametophyte are usually both functional and may enter two adjacent archegonia. The microgamete may become compressed as it passes between the neck cells, assuming a dumb-bell shape. In *Sequoia sempervirens* the microgamete nucleus is slightly smaller than that of the archegonium. The latter is centrally located. The fusing nuclei are surrounded by a dense mass of cytoplasm, but evidently no foreign cytoplasm is involved.

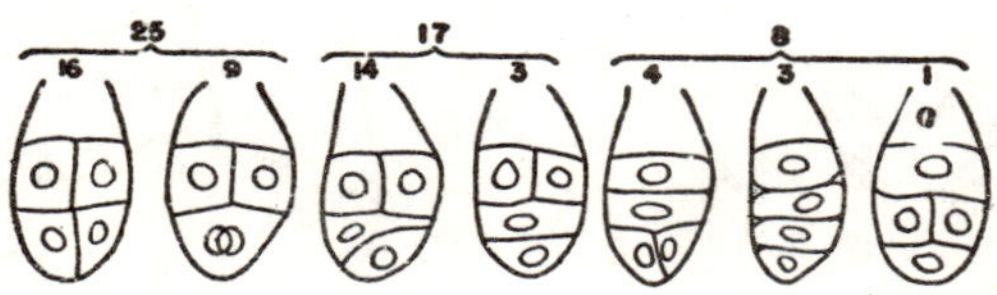

FIG. 14. — *Sequoia sempervirens.* Diagrams indicating frequency of distribution of variations in cell arrangement in fifty four-celled proembryos. (*After* BUCHHOLZ 1939).

Proembryo and Embryo:- The three genera differ so much that separate accounts of subsequent events are necessitated.

Sequoia sempervirens.— The first division is transverse to the long axis of the archegonium and is followed by cell wall formation. There is thus no free-nucleate stage in this species. A fertilized archegonium normally produces a

* At first the author accepted BUCHHOLZ's segregation of *Sequoiadendron* from *Sequoia*, as the arguments appeared to be based on valid morphological grounds. After the present account was set in type, however, there appeared a serious criticism of the separation of the two genera: DOYLE, J. 1945. Naming of the redwoods. Nature 155: 254 *et seq.* The segregation now appears to be unwarranted; consequently, for *Sequoiadendron* read *Sequoia* wherever the former name appears.

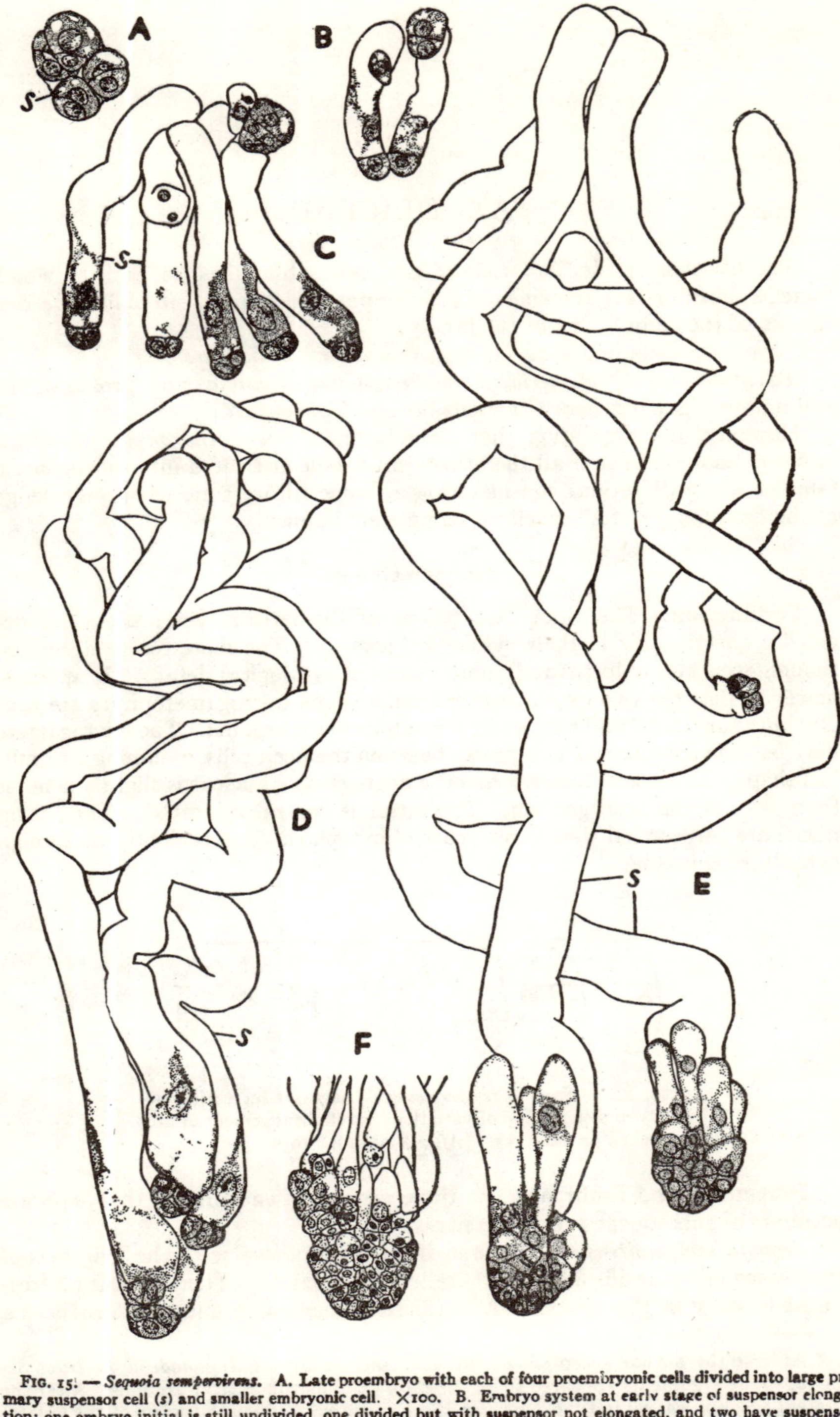

FIG. 15. — *Sequoia sempervirens.* A. Late proembryo with each of four proembryonic cells divided into large primary suspensor cell (*s*) and smaller embryonic cell. ×100. B. Embryo system at early stage of suspensor elongation: one embryo initial is still undivided, one divided but with suspensor not elongated, and two have suspensor elongated but embryonic cell undivided. ×100. C. Eight embryo units derived from the fertilization of two adjacent archegonia, some of which are aborting. ×100. D. Four embryos from the same archegonium with well developed primary suspensors (*s*). ×100. E. Later stage; many-celled embryo with embryonal tubes beginning to form the secondary suspensor. ×100. F. Multicellular older embryo on a secondary suspensor composed of numerous elongated embryonal tubes. ×100. (*From* BUCHHOLZ 1939).

four-celled proembryo, each cell of which is a potential embryo initial. That this is an actuality is indicated by FIG. 15A, which reveals what appears to be an eight-celled proembryo in linear arrangement but which is actually four two-celled embryo units. The type of cell arrangement in the four-celled proembryo is quite variable, probably being conditioned first by the shape of the archegonium and its position with reference to the microgametophyte and second by the axis of the archegonium. In FIG. 14 are depicted the various cell arrangements found in fifty four-celled proembryos, the numbers above each representing the frequency of distribution.

As the proembryo enlarges it fills all of the space formerly occupied by the archegonium and may even encroach upon the surrounding nucellar tissues. Each embryo unit consists of an embryonic initial and a vacuolated primary suspensor cell (FIG. 15B, 16A). The suspensor initial soon begins to elongate, but not all embryonic units develop as they may become aborted at any stage (FIG. 15C). The first division in the embryo initial is parallel to the axis of elongation (FIG. 15C). Subsequent divisions are somewhat irregular, but lobes corresponding to the position of the individual cells of the three- or four-celled stage can be distinguished. Some of these lobes may separately develop into independent small embryos or they may disintegrate. The irregularities tend to disappear and the mass of cells becomes a single embryo on the end of the primary suspensor (FIG. 15D). As the embryo becomes multicellular a fringe of embryonal tubes elongates around the primary suspensor (FIG. 15E, F). These numerous embryonal tubes abruptly become a multicellular secondary suspensor as the primary suspensor collapses. Rosette cells are ordinarily absent.

An apical cell is present for a limited period during the earlier developmental stages; it apparently does not persist as long as it does in *Pinus*.

Usually only one embryo attains maturity and it possesses two cotyledons. The time elapsing between fertilization and maturity is one year.

Sequoiadendron giganteum.— There are three successive free-nuclear divisions and the first walls are erected after eight free nuclei have been formed (FIG. 16B, C). This takes place in the lower half of the archegonium and three or four tiers of cells are organized. The lowest tier is commonly two- or three-celled and the upper tier, which divides again, is five- or six-celled (FIG. 16D, E). Subsequent divisions may occur at both ends of the proembryo, forming an additional tier of cells, as shown in FIG. 16F, G, thus producing a group of rosette cells in one or two tiers above the tier of five or six elongating cells. The latter thrust the embryonic group of cells deep into the nucellar tissues. These elongating cells constitute a prosuspensor (FIG. 16H). The latter soon becomes long and twisted but collapses shortly after attaining its maximum length.

The young embryo grows by means of an inconspicuous, transitory apical cell. The primary embryo initial divides to form a primary suspensor cell and an embryonic cell. The latter is considered to be a hemispherical apical cell and the primary suspensor cell as its first segment, which elongates and thrusts the embryonic cell ahead. In the meantime the hemispherical apical cell divides by a vertical partition, one of the daughter cells becoming a segment and the other the new apical cell. The segment presently elongates into an embryonal tube, while the apical cell divides repeatedly on several sides (FIG. 16I). These segments, which are not readily recognized, may become elongated embryonal tubes. The apical cell appears to persist until the end of the first growing season.

The primary suspensor cells segregated by the embryo initials elongate

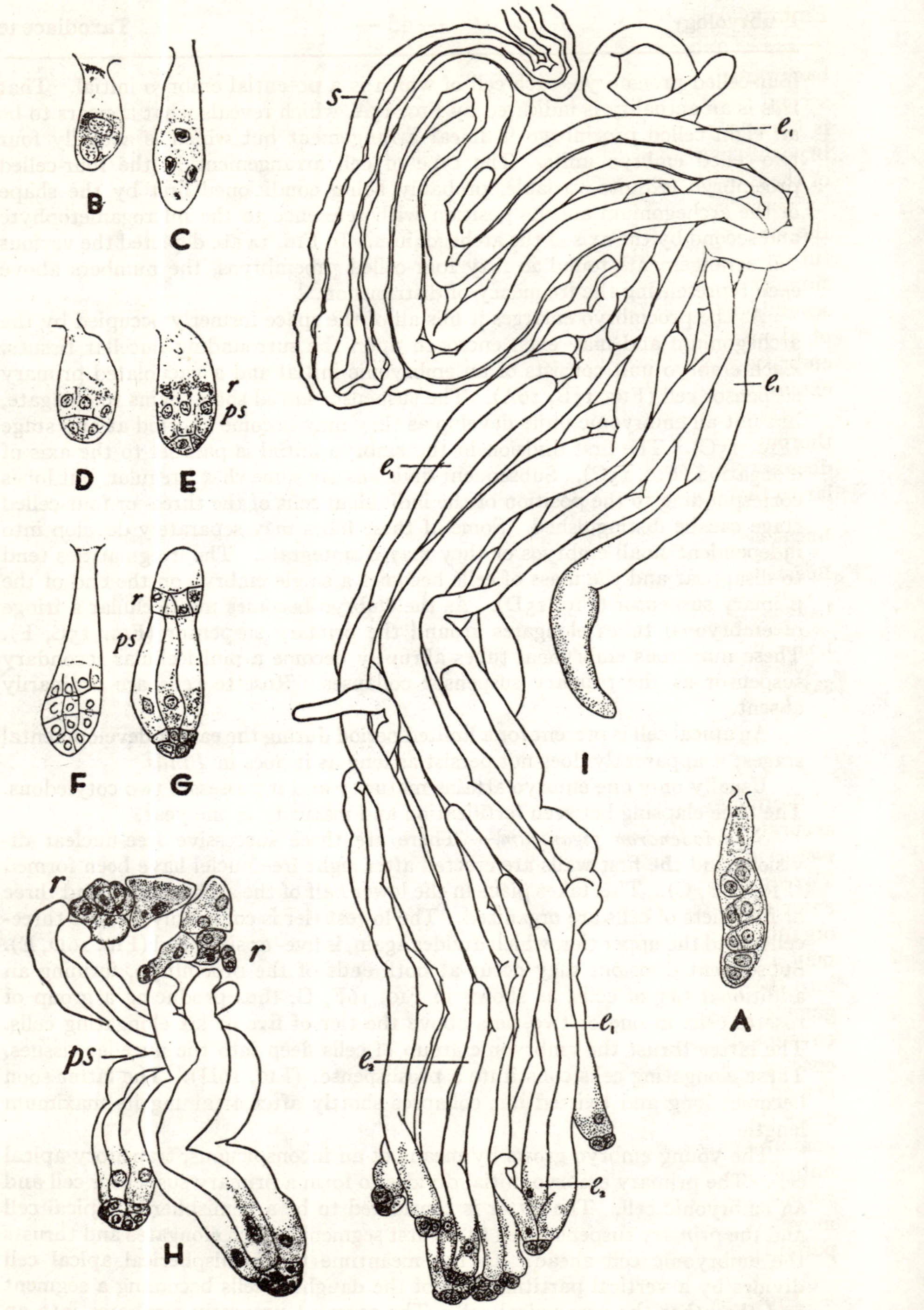

Fig. 16. — *Sequoia sempervirens.* A. Proembryo with four embryo units, each composed of a small embryonic cell and a primary suspensor cell. ×100. *Sequoiadendron giganteum* B–G. Stages in the development of the proembryo. ×100. H. Two neighboring embryo systems, with twisted prosuspensors with increased number of cells in terminal embryonic group. The rosette groups have divided, and free nuclei still remain in upper part of archegonium. ×100. I. Lower portion of an embryo complex after primary suspensors have begun to collapse. Each embryo unit is now borne on one or more embryonal tubes, whose order of origin and elongation (e_1, e_2, etc.) are indicated by their respective lengths. ×100. Symbols: *r*, rosette; *ps*, prosuspensor; *s*, primary suspensor. (A *from* Buchholz 1929; B–G *from* Looby and Doyle 1937; H–I *from* Buchholz 1939).

backwards rapidly and crush the prosuspensor. Embryonal tubes of the first second and third order (e_1, e_2, e_3, etc.) also appear; thus each embryonic group is borne terminally on a secondary suspensor composed of two or three embryonal tubes (FIG. 16I). Embryos may become aborted or dissociated at any of these developmental stages.

Growth of the embryo is resumed the following spring and it matures at the end of that season (*i.e.*, two years elapse between fertilization and maturity). The primordium of the stem apex becomes recognizable, in midsummer of the second year, at the tip of a cylindrical embryo composed of several hundred cells. The primordium soon becomes surrounded by a circle of about four cotyledon primordia. The number of cotyledons in mature embryos is not less than three nor more than five. Despite the prevalence of extensive cleavage polyembryony, only one embryo survives to maturity.

Athrotaxis selaginoides.— While a paucity of material has prevented a thorough investigation of any species included in the genus *Athrotaxis*, it does appear probable that the development of *A. selaginoides* is decidedly similar to that of *Sequoia sempervirens.*

BUCHHOLZ, J. T. 1939. The morphology and embryogeny of *Sequoia sempervirens.* Amer. Jour. Bot. 26: 93–101.
BUCHHOLZ, J. T. 1939. The embryogeny of *Sequoia sempervirens* with a comparison of the *Sequoias.* Amer. Jour. Bot. 26: 248–257.
LAWSON, A. A. 1904. The gametophytes, archegonia, fertilization, and the embryo of *Sequoia sempervirens.* Ann. Bot. 18: 1–28.
LOOBY, W. J., and J. DOYLE. 1937. Fertilization and proembryo formation in *Sequoia.* Sci. Proc. Royal Dublin Soc. 21: 457–476.
SAXTON, W. T., and J. DOYLE. 1929. The ovule and gametophytes of *Athrotaxis selaginoides.* Ann. Bot. 43: 833–840.

Taxodioideae

Fertilization:- All detailed accounts are rather old, but are surprisingly accurate from the modern standpoint. In *Cunninghamia sinensis* and *Cryptomeria japonica* only a single microgamete enters an archegonium, but in *Taxodium distichum* both microgametes from a given microgametophyte may enter the same archegonium. In the last species prothallial nuclei, the disorganized remains of the stalk and tube nuclei and the demolished neck cells may also be swept into the upper part of the archegonium but not actually into the cytoplasm of the latter. In at least the first two species the microgamete nucleus is surrounded by a dense sheath of starch which presently envelops the archegonial nucleus when the two come into contact. The presence of a similar sheath in *Taxodium distichum* has been both affirmed and denied. Fusion of the nuclei occurs in the center of the archegonium, and in all cases the microgamete nucleus is at first much smaller than that of the archegonium but rapidly increases in bulk until the two are nearly equal. The first mitosis occurs in the same position in the archegonium in *Cunninghamia sinensis* and *Cryptomeria japonica* but in *Taxodium distichum* the fusion nucleus moves to the base of the archegonium before dividing. The first mitosis apparently proceeds exactly as in *Pinus* and other carefully investigated genera.

Proembryo:- Two free nuclei result from the first division; except in *Taxodium distichum*, they then move to the base of the archegonium while still surrounded by the starch sheath. The two nuclei divide again about as soon as they reach their basal positions. The next division is also free-nuclear and the eight nuclei which are thus produced are arranged in two tiers. Wall formation now takes place, but the upper tier remains open toward the apex of the archegonium (FIG. 17A). The number of cells in each tier is not constant; the lower group of completely walled cells may even consist of two cells one above the other instead of the usual side by side position. Customarily there are two

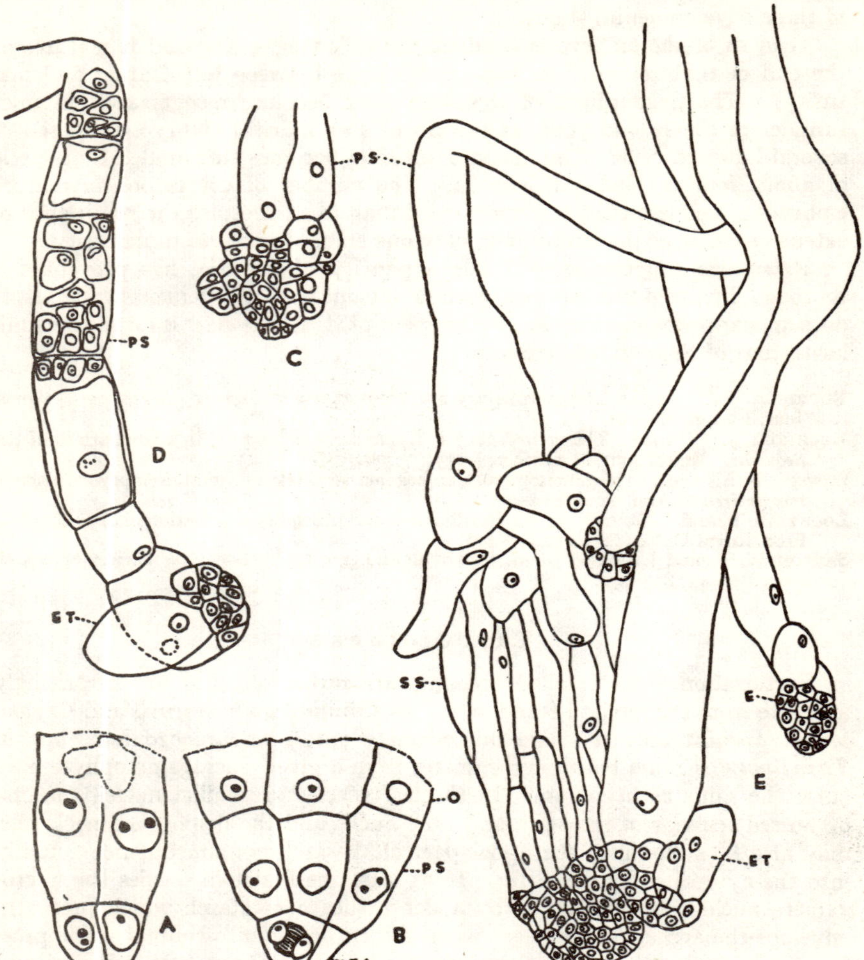

FIG. 17. — *Taxodium distichum.* A. Four of the eight nuclei in an eight-nucleate proembryo. ×379. B. Sixteen-nucleate proembryo, with two of the four in lowest group in telophase; open group above shows three of the six cells. ×379. C. Embryo system with multicellular embryo attached to prosuspensor; each unit has divided independently of the others ×172. D. Lower portion of a single prosuspensor cell which has become multicellular; embryo system at tips has one embryonal tube. ×172. E. One embryo system from a single archegonium, with only a portion of the extensive prosuspensor system depicted. Cleavage polyembryony is apparent. Secondary suspensor system is well developed in embryo unit at lower left. ×138. Symbols: *e*, embryo; *et*, embryonal tube; *ei*, embryo initials; *o*, open tier; *ps*, prosuspensor; *ss*, secondary suspensor. (*Redrawn from* KAEISER 1940).

cells in the basal tier with six above, or there may be three basal and five upper cells or even one basal and seven upper cells (FIG. 17B). The next divisions may be simultaneous in all cells, as in *Taxodium distichum*, but they are generally not. The nuclei of the upper tier may divide simultaneously to form a tier of prosuspensor cells and a tier of cells open above. The latter may persist for some time or disintegrate promptly. In *Cryptomeria* no walls have been discerned in the upper tier, but they may be merely transitory. The number of cells formed in the lower tier depends on how many were originally present: there may be a tier of four cells if there were two original cells, six if there were three or only two if there was but one terminal cell. A tier of four cells is most commonly encountered.

The upper tier of open cells should be designated as the open tier rather than the rosette tier, as has been done by the older investigators, since no rosette embryos are later produced. The middle tier gives rise to the prosuspensor, and the lower tier consists of embryo initials. In at least some species the cells of the prosuspensor are potential embryo initials since many of them may become embryonic.

Embryo:- All species so far investigated reveal extensive and constant cleavage polyembryony. This feature becomes apparent as soon as the embryo initials commence growth.

Growth of the embryo begins with extensive elongation and coiling of the prosuspensor cells, during which period the embryo initials remain quiescent, but in *Cryptomeria japonica* the embryo initials may become multicellular by the time the cells of the prosuspensor have become fully elongated. There are usually four embryo complexes, each complex being derived from a single archegonium, although as many as six have been observed. Each embryo initial develops independently of the others. The first division in an embryo initial usually forms a wall oriented in an oblique position. The second wall is generally formed perpendicular to the first and in the lower cell. One of these daughter cells functions as an apical cell which may persist into the multicellular embryo. However, the difficulties in distinguishing an apical cell in a well developed embryo are so great that reliance must be placed upon the arrangement of the successive cells produced at the apex of the embryo. The initials of a given embryo system may remain connected together for a short period but soon become separated, possibly because of the differences in the rate of elongation of the prosuspensor cells behind the initials. This separation, almost invariably found, indicates the commencement of cleavage polyembryony.

When the embryo has become multicellular, initial cells of embryonal tubes are delimited, generally at one level at a time close to the junction of embryo and prosuspensor. (These embryonal tubes have been called "supernumerary suspensors" by the older writers.) By extensive proliferation and enlargement these cells become the massive secondary suspensor of the maturing embryo. They surround the prosuspensor and replace the latter as it collapses.

The prosuspensor cells may behave in peculiar ways. They usually become more or less separated from one another and in *Taxodium distichum* there is a distinct tendency for them to become multicellular by internal cell proliferation, to cut off additional cells which may elongate slightly, or rarely to branch without forming transverse walls with small end cells forming at the tips of the branches. These end cells are potential embryo initials, but the embryos produced are abnormal.

The embryo initial which has the greatest chance of becoming eventually the one successfully to attain maturity is that borne on a vigorously elongating prosuspensor and which divides most rapidly.

The number of cotyledons varies somewhat according to the species: in

Cryptomeria japonica there are from two to four cotyledons, from two to nine in *Taxodium distichum* and from three to five in *T. mucronatum*, from two to four in *Cunninghamia lanceolata*, and two or three in *Taiwania cryptomanioides*.

BUCHHOLZ, J. T. 1932. The suspensor of *Cryptomeria japonica*. Bot. Gaz. 93: 221–226.
BUCHHOLZ, J. T. 1940. The embryogeny of *Cunninghamia*. Amer. Jour. Bot. 27: 877–883.
COKER, W. C. 1903. On the gametophytes and embryos of *Taxodium*. Bot. Gaz. 36: 1–27; 114–140.
KAEISER, M. 1940. Morphology and embryogeny of the Bald Cypress, *Taxodium distichum* (L.) Rich. Thesis, Univ. of Illinois.
LAWSON, A. A. 1904. The gametophytes, archegonia, fertilization, and the embryo of *Cryptomeria japonica*. Ann. Bot. 18: 417–444.
MIYAKE, K. 1910. The development of the gametophytes and embryogeny of *Cunninghamia sinensis*. Beih. Bot. Centralbl. 27: 1–25.

Chapter VII

CUPRESSACEAE

However coherently the *Cupressaceae* may be treated taxonomically, there are sharp embryological differences between either single genera or small groups of genera. The genus *Thuja*, for instance, stands in a class by itself, whereas *Libocedrus*, *Biota* and *Chamaecyparis* agree so substantially in their embryogeny that they form a group by themselves. In *Thuja* only a single embryonal initial develops further and cleavage polyembryony does not occur; in the other group of three genera practically all embryonal initials are functional and cleavage polyembryony is a characteristic feature. These embryonal manifestations, together with other morphological differences, render it desirable for at least the time being to divide the family into four subfamilies and to discuss them separately. A few genera can be assigned only provisionally to the most likely subfamily since too little is known of their embryogeny.

Cupressoideae

The genera *Cupressus*, *Chamaecyparis*, *Libocedrus*, *Biota* and *Thujopsis* are definitely included in this group; *Fokienia*, *Fitzroya*, *Pilgerodendron* and *Diselma* may possibly also belong here, but very little is known of their life history.

Fertilization:- The archegonia in all known instances are grouped together in a complex at the apex of the ovule, with their necks exposed at the base of a large common archegonial chamber. They may number from seven to twenty-two. Fertilization has been most closely followed out in *Thujopsis dolabrata* var. *hondai*, and in general is typical for the Coniferophyta. In *Chamaecyparis pisifera* the microgamete nucleus escapes from its cytoplasm when it moves toward the archegonial nucleus, but in *T. dolabrata* the microgamete nucleus remains surrounded by its cytoplasm which presently envelops the fusing nuclei. The contents of the microgametophyte are discharged into the archegonial cavity; it is rare for more than one microgamete to enter a given archegonium, although other nuclei may occasionally enter. The microgamete and archegonial nuclei come into contact in the upper part of the archegonium. The latter is approximately twice the bulk of the former. They then move toward the base of the archegonium and the first division occurs near the center or slightly below. Two separate sets of chromosomes are next resolved and the spindle is intranuclear but its orientation is variable (FIG. 18A).

Proembryo:- Two free nuclei result and move to the base of the archegonium while still surrounded by the cytoplasmic sheath derived from the microgamete cytoplasm. A second simultaneous free-nuclear division next occurs and the resulting four nuclei are disposed in one of three different ways according to the dimensions of the archegonium: (*1*) in a single row parallel to the long axis of the archegonium, (*2*) in two tiers with three nuclei above and one below, or (*3*) in two tiers with two nuclei in each tier. The third mitosis

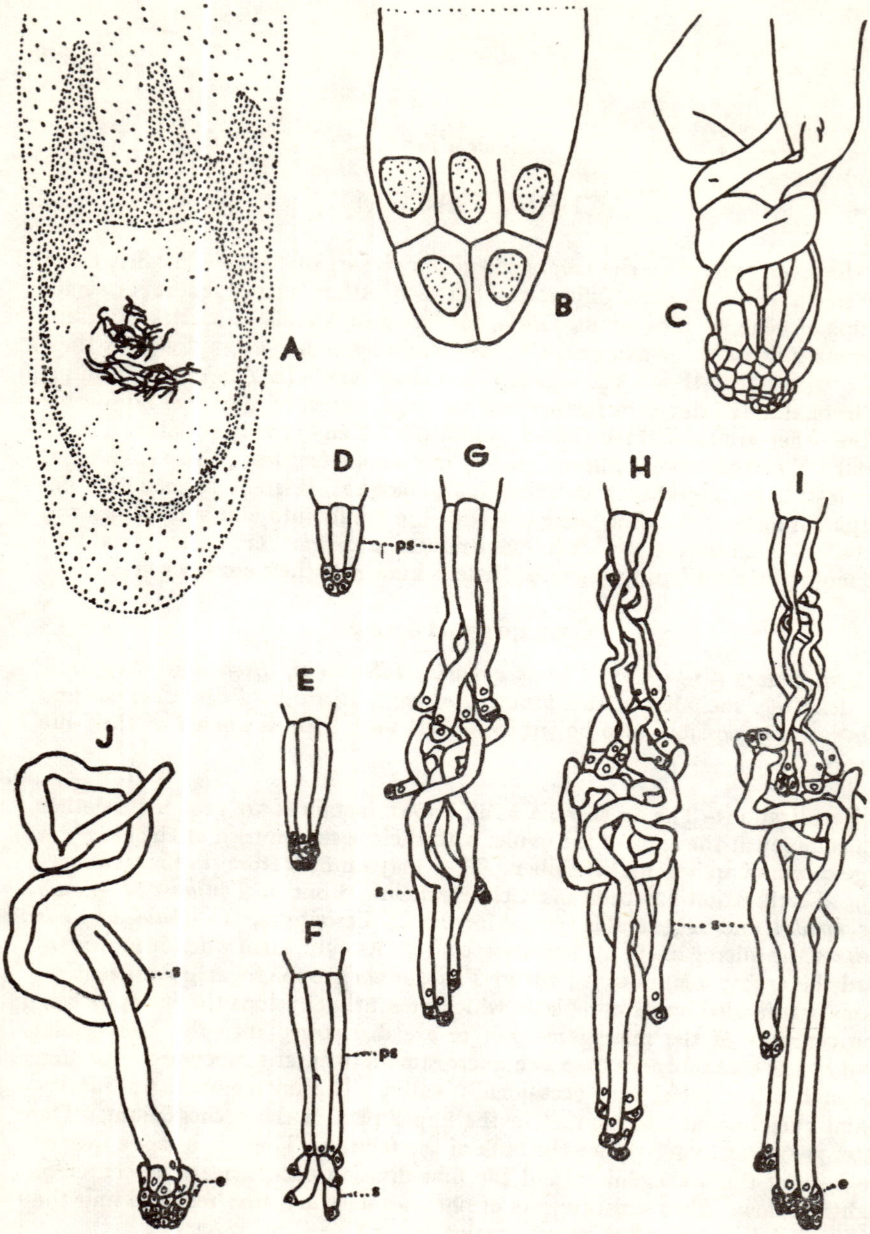

FIG. 18. — *Thujopsis dolabrata*. A. First division in proembryo with male and female chromosomes in separate groups. ×630. B. Eight-nucleate proembryo after wall formation. ×630. C. Early stage of cleavage polyembryony; two of the three embryos are depicted. ×630. *Chamaecyparis obtusa*. D–J. Diagrams illustrating successive steps in the development of a single embryo system. J represents a single primary embryo of a slightly older stage than in I. Symbols: *e*, embryonal tubes forming secondary suspensor; *ps*, prosuspensor cells; *s*, primary suspensors. (A–C *redrawn from* SUGIHARA 1939; D–J *redrawn from* BUCHHOLZ 1932).

is also simultaneous but the resulting nuclei become arranged in two tiers, with four, five or six nuclei in the upper tier. Wall formation occurs after this division; the cells of the upper tier always remain open above (FIG. 18B). In the following division the cells of the upper tier divide transversely to form a three-tiered proembryo. The upper cells, as before, remain open above. The uppermost tier later degenerates while the second tier becomes the prosuspensor and the lowest tier forms the embryo proper and the embryonal tubes (FIG. 18D–F).

In *Fitzroya patagonica* the proembryo is very variable in the arrangement of its cells and when fully developed is completely septate and fills the archegonium. *Fitzroya* is apparently in an intermediate position between the *Cupressoideae* and *Callitrichoideae.*

Embryo :- The cells of the lowest tier of the proembryo usually divide once before the prosuspensor elongates, and each cell appears to possess the capacity for developing into an embryo — that is, they are embryonic initials. However, only the initial at the apex of the group actually begins development and the others later become embryonal tubes. Nevertheless, two or more embryos may be formed at the tip of a prosuspensor.

Some of the prosuspensor cells have been observed to become separated from the main portion of the embryo and lack embryonic initials. They form irregularly shaped embryos which appear to correspond with the rosette embryos characteristic of *Pinus* and related genera. Apparently these never attain maturity.

In most of the genera, especially when there are a large number of archegonia each of which has been fertilized, it is difficult to trace out the relationship of embryonic initials, prosuspensors and suspensors; consequently, the development of the embryo complex may be delineated more satisfactorily by means of diagrams in which only a single embryo system is concerned. FIG. 18D–J shows such a series, based upon *Chamaecyparis obtusa*, but with which the other genera are in substantial agreement.

The elongating prosuspensor cells thrust the embryo deep into the gametophytic tissues and at the same time the former are pushed back toward the apex of the former archegonium. The prosuspensor cells, it should be noted, have an origin different from that of the primary suspensor cells. Sometimes the prosuspensor cells may collapse or shrivel completely; again, they may enlarge and become multinucleate, or become filled with cellular tissue or cut off cells near the tip. Since the prosuspensors are derived from embryonic initials, they are actually potential embryos whose development was more or less suppressed.

In *Biota orientalis* the activity of an apical cell in each embryo has been detected at an early stage; in *Chamaecyparis obtusa* one is still observable when the embryo has become massively multicellular. One has not definitely been found in the other genera.

After the embryos of *Biota orientalis* have become multicellular some of them may become broken up into fragments. This results in an irregular secondary cleavage stage in the embryo. This cleavage occurs at the stage when embryonic selection has been interrupted, when the primary suspensors begin to collapse and before the secondary suspensors commence development. These embryonic masses do not immediately produce additional embryonal tubes to form suspensors but may later do so.

The phenomenon of cleavage polyembryony is prevalent among the *Cupressoideae* and can be detected at an early stage. Simple polyembryony has been described for *Thujopsis dolabrata.* The mature embryo is dicotyledonous.

BUCHHOLZ, J. T. 1932. The embryogeny of *Chamaecyparis obtusa.* Amer. Jour. Bot. 19: 230–238.
DOAK, C. C. 1937. Morphology of *Cupressus arizonica:* gametophytes and embryogeny. Bot. Gaz. 98: 808–815.
LAWSON, A. A. 1907. The gametophytes and embryo of the *Cupressineae* with special reference to *Libocedrus decurrens.* Ann. Bot. 21: 281–301.
SUGIHARA, Y. 1938. Fertilization and early embryogeny of *Chamaecyparis pisifera* S. et Z. Sci. Rep. Tokohu Imp. Univ., Ser. IV, Biol. 13: 9–14.
SUGIHARA, Y. 1939. Embryological observations of *Thujopsis dolabrata* var. *hondai* Makino. Sci. Rep. Tokohu Imp. Univ., Ser. IV, Biol. 14: 291–303.

Callitroideae

The three genera, *Actinostrobus*, *Callitris* and *Widdringtonia*, exhibit a remarkable similarity in their early embryogeny, and differ from all other Coniferophyta in the complete absence of a prosuspensor. The embryogeny of *Tetraclinis* and *Callitropsis* has not yet been worked out with sufficient completeness but these genera may provisionally be assigned to the *Callitroideae.*

Fertilization:- In *Actinostrobus* the archegonia are lateral in groups of twenty-five to thirty; in *Tetraclinis articulata* they may be apical or both apical and lateral and number nine or ten to fifteen; in *Widdringtonia* the archegonia occur in lateral groups and may total around two hundred; in *Callitris verrucosa* there is a single lateral group of from seventeen to twenty archegonia and in *C. rhomboidea* there are even more. In practically all recorded instances the microgametophytes traverse the nucellus laterally, in sharp contrast to all other conifers except *Sequoia sempervirens.*

In general, two adjacent archegonia are fertilized from a given microgametophyte, one of the two microgametes going into each archegonium. However, cases are known where all the nuclei from one microgametophyte were discharged into a single archegonium. When there are numerous archegonia, not all are fertilized; the unfertilized archegonia quickly degenerate.

In *Actinostrobus pyramidalis* and *Callitris verrucosa* both the sperm and archegonial nuclei are of approximately equal size and similar structure. Those of *A. pyramidalis* and *Callitris rhomboidea* are almost invariably surrounded by a starch sheath while in contact; the origin of the starch is obscure but it perhaps is derived from the microgametophyte. The spindle of the first division has been described as being entirely intranuclear, but the segregation of the chromosomes into two groups has not been observed.

Proembryo:- After the first division in *Actinostrobus pyramidalis* one free daughter nucleus is located above the other in the longitudinal axis of the archegonium, no matter what the orientation of the latter (FIG. 19B). The basal nucleus next divides to form two daughter nuclei located one above the other, and is quickly followed in division by the upper nucleus which divides in the transverse plane (FIG. 19C–F). Four such embryonic units are formed from each fertilized archegonium; the remainder of the cells in the upper part of the proembryo degenerate and disappear. No region of the proembryo elongates as a whole. There is no definite orientation of the embryonic initials and they point in various directions.

In *Callitris rhomboidea* the first division of the zygote nucleus inferentially results in two free nuclei; a second simultaneous division then follows and walls are formed to produce a four-celled proembryo. The disposition of the cells in such proembryos depends entirely on the orientation of the spindles at the second division. The type most regularly formed consists of three tiers which are respectively 1-, 2- and 1-celled (FIG. 19J). Or there may be two juxtaposed

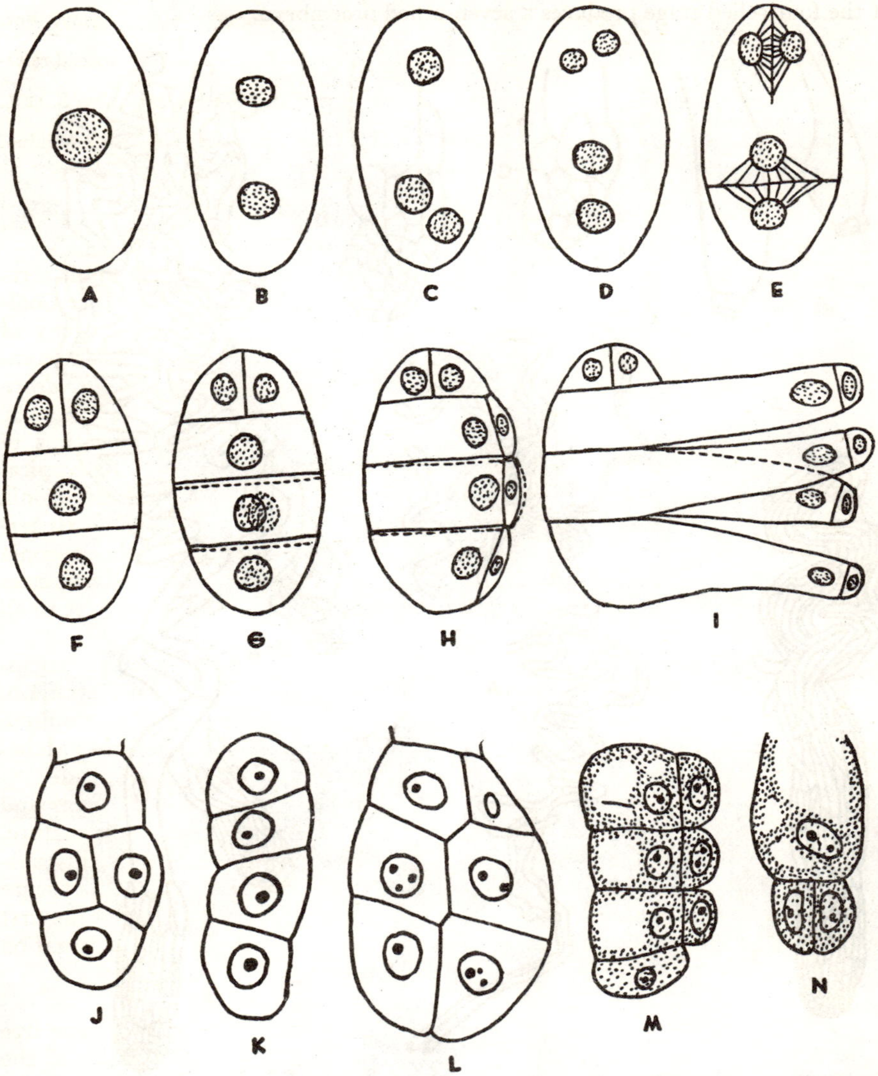

FIG. 19. — *Actinostrobus pyramidalis*. A–I. Series of diagrammatic figures to show proembryo development, formation of embryonic initials and early elongation of suspensor cells. (Since the archegonia are lateral, the apex of the gametophyte is to the left.) A. Zygote. B. Two-nucleate proembryo. C. Three-nucleate proembryo. D. Four-nucleate proembryo. E. Origin of walls after four-nucleate stage. F. Wall formation complete. G. Six-celled proembryo. H. Embryonic initials delimited. I. Early elongation of suspensors. All × ca. 350. *Callitris rhomboidea*. J. Four-celled proembryo. K. Four-celled proembryo, linear type. L. Symmetrical six-celled proembryo of which the upper two cells play no further role. M. Linear proembryonic group in three of which the embryo cells are cut off on the same side. N. Typical two-celled embryo. All × 375. (A–I *redrawn from* SAXTON 1913; J–N *redrawn from* LOOBY and DOYLE 1940).

and two superposed cells; again, a linear proembryo may be produced (FIG. 19K). Further divisions commonly occur in any two cells of the four-celled proembryo, to create a six-celled stage (FIG. 19L). There is, however, no real regularity at this period of development. It is doubtful whether most of the irregularly constructed proembryos develop further. Rarely another division at the four-celled stage produces a seven-celled proembryo.

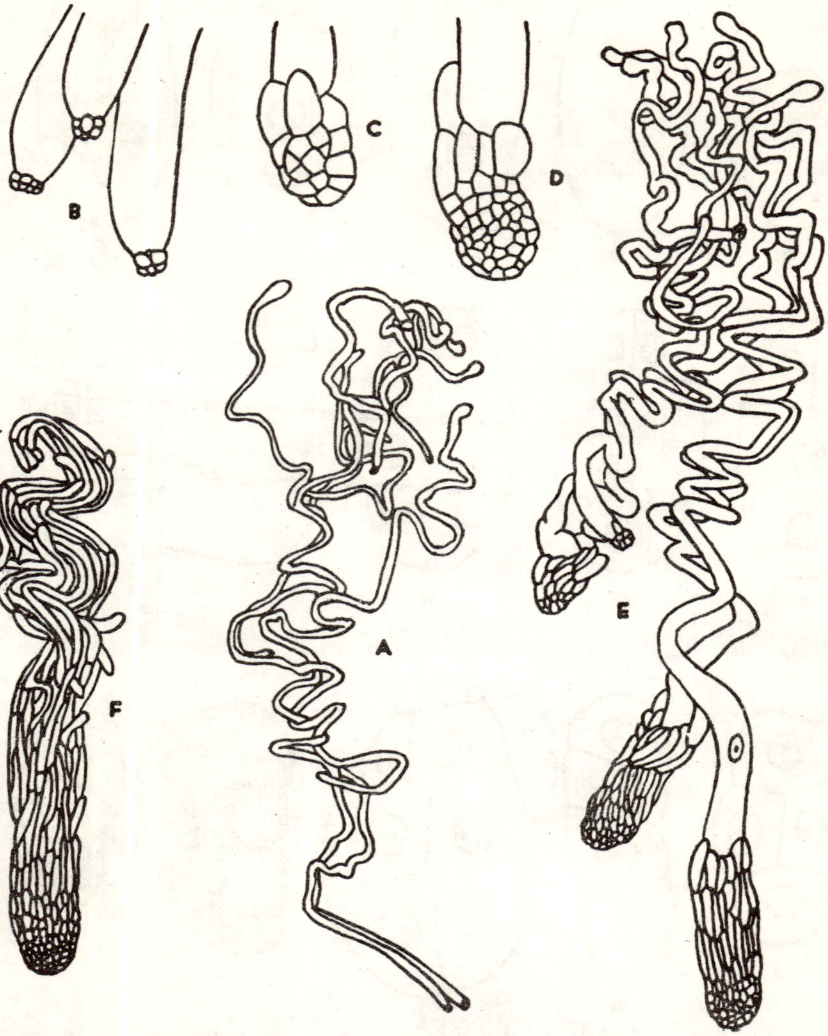

FIG. 20. — *Actinostrobus acuminatus* (A, B, D, F) and *A. pyramidalis* (C, E). A. Embryo system with embryos at two-celled stage. B. Terminal embryos of a group. C. Early formation of embryonal tubes. D. Later stage than the preceding. E. Embryonal system with primary and secondary suspensors. F. Fully developed secondary suspensors. A–C × 25; D × 130; E, F × 30. (*Redrawn from* BAIRD 1936–37).

Embryo :- Each of the four larger cells in the final proembryonic stage of *Callitris rhomboidea* divide to form a small outer embryonic cell and a larger inner cell which elongates to produce a suspensor (FIG. 19M). There is a complete absence of regularity in the direction of growth. The upper sterile cells usually disintegrate rapidly. It appears to be an invariable rule that four embryos and suspensors are formed from a given proembryo.

In *Actinostrobus pyramidalis* the embryo initials remain undivided for about two weeks. The first division is vertical (FIG. 20A) and is quickly followed by a second vertical division in each of the resulting cells but in a plane perpendicular to the first wall. There thus results a single tier of four cells at the tip of the suspensor (FIG. 20B). The next set of divisions are more or less horizontal (transverse), and the later ones quite irregular. A recognizable apical cell has never been detected. After a considerable number of cells have been produced the cells immediately adjacent to the suspensor elongate to form embryonal tubes (FIG. 20C, D). There is great variation in the number and size of these cells, the position of the walls and relative size of suspensor and embryos. Growth of the embryo, hitherto rather slow, becomes very rapid after initiation of the embryonal tubes and a massive multicellular structure is formed (FIG. 20E, F). The apex of the embryo becomes rounded and the cotyledons presently become recognizable as protuberances at the sides of the mass. The cotyledons usually number two, but three have been reported once in *Actinostrobus* and are apparently frequent in *Widdringtonia*. The stem apex is broad and conspicuous.

Summarizing, the *Callitroideae* differ from all other Coniferophyta whose embryology has thus far been clearly elucidated in the complete absence of a prosuspensor and exhibit a remarkable similarity among themselves. The affinities of the *Callitroideae* are with the *Cupressaceae* (some *Fitzroya*-like type) rather than with the *Taxodiaceae*.

BAIRD, A. M. 1936–37. The suspensor and embryo of *Actinostrobus*. Jour. Roy. Soc. West Australia 23: 89–95.
LOOBY, W. J., and J. DOYLE. 1940. New observations on the life-history of *Callitris*. Sci. Proc. Royal Dublin Soc. 22: 241–255.
MOSELEY, M. F., Jr. 1943. Contributions to the life history, morphology and phylogeny of *Widdringtonia cupressoides*. Lloydia 6: 109–132.
SAXTON, W. T. 1910. Contributions to the life-history of *Widdringtonia cupressoides*. Bot. Gaz. 50: 31–48.
SAXTON, W. T. 1910. Contributions to the life-history of *Callitris*. Ann. Bot. 24: 557–569.
SAXTON, W. T. 1913. Contributions to the life-history of *Actinostrobus pyramidalis*. Ann. Bot. 27: 321–345.
SAXTON, W. T., 1934. Notes on conifers. IX. The ovule and embryogeny of *Widdringtonia*. Ann. Bot. 48: 429–431.

Thujoideae

Apparently only the genus *Thuja*, whose various species and varieties seem to be alike embryologically, belongs in this subfamily. *Thuja* is distinguished from the other *Cupressaceae* in that only one of all the embryo initials proliferates rapidly to produce an embryo. (The genus *Biota*, merged by some taxonomists with *Thuja*, differs sharply in its embryogeny and should be kept separate.)

Fertilization:- There are usually six archegonia, arranged in an apical group. All that has been recorded of the fertilization process is the contact of the respective nuclei in *Thuja occidentalis*. The microgamete nucleus escapes from its sheath in the upper part of the archegonium and comes into contact with the archegonial nucleus laterally.

Proembryo:- The first division, which appears to be intranuclear as in other Coniferophyta, takes place a little below the center of the archegonium. The two free nuclei then move to the base of the archegonium and there divide simultaneously (FIG. 21A, B). The third division is also simultaneous and is followed by the formation of cell walls (FIG. 21C). The wall first laid down is transverse and separates the eight nuclei into two groups of four nuclei each. Longitudinal walls are quickly formed in the lower tier; the cells in this tier are completely walled in. Similar walls are erected at the same time in the upper tier, but the cells here remain open toward the upper end of the archegonium (FIG. 21D). Events in the *Thujoideae* thus far closely resemble the corresponding stages in the *Cupressoideae.* The four nuclei in the upper tier next divide simultaneously and walls transverse to the long axis of the proembryo are formed. The proembryo thus consists of twelve cells arranged in three tiers of four cells each. The upper tier of four cells soon disintegrates; the second tier rapidly elongates to form the four suspensor cells; and the lower tier develops a single embryo.

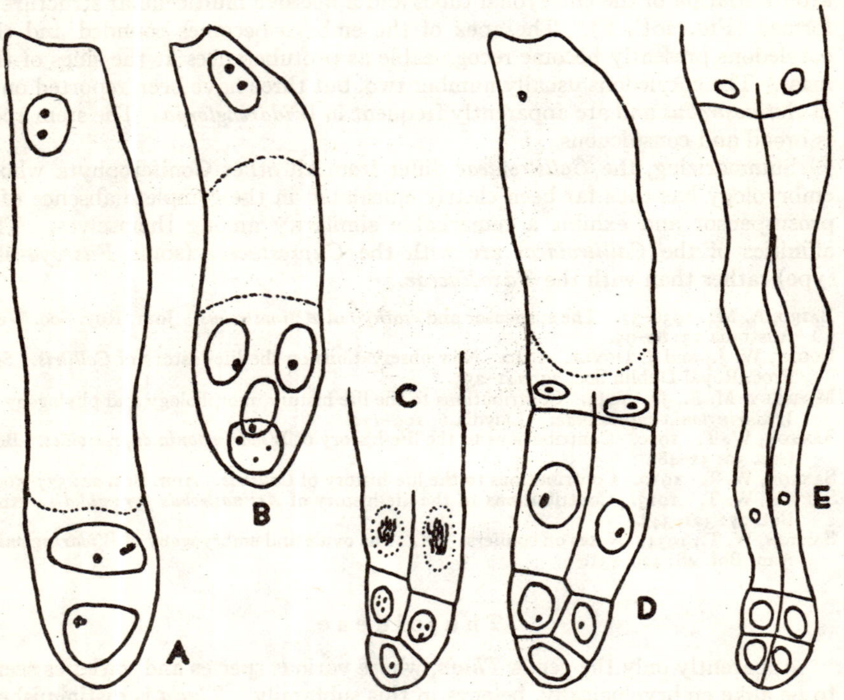

FIG. 21. — *Thuja occidentalis.* A. Two-nucleate proembryo (ventral nucleus remains in upper portion). B. Four-celled proembryo. C. Upper tier of nuclei dividing to form suspensors. D. Second tier elongating into suspensors. E. Suspensors and young embryo. (*Redrawn from* LAND 1902).

Embryo :- The embryo of *Thuja* is characterized by the activity of a prominent apical cell, which can be found even in well developed multicellular embryos. As in *Pinus,* the apical cell disappears before the tissue regions of the embryo are organized.

In the earlier stages of embryo development a tendency to form two groups of cells is apparent, especially in transverse sections of the embryo, but the line of division disappears as the embryo becomes older. There is no evidence that rosette cells occur in *Thuja.* Cleavage polyembryony may possibly occur occasionally, but this is predicated upon inference rather than on actual observation.

LAND, W. J. G. 1902. A morphological study of *Thuja.* Bot. Gaz. 36: 249–259.

Juniperoideae

The subfamily comprises the genera *Juniperus, Sabina* and *Arceuthos*. The last two have not been investigated.

Fertilization:- The two microgametes in *Juniperus* are equal in size and each apparently may enter a different archegonium. No other nuclei from a microgametophyte pass into an archegonium, nor is starch present. From one to occasionally more than three archegonia in each complex may be fertilized. The microgamete nucleus is either the same size or slightly smaller than that of the archegonium. The cytological aspects of fertilization have not been described.

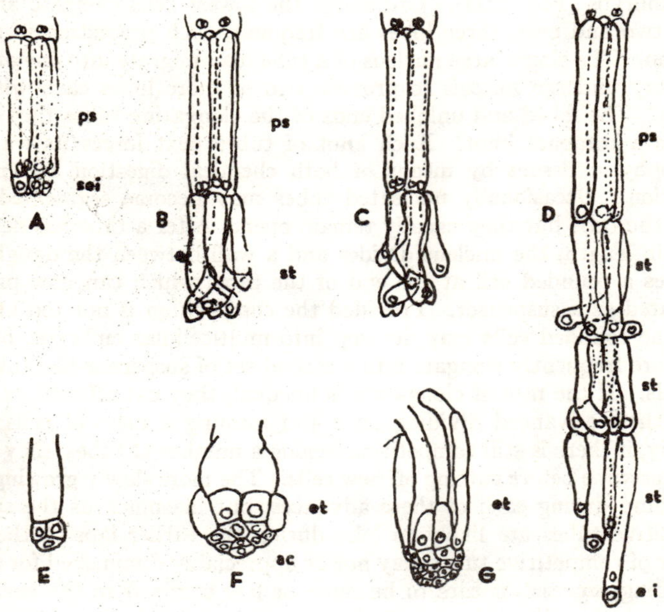

FIG. 22. — *Juniperus communis.* Diagrammatic representation of the embryogeny of the product of one fertilized archegonium. A. Proembryo of three tiers. B. Prosuspensor and terminal cells elongating. C. Terminal cells elongating, lobing and intertwisting; two tubes being left behind. D. Prosuspensor tier and three tiers of elongated tubes, competition reduced in last tier, terminal cells cut off and about to produce multicellular embryo. E. Most advanced cell of preceding becoming multicellular; apical cell prominent. F. Further development of preceding, suspensor tubes swollen and embryonal tubes forming secondary suspensor. G. Still further development, embryo becoming massive. Symbols: *e*, embryo; *ei*, embryo initial; *ac*, apical cell; *sei*, suspensor-embryo initial; *ps*, prosuspensor; $st^1, ^2, ^3$, suspensor tubes. (*Redrawn from* Cook 1939).

Proembryo:- The first division occurs in the middle of the lower half of the archegonium and two free nuclei are produced. Two more successive free-nuclear mitoses now occur. Cell walls are erected after the subsequent division, but not all nuclei divide so that the complete proembryo consists of about twelve cells arranged in three more or less definite tiers. The upper tier of cells remains open toward the apex of the archegonium and this tier does not participate in the further development of the proembryo. The cells of the middle tier elongate to produce the prosuspensors, and those of the lowest tier eventually give rise to embryos (FIG. 22A).

Embryo:- In *Juniperus communis* the two lower tiers of proembryonic cells now commence a developmental process which is unique among the Coniferophyta and approximated elsewhere only among the Ephedrophyta. Both tiers elongate similarly, thus indicating that each cell in each tier is a potential embryo initial (FIG 22B–D). The rate of elongation, however, may vary, so that the cells of the middle tier may sometimes push aside those of the lower tier. Nevertheless, the cells of the lowest tier, because of their more favorable position, reach more deeply into the nutritive tissue of the gametophyte. Those cells furthest advanced may immediately give rise to multicellular embryos (FIG 22D). Ordinarily the lowest cells elongate and become lobed; two and even three lobes are frequent, each projecting in a different direction. The single large nucleus of a tube may migrate into one of the lobes, or it may undergo mitosis to provide two or more lobes each with its own nucleus. The lobed and unlobed ends of the elongating tubes then intertwine to form a compact knot. Each knot of tubes next forces its way into the gametophytic tissues by means of both chemical digestion and mechanical elongation. Occasionally nucleated tubes may become segmented from the rest of the tube but they usually remain open. After a tube has developed to a certain length, the nucleus divides and a wall between the daughter nuclei produces a rounded cell at the end of the tube, which can now probably be designated as a suspensor. Provided the competition is not too keen at this time, the rounded cells may develop into multicellular embryos, but they in turn more frequently elongate into a second set of suspensor-like tubes lacking end cells. If the rate of elongation is unequal, they cut off cells at their tips, the farthest advanced dividing once and forming a mass of cells; but, customarily, if there is still competition among a number of tubes they again lobe and intertwine before cutting off new cells. The more slowly growing tubes are left behind during each of these advances, but inasmuch as the successfully competitive tubes are likely to lobe during the earlier laps of the race, the number of competitive tubes may not be appreciably diminished for some time. The lobing process appears to be more or less confined to the earlier sets of tubes, but the elongating-intertwining process may be repeated for an indefinite number of times, forming a fairly recognizable tier of tubes for each such occurrence.

After each set of new suspensor-tubes has developed, those preceding that set begin to collapse. The elongated cells that originally constituted the middle tier in the proembro and the next to the last formed tier of tubes remain fairly turgid, while the tubes last formed are extremely rounded and plump.

Eventually the keenness of competition becomes lessened and the tubes that have penetrated the farthest cut off terminal cells that grow into multicellular embryos. An apical cell soon becomes established, and the single suspensor-tubes become enormously swollen (FIG. 22F). When the embryo consists of about twelve cells, those cells on the outside and farthest from the apex elongate backward to form a massive secondary suspensor which pushes back any competitive embryos which may be behind it (FIG. 22G). Since the number of tiers of suspensor-tubes is indefinite, the time of formation of the multicellular embryo also varies; in a series of ovules collected from the same plant six months after fertilization it is therefore possible to find embryo complexes with embryos in all save the final stages of differentiation. There is thus no definite overwintering condition, although the embryo requires two growing seasons for attainment of maturity. The mature embryo normally has two cotyledons.

Cook, P. 1939. A new type of embryogeny in the conifers. Amer. Jour. Bot. 26: 138–143.
Nichols, G. E. 1901. A morphological study of *Juniperus communis* var. *depressa*. Beih. Bot. Centralbl. 25: 201–241.
Ottley, A. M. 1909. The development of the gametophytes and fertilization in *Juniperus communis* and *Juniperus virginiana*. Bot. Gaz. 48: 31–46.

Chapter VIII

SAXEGOTHAEACEAE

Both genera included in the family, *Saxegothaea* and *Microcachrys*, have been investigated.

Fertilization:- The archegonia of *Saxegothaea conspicua* are uncommonly long and narrow; they are broadest at the apex and become attenuated toward the base. Those of *Microcachrys tetragona* are short and not attenuated. In the latter species the body cell does not divide until just before fertilization; since the archegonia are grouped together within a common jacket, male gametes from the same microgametophyte may each enter a different archegonium. In *Saxegothaea conspicua* the archegonia are surrounded by individual jackets; one or both microgametes may enter an archegonium.

The actual fusion has not been observed in *Microcachrys tetragona*; in *Saxegothaea conspicua* the microgamete and the archegonial nucleus fuse completely. It is stated for the latter species that separate spindles could not be observed during the first mitosis.

Proembryo:- In *Saxegothaea conspicua* fertilization occurs near the upper and wider end of the archegonium and the two successive mitoses give rise to four free nuclei (FIG. 23A). These free nuclei move gradually to the base of the archegonium and there undergo a third free-nuclear division (FIG. 23B). The eight free nuclei are not arranged in definite positions, although one is located near the very base of the pointed archegonium. Each nucleus divides again, though it is not known whether the divisions are simultaneous, and walls are then formed. A certain degree of tier formation occurs, resulting in (*a*) an open tier of nuclei, numbering from three to five, (*b*) a tier of three to five cells which will presently form the prosuspensors, and (*c*) a conical group of embryonic cells numbering from six to ten (FIG. 23C). There is a single apical cell at the apex of the cone. The celled portion occupies but a very small part of the basal end of the archegonium. Some of the cells may later become binucleate (FIG. 23C, D).

For *Microcachrys tetragona* all that is known concerning the proembryo is that there are two tiers of cells and one tier with the upper ends open. The lowest tier consists of embryonal initials and the middle tier forms the prosuspensor.

Embryo:- Either normal embryogeny or various types of lobed embryos, embryonic buds and even occasionally double embryos may occur in *Saxegothaea conspicua*.

Normal embryogeny will be described first. When this occurs, cleavage is absent but simple polyembryony is fairly common. Not more than one of the embryos attains maturity.

At an early stage all the nuclei in the embryonic group, but not those in the prosuspensor tier, divide more or less simultaneously and as no walls are developed the cells become binucleate as in species of *Podocarpus* (FIG. 23D).

The prosuspensor cells at this time elongate rapidly; the cells commonly number four, but three and five have been observed. One or even two of the cells appear to be unable to keep pace with the others, then they cease to elongate

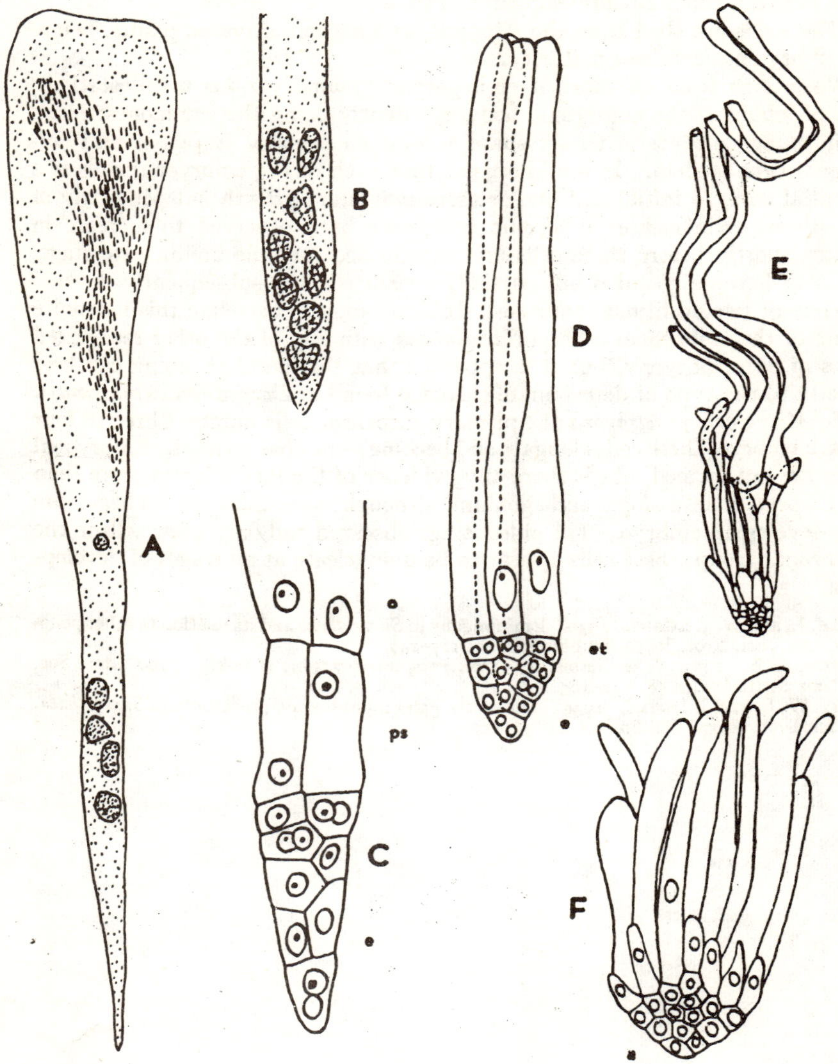

FIG. 23. — *Saxegothaea conspicua.* A. Four free nuclei moving to base of archegonium. ×90. B. Eight free nuclei in proembryo. ×320. C. Walled proembryo with open tier at top. ×320. D. Typical young proembryo. ×135. E. Typical older stage. ×55. F. Embryo cells and embryonal tubes. ×135. Symbols: *a*, apical cell; *e*, embryonic cells; *o*, open tier; *ps*, prosuspensor. (*Redrawn from* LOOBY and DOYLE 1939, and DOYLE and LOOBY 1939).

and tend to collapse. The nuclei of the open tier of cells degenerate early.

After the prosuspensor cells have become so long that they have begun to bend, a membrane makes its appearance between each of the pairs of nuclei in the binucleate embryonic cells and then further divisions are initiated

(FIG. 25F). At this stage an apical cell is generally recognizable. Embryonal tube formation commences at a comparatively early stage: a single belt of elongated tubular cells is formed, and to this are added other cells which grow between those already elongated (FIG. 25F).

The seeds are shed from the strobilus, at least in cultivated plants, before the embryo has attained maturity.

Variations from the normal embryogeny assume various manifestations, that of lobing of the embryonic body apparently being the commonest form. Such young embryos generally share a common primary suspensor and embryonal tube system. It would appear that each of the embryonal cells is a potential embryo initial and one or more may start growth independently of the others. Binucleate embryonic cells have been observed to increase in volume shortly before these cells are transformed into the uninucleate stage, and may become rounded off, partially separated, and subsequently develop into one or two additional embryos. It is not easy to correlate this indefinite lobing of the embryonal cells in *Saxegothaea* with any of the other recognized types of polyembryony, but it is apparent that this peculiar lobing can lead directly to the type of determinate cleavage found in *Dacrydium cupressinum*.

In *Microcachrys tetragona* the primary suspensor cells number three to four to each embryo; these cells elongate and become somewhat twisted. Embryonal tubes are not formed, nor is there any evidence of the formation of more than one embryo from a single archegonium, although more than one archegonium may develop an embryo. The oldest stage observed had only a few cells in the embryonal group; these cells appear to be uninucleate at all stages of development.

DOYLE, J., and W. J. LOOBY. 1939. Embryogeny in *Saxegothaea* and its relation to other podocarps. Sci. Proc. Royal Dublin Soc. 22: 127–147.

LAWSON, A. A. 1923. The life-history of *Microcochrys tetragona* (Hook.). Proc. Linn. Soc. New South Wales 48: 177–193.

LOOBY, W. J., and J. DOYLE. 1939. The ovule, gametophytes and proembryo in *Saxegothaea*. Sci. Proc. Royal Dublin Soc. 22: 95–117.

Chapter IX

PODOCARPACEAE

Podocarpus, *Dacrydium*, *Acmopyle* and *Phyllocladus* make up the family *Podocarpaceae*. All genera have been investigated, *Acmopyle* the least satisfactorily. *Podocarpus* is a rather large genus which exhibits remarkably diverse types of embryogeny.

Fertilization:- In *Phyllocladus* there are normally two archegonia, occasionally three or four; the number averages from three to five in some species of *Podocarpus*, but becomes as many as twenty in other species (*e.g.*, *P. nagei*). The peculiarity about fertilization in many species, notably *Podocarpus totarri*, is that it occurs after the seeds are shed from the plants. In most species it appears that only one of the two microgametes is functional; the other and smaller one soon degenerates. The microgametophyte also contains a variable number of prothallial and vegetative cells — usually six to eight in *Podocarpus* and from four to six in *Dacrydium* — as well as the tube and stalk nuclei.

In *Phyllocladus* the microgametophyte does not demolish the neck cells of the archegonium when its contents are discharged into the latter. The archegonial nucleus is surrounded by a dense mass of cytoplasm and more is contributed from the microgamete. The functional microgamete nucleus is about one-third the bulk of that of the archegonium and becomes appressed laterally. Fusion occurs either in the upper part of the archegonium or near the basal end. In most species of *Podocarpus* the functional microgamete, surrounded by a dense layer of cytoplasm, is about half the size of the archegonial nucleus; it becomes very much elongated as it flattens laterally against the latter.

Some details of the first mitosis are available for *Podocarpus andinus*. It is intranuclear, but the chromosomes and spindles are so extremely minute that it apparently could not be ascertained whether the spindles are separate. The two nuclei which result slip out from the remains of the fusion nucleus and move some distance toward the center of the archegonium.

Proembryo:- Eight free nuclei are formed in all species that have been investigated and walls are erected following the succeeding division. The first four nuclei are more or less linearly arranged in the lower half of the archegonium (FIG. 24A); they then move further toward the base and the eight nuclei next formed are irregularly disposed in the basal region of the archegonium (FIG. 24B). The organization of the proembryo after the appearance of walls is extremely variable and in *Podocarpus* each type is characteristic of a species or a group of a few species rather than of the genus as a whole. Separate discussions of the genera and of the species of *Podocarpus* are therefore necessitated.

In *Phyllocladus alpinus* the sixteen free nuclei may be arranged more or less in tiers of 5, 4, 3, 3, 1 or 6, 4, 3, 2, 1 cells, with the upper five- or six-celled tier remaining unwalled above. A nuclear division follows in each cell, the upper tier producing five or six prosuspensor cells and an equal number of relict nuclei (which soon become disorganized) in an unwalled tier. The nine to eleven walled cells below the prosuspensor tier each become binucleate and develop into an embryo after the prosuspensor elongates.

Podocarpus, sect. *Dacrycarpus:* Details of proembryo formation have not been worked out.

Podocarpus, sect. *Eupodocarpus:* (The species which COKER investigated is probably *P. macrophyllus* ssp. *maki* rather than the true *P. coriaceus*, a synonym under *P. urbanii.*) In *P. macrophyllus* ssp. *maki* the sixteen-nucleate proembryo is arranged in three tiers after wall formation is completed. The upper tier is not a definite rosette tier, although it has been designated as one: the cells are usually open above, more or less isolated, but an occasional cell may be completely but irregularly walled. There are about fourteen nuclei or cells in this tier. The next tier is that of the prosuspensor and usually consists of fourteen cells. The last tier ordinarily has only one binucleate cell but may

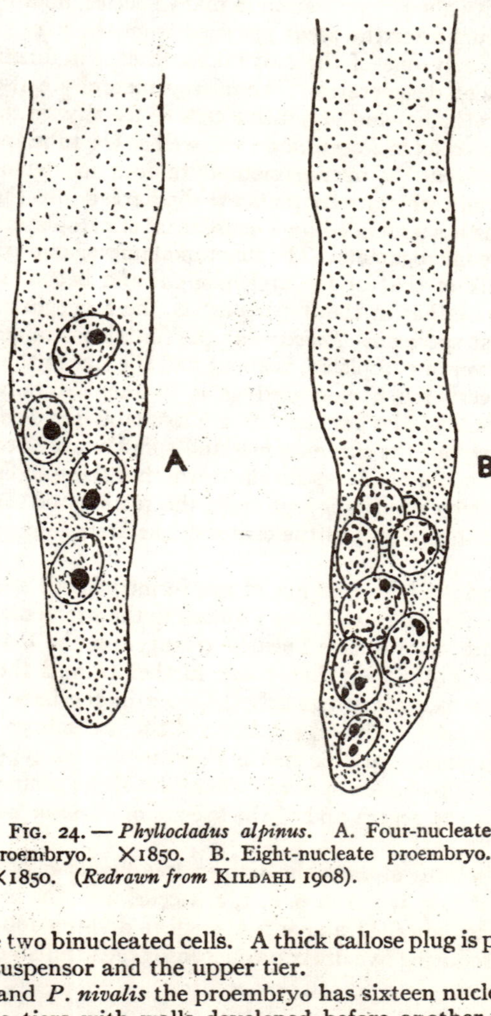

FIG. 24. — *Phyllocladus alpinus*. A. Four-nucleate proembryo. ×1850. B. Eight-nucleate proembryo. ×1850. (*Redrawn from* KILDAHL 1908).

occasionally have two binucleated cells. A thick callose plug is presently formed between the prosuspensor and the upper tier.

In *P. totarra* and *P. nivalis* the proembryo has sixteen nuclei which become arranged in three tiers with walls developed before another division occurs. The terminal embryonic unit is also usually a single binucleate cell; there are from seven to nine prosuspensor cells in the next tier; and the remaining nuclei, numbering less than those forming the prosuspensor, constitute a relict tier above the prosuspensor and soon disintegrate. There is no walled tier of rosette cells and no isolated nuclei become completely walled.

South American (*P. glomeratus*, *P. coriaceus*, *P. purdeanus*) and Mexican (*P. matudai*) species usually have twelve cells in their prosuspensors.

Podocarpus, sect. *Stachycarpus*. Species placed in this section (*P. gracilior*, *P. amarus*) have about twenty cells in their prosuspensors.

In *P. andinus* the four free nuclei migrate to the base of the archegonium where successive simultaneous divisions produce thirty-two free nuclei. Of these, as many as ten are not included in the subsequent cleavage formation of the embryo; they are designated as relict nuclei.

Podocarpus, sect. *Nageia*. In *P. nagei* the proembryo contains thirty-two nuclei before walls are formed. From seven to nine of the nuclei form a completely walled tier in the lower end of the archegonium; the rest of the nuclei become organized into another tier above the first tier, with the upper ends of the cells remaining open above. The next mitosis results in a tier of 23–25 prosuspensor cells, with an open tier of relict nuclei above. A rosette tier is not formed, nor are isolated rosette cells found. In the uninucleate cells of the lower tier, each nucleus divides into two free nuclei. This process has been followed out cytologically: the two nuclei in a binucleate cell are not derived from two of the thirty-two nuclei of the free-nucleate proembryo but originate from the single nucleus in an uninucleate cell.

A series of diagrams (FIG. 25) of the various species of *Podocarpus* and of *Dacrydium* and *Phyllocladus* will serve better to illustrate the comparative arrangements of the component cells of the proembryo. A diagram of the proembryo of the related *Saxegothaea conspicua* (*Saxegothaeaceae*) is also included.

Embryo:- Descriptions of the late proembryo and early embryo of *Phyllocladus* are somewhat unsatisfactory. The number of binucleate embryonic cells appears to vary from nine to twelve. The pair of nuclei in each cell divide nearly simultaneously and walls appear after the telophase stage. The binucleate stage thus passes directly into a four-celled condition, to produce a total of from forty to forty-four cells in the embryonic mass. It is extremely doubtful, despite apparent appearances to the contrary, whether an apical cell is present in *Phyllocladus*. The prosuspensor is comparatively short. Later embryonal tubes form a secondary suspensor. The mature embryo is dicotyledonous.

In *Dacrydium cupressinum* the prosuspensor of the young embryo is composed of seven to nine cells, and the embryonic initials at the tip number five to nine binucleated cells (FIG. 26A). Their arrangement varies somewhat. No rosette tier or cells are formed. The prosuspensor presently becomes greatly elongated, coiled and interwoven. A large deposit, described as a callose plug, is usually found immediately above the prosuspensor cells. At first the embryonic units do not increase in number. The transition from the binucleate to the uninucleate condition has not been described in detail, but presumably occurs by quadripartitioning after each nucleus in a given cell divides simultaneously. Quadripartitioning is plainly not simultaneous in all cells of an embryo since lobes with uninucleate cells are found in some regions of an embryo while other regions remain binucleate (FIG. 26C). The evidence as to the activity of an apical cell is somewhat inconclusive, but one may be present for a brief period. The largest terminal embryo presently develops a secondary suspensor by the elongation of embryonal tubes. These tubes are apparently formed in only a few layers (seemingly not more than two) and are

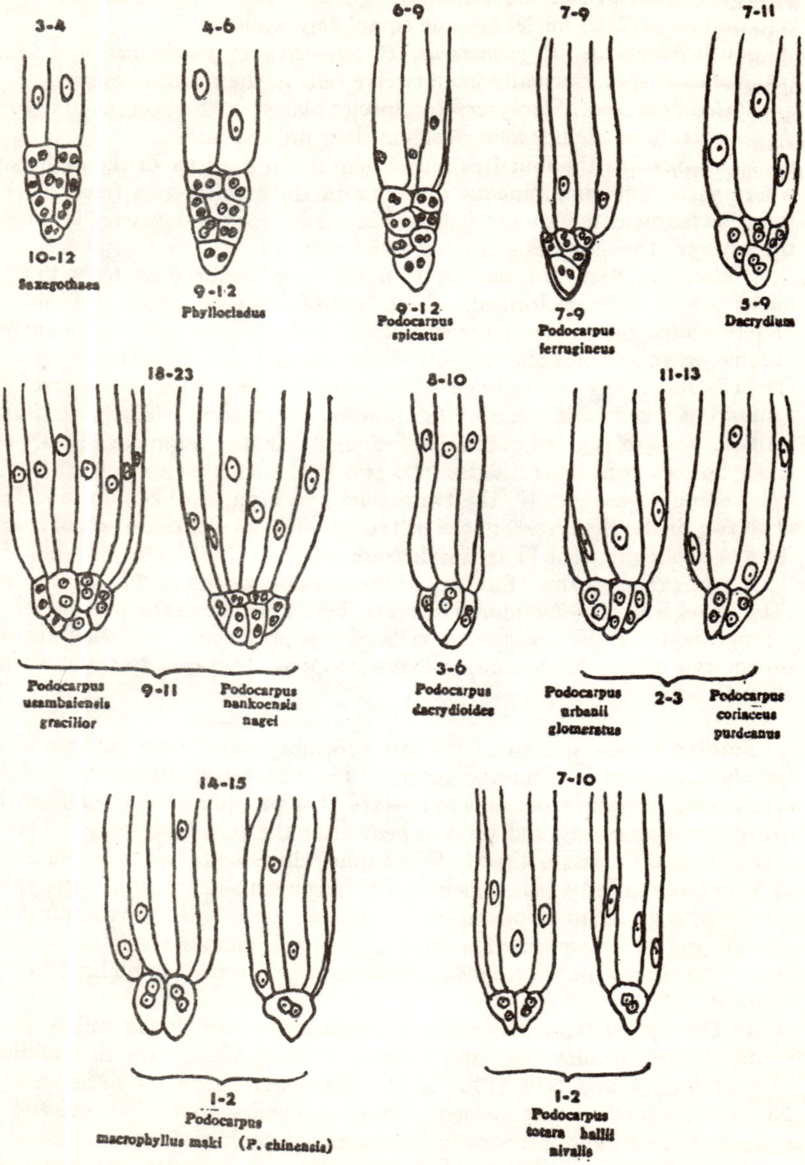

FIG. 25. — Diagrammatic comparison of the *Podocarpaceae*, and of *Saxegothaea*, *Phyllocladus* and *Dacrydium*, to show, by numerals placed above each figure, variations in the number of prosuspensor cells and, by numbers below each figure, variations in the number of binucleate cells. All diagrams are on approximately the same scale. Further description in text. (*Redrawn from* BUCHHOLZ 1941).

not as numerous as in many other Coniferales (FIG. 26D, E). Determinate cleavage polyembryony is the general rule in *Dacrydium*.

Podocarpus, sect. *Dacrycarpus*: *P. dacrydioides* and *P. imbricatus* are essentially similar in their embryogeny. The early embryo shows the seven or more prosuspensor cells well elongated, with a tier of relict nuclei above, and with usually four binucleate terminal cells which are considered to be actual or potential embryo initials. Rosette cells have been observed only rarely. The binucleate condition of the terminal cells, as indicated by the persistent spindle fibers commonly observed, is a long-delayed telophase. The prosuspensors elongate greatly, become much twisted, and bury the tips deeply in the gametophyte. Not all prosuspensor cells elongate equally and some may even cease development. After prosuspensor growth is nearly completed, walls are erected

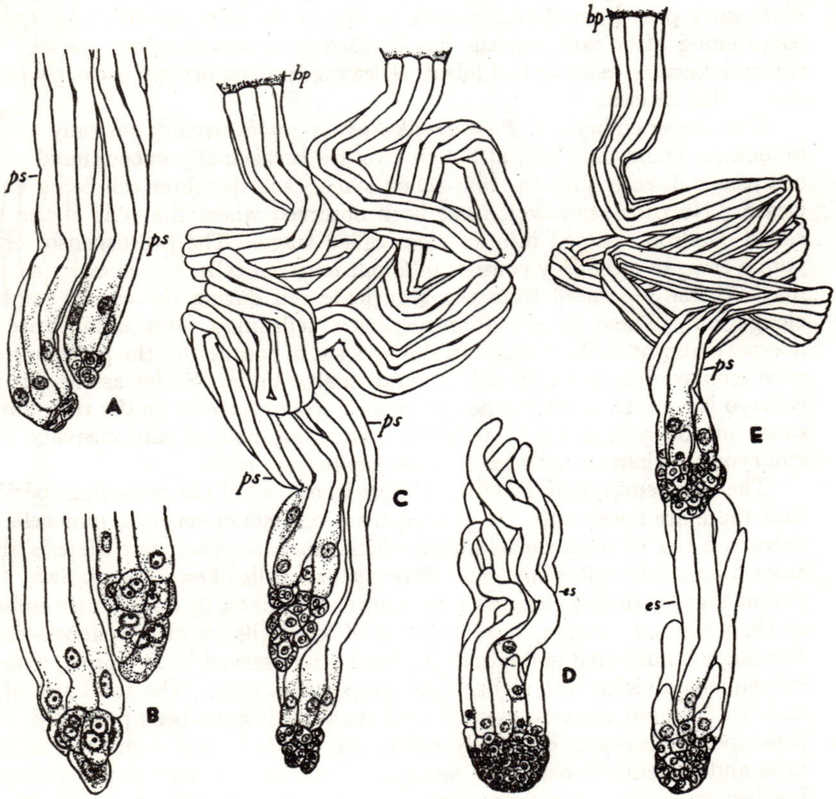

FIG. 26. — *Dacrydium cupressinum*. A. Young embryo system. ×75. B. Two adjacent embryo systems, with binucleate embryo initials. ×150. C. Two embryo systems showing characteristic convolution of prosuspensor. Some embryo initials are binucleate, others uninucleate. ×75. D. Early stage in formation of embryonal tubes. Embryo cells are all uninucleate. ×75. E. Later embryo with well developed embryonal tubes. ×75. Symbols: *ps*, prosuspensor; *es*, embryonal tubes. (*From* BUCHHOLZ 1933).

in the binucleate cells. Some of the embryonic units may become aborted, leaving five or less units to enter into proliferation of the individual embryos. They do not develop at the same rate. The later embryology has not been followed out, but the general situation resembles that in ***Dacrydium*** and cleavage polyembryony occurs.

Podocarpus, sect. *Eupodocarpus*: In ***P. urbanii*** the young embryo has a prosuspensor of twelve to fourteen cells, a terminal group of two or three binucleated cells, and rosette cells at the upper end which may develop into small embryos. Although a true primary suspensor is not formed in this species, many of the prosuspensor cells do not elongate fully and they may emerge from the tangle of elongated cells and later cut off at their tips one or more cells which develop into small secondary embryos. The main group of prosuspensor cells become greatly elongated, much convoluted and intertwisted and irregularly swollen. The embryonic initials remain in the quiescent binucleate condition during the period of prosuspensor elongation. On each embryo system derived from a single fertilized archegonium an average of fifteen embryos are to be found; *i.e.*, about one for each prosuspensor cell plus two or three derived from terminal binucleate embryonic cells. The embryonic initials early pass from the binucleate to the uninucleate condition by quadripartitioning after each nucleus has divided into two daughter nuclei. The embryos become massive and lobed. Cleavage polyembryony is clearly prevalent in this species.

The young embryo of *P. macrophyllus* ssp. *maki* usually has only a single binucleate embryonic cell, although two are occasionally encountered. This cell passes directly into the four-celled stage, each daughter cell being uninucleate. No cases, however, have been observed where the cells of the two-celled condition passed into an eight-celled stage. The prosuspensor of the young embryo ordinarily consists of about fourteen cells, but as many as eight may become separated from the binucleate cell during the earlier stages of elongation. These detached prosuspensor cells may form uninucleate abnormal embryos at their tips. In this species, in contrast to the preceding one, each embryo does not split into several smaller units. So far as the terminal embryo is concerned, simple polyembryony would seem to be the rule, but because of the occurrence of secondary embryos, determinate cleavage polyembryony is characteristic of *P. macrophyllus* ssp. *maki*.

The early embryo of *P. totarra* differs from that of the preceding species in that there are fewer cells; the prosuspensor consists of seven to nine cells and there is a tier of relict nuclei above which soon disappear, but there is also a single binucleate embryonic cell. Prosuspensor cells likewise break away from the main group during elongation but here they apparently do not form embryos at their termini. Again, the formation of four cells by quadripartitioning of the single binucleated embryonic cell has been observed in *P. totarra*, with the subsequent division into eight, then into sixteen, cells. The total size of the embryo does not increase greatly until many cells have been produced. The prosuspensor becomes extremely coiled and twisted. Later embryonal tubes arise and presently constitute a secondary suspensor. Cleavage polyembryony has become more or less suppressed in *P. totarra* so that simple polyembryony prevails.

The embryogeny of *P. glomeratus* is very similar to that of *P. urbanii*.

Podocarpus, sect. *Stachycarpus*: In *P. gracilior* and *P. amarus* the young embryo has about twenty cells in the prosuspensor and about ten binucleated cells at the tip. Each of the binucleate cells in species belonging to this section gives rise to a separate embryo provided it has not earlier aborted. This therefore indicates that cleavage polyembryony is strongly prevalent.

After cleavage in *P. andinus* there is an upper tier of from nine to fourteen cells open above to the archegonium and an irregularly tiered group of from eleven to fifteen cells with a large terminal cell. The upper tier segments to produce a prosuspensor tier and an impermanent open-cell tier. The single nuclei of the other cells each divide to present the typical binucleate stage which then enters into a peculiar resting stage during prosuspensor elongation. At a later stage double divisions in each cell followed by wall formation give rise to groups of four cells throughout the embryo. This is designated as the "tetrad stage." Subsequently, the embryo becomes homogeneous and develops further without cleavage polyembryony.

Podocarpus, sect. *Nagei*: Embryogeny in this section is similar to that in sect. *Stachycarpus*. The prosuspensor of the young embryo possesses from nineteen to twenty-five cells and the embryonic group has ten or eleven binucleated cells. Generally several rosette cells are present. One of the largest cells in the embryo, centrally located, becomes aborted; there are two very small cells in the interior, surrounded by prosuspensor cells above and by neighboring binucleate cells below.

BUCHHOLZ, J. T. 1933. Determinate cleavage polyembryony, with special reference to *Dacrydium*. Bot. Gaz. 94: 579–588.

BUCHHOLZ, J. T. 1936. Embryogeny of species of *Podocarpus* of the subgenus *Stachycarpus*. Bot. Gaz. 98: 135–146.

BUCHHOLZ, J. T. 1941. Embryogeny of the *Podocarpaceae*. Bot. Gaz. 103: 1–37.

COKER, W. C. 1902. Notes on the gametophyte and embryo of *Podocarpus*. Bot. Gaz. 33: 89–107.

HOLLOWAY, J. T. 1937. Ovule anatomy and development and embryogeny in *Phyllocladus alpinus* (Hook.) and in *P. glaucus* (Carr.). Trans. Proc. Roy. Soc. New Zealand 67: 149–165.

KILDAHL, N. J. 1908. The morphology of *Phyllocladus*. Bot. Gaz. 46: 339–348.

LOOBY, W. J., and J. DOYLE. 1944. Fertilization and early embryogeny in *Podocarpus andinus*. Sci. Proc. Royal Dublin Soc. 23: 257–270.

SINNOTT, E. W. 1913. The morphology of the reproductive structures in the *Podocarpaceae*. Ann. Bot. 27: 39–82.

STILES, W. 1912. The *Podocarpaceae*. Ann. Bot. 26: 443–514.

TAHARA, M. 1941. Embryogeny of *Podocarpus macrophyllus* and *Podocarpus nagei*. Sci. Rept. Tokohu Imp. Univ., Ser. IV, Biol. 14: 91–98.

YOUNG, M. S. 1910. The morphology of the *Podocarpineae*. Bot. Gaz. 50: 81–100.

Chapter X

PHEROSPHAERACEAE

The one genus, *Pherosphaera*, has often been placed under the *Podocarpaceae*, but it shows no close relationship to any species in that family and is best included in a family by itself.

Two species are known: *P. fitzgeraldi* and *P. hookeriana*. Both have been investigated more or less and are similar embryologically.

Proembryo:- Of the three or four archegonia, two or three may be fertilized and each produces an embryo. The preliminary stages in proembryo formation were not found. The earliest described stage is that of two embryos, apparently from separate archegonia, of about three cells each at the ends of long, twisted suspensors. A later stage reveals considerable growth in one of two embryos, with the other only slightly developed and apparently in process of degeneration.

LAWSON, A. A. 1923. The life-history of *Pherosphaera*. Proc. Linn. Soc. New South Wales 48: 499–516.

Chapter XI

CEPHALOTAXACEAE

Of the two genera which comprise the *Cephalotaxaceae*, *Cephalotaxus* has been well studied, but *Amentotaxus* remains uninvestigated.

Fertilization:- The archegonia of *Cephalotaxus*, which usually number four, are narrow, attenuated and sharply pointed at the lower end. Accounts differ as to the structure of the microgametes: those of *C. fortunei* have been described as uninucleate cells, while those of *C. drupacea* are naked nuclei which are discharged into an archegonium while still remaining within the original membrane of the body cell. These nuclei are presently released; one is larger than the other but about a third the bulk of the archegonial nucleus. The celled microgametes are also very unequal in size and only the larger one is functional. How the nucleus is released has not been described, but the cytoplasm of the microgamete envelops the fusing nuclei. In the case of the naked microgametes, nothing else is carried into the archegonial cytoplasm. In both species fusion occurs in the center of the archegonium. The first division has not been accurately described.

Proembryo:- Two free nuclei are produced as a result of the first mitosis (FIG. 27A); they travel a short distance toward the base of the archegonium and there divide into four free nuclei which thereupon move the remainder of the way to the base. A third mitosis next ensues (FIG. 27B), followed by a fourth after which cell walls are erected. The cells are not arranged in clearly defined tiers (FIG. 27C); the irregularity of disposition is most pronounced in the central regions. Sometimes there may be three indefinite tiers, sometimes four. The proembryo of *Cephalotaxus* differs from those of related Coniferophyta in that there are no free nuclei above nor cells with open upper ends, and further in that there is usually a single large terminal or tip cell (FIG. 27C–E). There are several cells above this terminal cell, which may have been produced by the latter, that remain large and mitotically inactive.

The cap, slightly reminiscent of that in the *Araucariaceae*, but scarcely as clearly organized, is probably in the nature of a secreting organ and becomes deciduous soon after the prosuspensor is developed. The nuclei are comparatively large. The terminal cell of the cap occupies the position of the apical cell in other genera: the embryo proper has no sharply defined apical cell, although one is depicted in several of STRASBURGER's figures. The upper tier of the cap consists of five or six cells not much larger than the cells of the embryo proper. Since cleavage polyembryony has been almost entirely suppressed in *Cephalotaxus*, it has been conjectured that "the organization of this cap was partially, if not largely, instrumental in the elimination of cleavage polyembryony during the early stages." The cap was probably not present in ancestral forms; hence its appearance may be associated with the suppression of both the apical cell and cleavage polyembryony.

The cells of the embryonic region, each of which is doubtless a potential embryo initial, multiply rapidly and as many as sixteen or thirty-two cells may

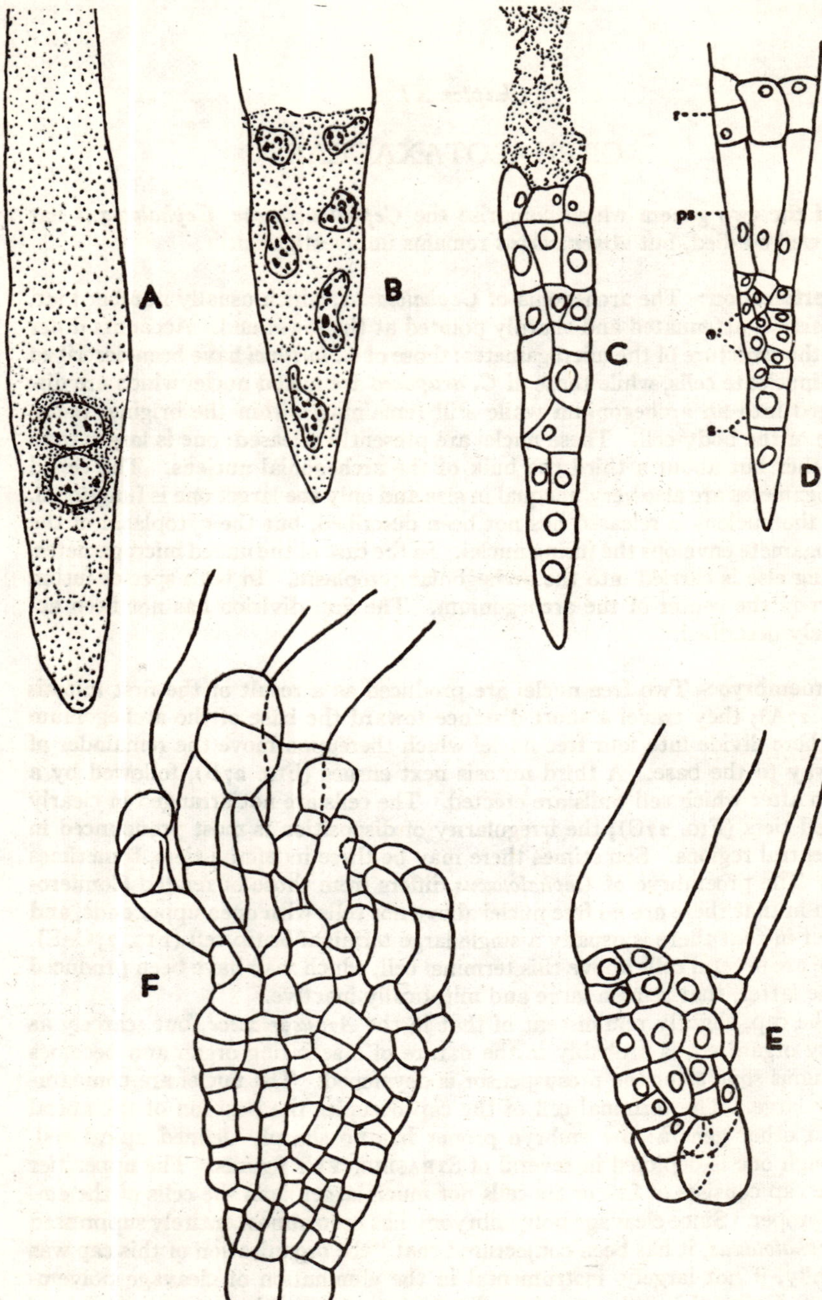

FIG. 27. — *Cephalotaxus drupacea*. A. Two-nucleate proembryo. B. Eight-nucleate proembryo. *Cephalotaxus fortunei*. C. Thirty-two-celled proembryo just before elongation of suspensor cells. ×185. D. Slightly older proembryo; suspensor beginning to elongate. ×150. E. Early embryo before formation of embryonal tubes. ×230. F. Young embryo with embryonal tubes. ×230. Symbols: *e*, embryo; *r*, rosette tier; *ps*, prosuspensor; *s*, sterile cells. (A, B *redrawn from* LAWSON 1907; C, D *redrawn from* COKER 1907; E, F *redrawn from* STRASBURGER 1879).

be formed before the cells of the prosuspensor region commence elongation.

The prosuspensor cells then enter into a period of rapid elongation. At first they elongate in a straight line but become gradually more and more curved or convoluted as growth proceeds.

Embryo :- The embryo soon becomes multicellular and presently embryonal tubes are developed (FIG. 27E, F). The layer of embryo cells immediately adjacent to the prosuspensor first elongate backwards, followed by tubes produced by the second layer of embryo cells, then by the third layer and so on in a regular series. Several of the early layers of embryonal tubes become very much elongated, but later series of tubes become progressively shorter. The tubes together constitute a massive secondary suspensor. This secondary suspensor is the means whereby one embryo gains the ascendancy over those from other archegonia and is the one to attain maturity.

Another feature distinguishing the embryo of *Cephalotaxus* is that the stem tip is the last embryonal region to be differentiated. The mature embryo possesses two cotyledons.

Rosette Embryos :- There is a meristematic rosette tier which is quite active, giving rise to a number of embryos. These embryos lack the terminal cap characteristic of the primary embryo and have no suspensor cells. After they have become multicellular embryonal tubes are developed, thus demonstrating that they are true embryos. It therefore appears that the embryo of *Cephalotaxus* is a compound structure.

BUCHHOLZ, J. T. 1925. The embryogeny of *Cephalotaxus fortunei.* Bull. Torrey Bot. Club 52: 311–423.

COKER, W. C. 1907. Fertilization and embryogeny of *Cephalotaxus fortunei* Bot. Gaz. 43: 1–40.

LAWSON, A. A. 1907. The gametophytes, fertilization and embryo of *Cephalotaxus drupacea.* Ann. Bot. 21: 1–23.

STRASBURGER, E. 1897. Die Angiospermen und die Gymnospermen. Jena, Fischer.

Chapter XII

TAXACEAE

Three genera are included in the *Taxaceae*: *Taxus*, *Austrotaxus* and *Torreya* (*Tumion*); all have been investigated but none too thoroughly.

Fertilization:- Accounts are not very satisfactory from the cytological standpoint and some of the older investigations require confirmation in details.

The microgametophytes of the several species differ considerably in their relative stages of development at the time they have reached the archegonia. This is most marked in *Torreya*: in *T. taxifolia* the microgametophyte is fully developed when it reaches the megagametophyte while the latter is still at the free-nuclear stage of growth, whereas in *T. californica* the microgametophyte has grown only about halfway through the nucellus at the time the archegonia are mature. The number of archegonia is small in all genera (usually three in *T. californica*, solitary in *T. taxifolia*, three in *T. nucifera*, about eight in *Taxus canadensis*, ordinarily four in *Austrotaxus spicata*) and they are always separate and not localized in complexes. The mature archegonia in *Torreya* are extraordinarily small, those of *Taxus* a trifle larger; but the archegonia of *Austrotaxus* are twice as large and comparable in size with those of *Cephalotaxus* and related genera. Many ovules in at least two species, however, are sterile; in *Torreya nucifera*, for instance, ten out of fifty ovules completely lacked megagametophytes. In certain species, again, more than one megagametophyte may be present; this means that more than one of the several megasporocytes may have become functional. In *Taxus canadensis* as many as five megagametophytes have been observed, but not all supernumerary megagametophytes contain archegonia. Fertilization is not restricted to the archegonia in the upper megagametophytes; they may be by-passed by the microgametophytes and archegonia lower down are fertilized.

The time of fertilization is apparently inconstant and covers a considerable range of time. Generally a given archegonium is fertilized by a single microgametophyte, only one of whose two microgametes may be functional. The condition in which the microgamete enters the archegonium is variable according to the species: in *Taxus canadensis* the microgamete is still invested with its cytoplasm, contacts the archegonial nucleus at the basal end of the archegonium, and the cytoplasm invests the fusion nucleus; conditions are similar in *Torreya taxifolia* save that the fusion nucleus is located in the center of the archegonium; *Torreya californica* resembles *T. taxifolia*; in *Torreya nucifera* the nucleus is extruded from the microgamete in the form of an hourglass and the fusion nucleus is slightly below the center of the small archegonium. In all cases the male nucleus is much smaller than that of the archegonium.

No accurate cytological details of the first mitosis have been described.

Proembryo:- The two free nuclei resulting from the first mitosis are, in most species, surrounded by a dense cytoplasmic sheath derived principally from the microgamete in some species, but formed in the archegonial cytoplasm

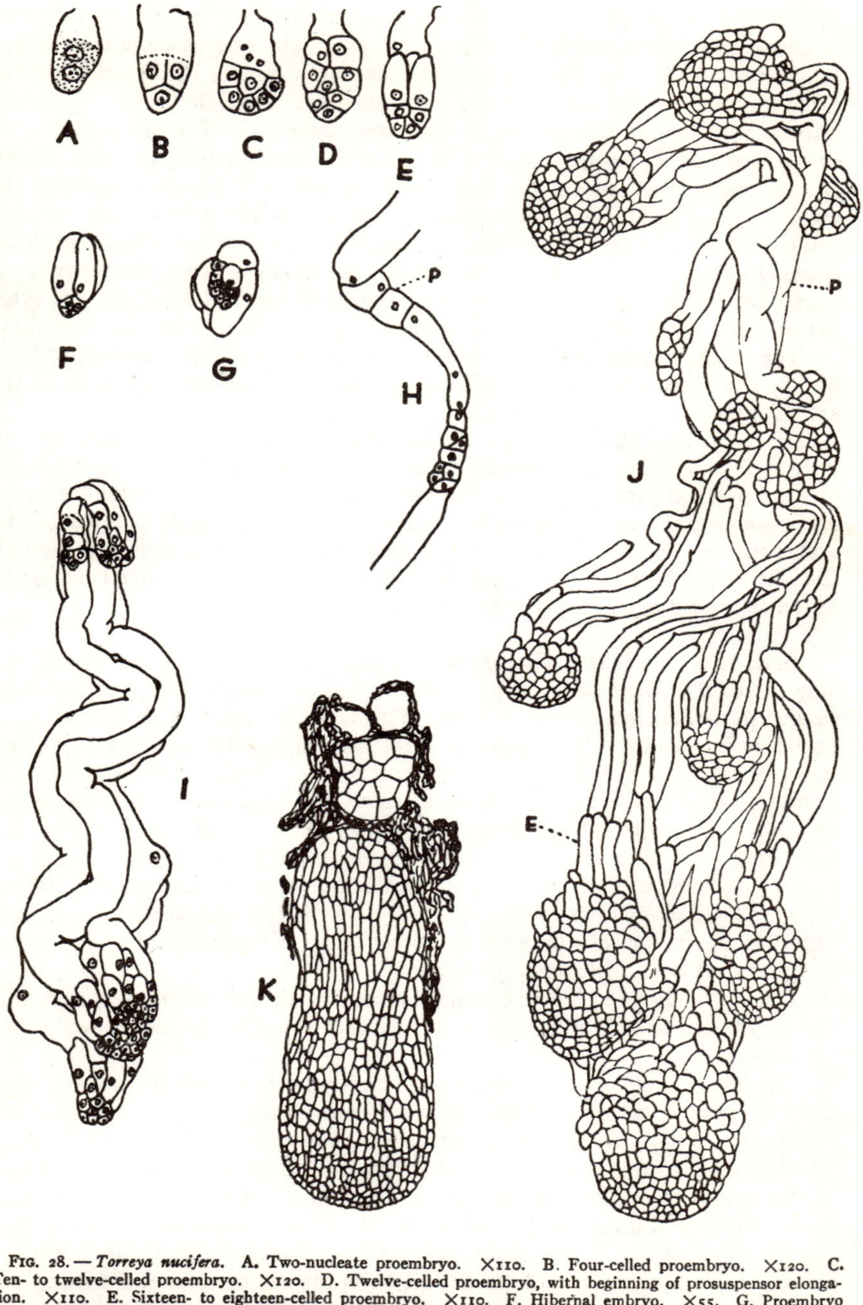

FIG. 28.—*Torreya nucifera*. A. Two-nucleate proembryo. ×110. B. Four-celled proembryo. ×120. C. Ten- to twelve-celled proembryo. ×120. D. Twelve-celled proembryo, with beginning of prosuspensor elongation. ×110. E. Sixteen- to eighteen-celled proembryo. ×110. F. Hibernal embryo. ×55. G. Proembryo with cell arrangement disturbed so that embryonic cells are lateral to prosuspensor cells. ×55. H. A single prosuspensor cell from an older embryo system which has given rise to embryonic masses. ×55. I. Early embryo system with prosuspensors elongated. ×45. *Torreya taxifolia*. J. Longitudinal section of embryo on secondary suspensor with remains of smaller embryos resulting from cleavage polyembryony. ×60. *Torreya californica*. K. A complete embryo system; the prosuspensors are becoming embryonic, and the embryonal tubes are well developed. ×55. Symbols: *e*, embryonal tubes; *p*, prosuspensor. (*Redrawn from* BUCHHOLZ 1940).

in those cases where the nucleus of the microgamete escapes some time before fusion (FIG. 28A). In *Torreya taxifolia* the nuclei remain in the center of the archegonium; in the other species they are near or at the base. A second mitosis next occurs, giving rise to four free nuclei. In *Torreya taxifolia*, *T. californica* and *T. nucifera* (FIG. 28B) walls are formed after this mitosis, although one account states that cell wall formation in *T. nucifera* begins at the eight-nucleate stage. In *Austrotaxus spicata* about half of the nuclei become walled during the transition from the eight-nucleate to the sixteen-nucleate stage (FIG. 29A). After the succeeding mitosis in this species, the lower seven nuclei are walled and the remaining nine nuclei lie free in a small mass of cytoplasm (FIG. 29B). A somewhat similar arrangement occurs in *Taxus baccata*.

The young proembryo completely fills the archegonium in *Torreya taxifolia*, but in the other species it occupies only the lower half or two-thirds of the archegonium.

After the four-nucleate stage, the divisions are not always regular, particularly in *Torreya*: there may occur six-celled, eight-celled and twelve- to fourteen- or sixteen-celled proembryos. If the proembryo does not already fill the entire archegonium, it now elongates to occupy the unfilled upper portion of the archegonium. Except in *Austrotaxus*, growth now becomes arrested through the late autumn, winter and early spring. This resting stage has been designated as the hibernal embryo (FIG. 28F).

Embryo:- In *Austrotaxus spicata* at the transition from the proembryo stage to that of the embryo the structure consists of about thirty-two cells arranged somewhat indefinitely in tiers. This arrangement may be typical but is probably inconstant. The upper tier has been designated as the rosette tier, but as rosette embryos have not been observed, it is doubtful if this is actually a rosette tier; it may be recalled that in other Coniferophyta — *e.g.*, *Podocarpus spicatus* — a rosette is variable in occurrence. The second tier gives rise to a very long prosuspensor (FIG. 29C). The remaining cells consist of embryo nitials, arranged in four tiers of 5, 5, 3 cells and a single terminal cell respectively. The latter is not in the nature of a sterile cap cell, in which respect it differs from similar cells in other Coniferophyta. The embryogeny of *Austrotaxus* has not been investigated further.

The embryogeny of *Taxus baccata* is much like that of *Austrotaxus* save that the upper tier of cells degenerates at an early stage in the elongation of the prosuspensors.

The hibernal embryo of *Torreya* is distinctly past the true proembryo stage since the prosuspensor cells have already elongated slightly when the structure begins the inactive period lasting until the following spring. In *T. nucifera* the hibernal embryo displays various arrangements of its component cells; many of the apparently abnormal cell orientations are due to accidental displacements suffered during the period when the gametophyte envelops the proembryo (FIG. 28G). Many of the irregularities observed during later developmental stages are attributable to variations in the cell number and orientations in the hibernal embryo. The early embryo shows the prosuspensor cells greatly elongated and unequally swollen. At the upper end is to be found a group of cells in the position of rosette embryos but they are not actually such since there is usually no rosette tier. They probably were derived from some of the cells of the hibernal embryo which did not elongate in the uppermost tiers of the prosuspensor. Sometimes these embryos appear to be oriented in the direction of the former apex of the archegonium. Again, single prosuspensor cells on older embryo systems may internally develop embryo-like masses (FIG. 28H).

Apical cells have not been discerned in any of these embryos. One account states that rosette embryos are very common, but this is doubtful. After the terminal embryo has produced a considerable number of cells embryonal tubes make their appearance immediately below the connection of the embryo to the prosuspensor (FIG. 28E, I). These embryonal tubes elongate backward, in several successive layers, and form massive secondary suspensors. Only one embryo reaches maturity in each ovule; it is not fully developed when the seed is ripe but undergoes considerable growth within the seed after it has fallen (FIG. 28K).

In *Torreya californica* the elongating prosuspensor fills the entire archegonium and crushes the upper free nuclei. The hibernal embryo has not been studied. There are from four to seven cells in the prosuspensor and these ex-

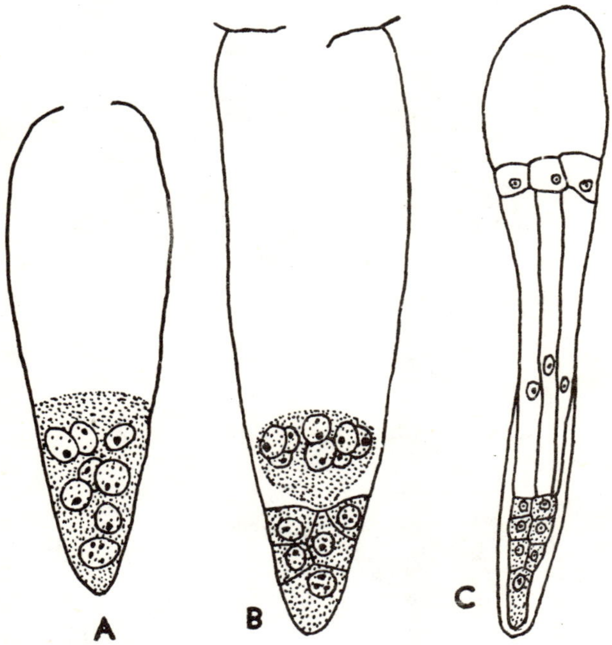

FIG. 29. — *Austrotaxus spicata.* A. Eight-nucleate proembryo. ×220. B. Sixteen-nucleate proembryo; the lower seven nuclei have formed walls and the upper nine are free in a small mass of cytoplasm. ×220. C. Young embryo with prosuspensor and upper tier. ×125. (*Redrawn from* SAXTON 1934).

hibit a peculiar nuclear behavior. There are a number of free-nuclear mitoses, followed by wall formation after a considerable number of such nuclei have been produced, then cell walls appear after each nuclear division. Misshapen and abnormal embryos are presently formed in this unusual manner. A series of embryonal tubes soon become produced on the terminal embryos to form massive secondary suspensors. Before the cotyledons appear, the developing embryo has the shape of a cylinder. Cleavage polyembryony is prevalent. It appears that each fertilized archegonium may produce at least as many embryos as there are cells in the young proembryo and occasionally as many as there are cells in the hibernal embryo. In addition, as mentioned above, the prosuspensor cells add more embryos. There are two cotyledons on the mature embryo.

Conditions in *Torreya taxifolia* are essentially similar to those in *T. nucifera*. There are likewise two cotyledons on the mature embryo.

BUCHHOLZ, J. T. 1940. The embryogeny of *Torreya*, with a note on *Austrotaxus*. Bull. Torrey Bot. Club 67: 731–754.
COULTER, J. M., and W. J. G. LAND. 1905. Gametophytes and embryo of *Torreya taxifolia*. Bot. Gaz. 39: 161–178.
DUPLER, A. W. 1917. The gametophytes of *Taxus canadensis* Marsh. Bot. Gaz. 64: 115–136.
JÄGER, L. 1889. Beiträge zur Kenntniss der Endospermbildung und zur Embryologie von *Taxus baccata* L. Flora 86: 241–288.
ROBERTSON, A. 1904. Studies in the morphology of *Torreya californica*. II. The sexual organs and fertilization. New Phytol. 3: 205–216.
SAXTON, W. T. 1934. Notes on conifers. VIII. The morphology of *Austrotaxus spicatus* Compton. Ann. Bot. 48: 411–427.
TAHARA, M. 1940. Embryogeny of *Torreya nucifera*. Bot. Mag. Tokyo 54: 469–472. (In Japanese.)

EPHEDROPHYTA

Chapter XIII

EPHEDRACEAE

The family consists of only one genus, *Ephedra*.

Fertilization:- When the pollen grain of *Ephedra trifurca* begins to germinate, the two prothallial cells soon disappear and the body cell divides. The tube nucleus and stalk cell at first lie at opposite ends of the pollen grain, then move through 90° and come to lie side by side. The tube nucleus comes to rest against the wall and soon afterward the pollen tube protrudes from this point of contact. The two microgametes are of equal volume and truly elliptical in shape. One of them almost invariably enters the microgametophyte ahead of the stalk cell, which takes a position in advance of the second microgamete. The microgametophyte forces its way between the neck cells of the archegonium, rarely destroying them. At about this time, the stalk nucleus and tube nucleus disappear. The second microgamete usually remains in the upper part of the archegonium and later the region it occupies is filled with a number of small cells which may have been derived either from that nucleus or jointly from the jacket cells. However, this explanation is probably erroneous: to judge from what happens in *E. campylopoda*, which was investigated at a much later time, it is probable that the ventral canal nucleus in *E. trifurca* fused with one of the microgametes and thereupon gave rise to a number of cells. This mass of cells is ephemeral and is quickly absorbed by the functioning proembryo.

As in *E. trifurca*, there are two microgametes, equal in size, in *E. altissima*. They increase in size as they migrate down the microgametophyte. No cytoplasm was observed to invest the microgamete which comes into contact with the upper end of the more or less pointed archegonial nucleus, although a zone of radiating strands of cytoplasm is described as appearing around the zygote nucleus.

In *E. campylopoda* the microgametophyte grows down through the micropylar canal and may force the neck canal cells ahead of it into the archegonium. Both microgametes enter an archegonium; one fuses with the ventral canal nucleus and the other with the archegonial nucleus. This phenomenon is the equivalent of "double fertilization" in the Anthophyta, although the equivalent of an endosperm is not produced. Moreover, nuclei from the cells surrounding the archegonium may become freed to migrate into the latter and the union of the archegonial nucleus with such a free nucleus was actually observed in an unfertilized archegonium. Whether such an occurrence can give rise to an embryo is not known.

That microgamete in *E. campylopoda* which fuses with the archegonial nucleus is from one-third to one-fourth the bulk of the latter. The former may come into contact with the latter at almost any point, but this is usually near the upper end. The actual details of fusion are obscure, though it appears to be probable that separate groups of maternal and paternal chromosomes are resolved.

Fusion of the ventral nucleus with the second microgamete has been described in *E. foliata*.

Proembryo:- The zygote of *E. trifurca* gives rise to eight free nuclei more or less unequal in size. From three to five of the largest of these nuclei become surrounded by walls. The cells at first are very irregular in shape but presently become more or less globular. The nuclei which remain unwalled are either arranged in a row down the center of the archegonium or are scattered throughout the latter. In general the lower walled cells successfully produce embryos, although it was noted that in a few instances the lower ones did not progress further than the one-celled stage, the micropylar ones alone being successful.

The nucleus of a proembryonal cell divides into two free nuclei which are equal in size at first but one soon becomes noticeably larger than the other (FIG. 30A). The nuclei migrate to the lower end of the cell, with the larger of the two invariably placed below the other. Next two papillae, presumably initiated by the nuclei in some inexplicable manner, begin to project from the cell at the points nearest the nuclei. Presently the papilla closest to the larger nucleus rapidly elongates and the other papilla becomes obliterated. The developing papilla is a suspensor tube; if it is from the cell that is located lowermost in the archegonium, it is immediately directed downward; but outward and then downward if it projects from any of the upper cells. The larger nucleus then passes down nearly to the end of the suspensor tube. Immediately behind it a cleavage ring, commencing at the wall of the suspensor tube and gradually extending inward, appears in the cytoplasm and separates the tube into two portions, each with a nucleus (FIG. 30B). A wall is established at the cleavage plane. The nucleus left behind passes down the suspensor tube to the lower end and there eventually disintegrates. The suspensor elongates rapidly, thrusting the proembryonic initial deeply into the gametophytic tissues, and remains turgid until the secondary suspensor is produced, after which it collapses (FIG. 30C–E).

In *E. foliata* there is but one suspensor tube and the nucleus of the proembryonal cell does not divide until after the tube has been formed and the former moves down to the tip of the latter.

In *E. foliata* the suspensor tube is not very long; in *E. trifurca* and *E. distachya* it elongates considerably, and is so long in *E. altissima* that its course is difficult to trace. In the latter species a second transverse wall may occasionally be erected in the suspensor tube. The cell at the tip of the suspensor tube segments transversely into a basal cell and a terminal cell. In *E. trifurca* the basal cell segments first, by a longitudinal wall (FIG. 30E); in *E. foliata* the second wall is longitudinal in the terminal cell, and in *E. distachya* the terminal cell divides by two transverse walls. These features appear to be specific diagnostic characters of prime importance.

Subsequent events have been followed out most closely in *E. trifurca*. The third division is longitudinally diagonal in the terminal cell (FIG. 30F) and divides it into two unequal cells (FIG. 30G). The succeeding division is oblique in the larger of the two daughter cells (FIG. 30H), resulting in a lower cell and an upper triangular cell which has every appearance of being an apical cell. This procedure is decidedly reminiscent of the origin of the epiphysis initial in certain angiospermic genera. The apical cell can be traced through favorable

FIG. 30. — A–L. *Ephedra trifurca*. A. Proembryonic cell with two free nuclei and commencement of suspensor tube development. B. After wall formation in suspensor tube. C, D. Transverse segmentation of proembryonic initial and elongation of suspensor tube. E. Vertical partitioning of upper daughter cell. F. Mitosis after which oblique wall will be formed in lower daughter cell. G. Formation of oblique wall completed. H Oblique wall formed in larger cell resulting from preceding division (at lower left); the triangular cell at apex is the apical cell. I. After transverse segmentation of two upper cells (apical cell not yet delimited). J. Further growth of proembryo; apical cell is still visible. K. Slightly oblique section of developing proembryo; apical cell not apparent. L. Formation of periclinal walls in apical cell and adjoining cell (a secondary suspensor cell, without nucleus, is at the top). A–C, × 950; D–L, × 470. (*All figures original*).

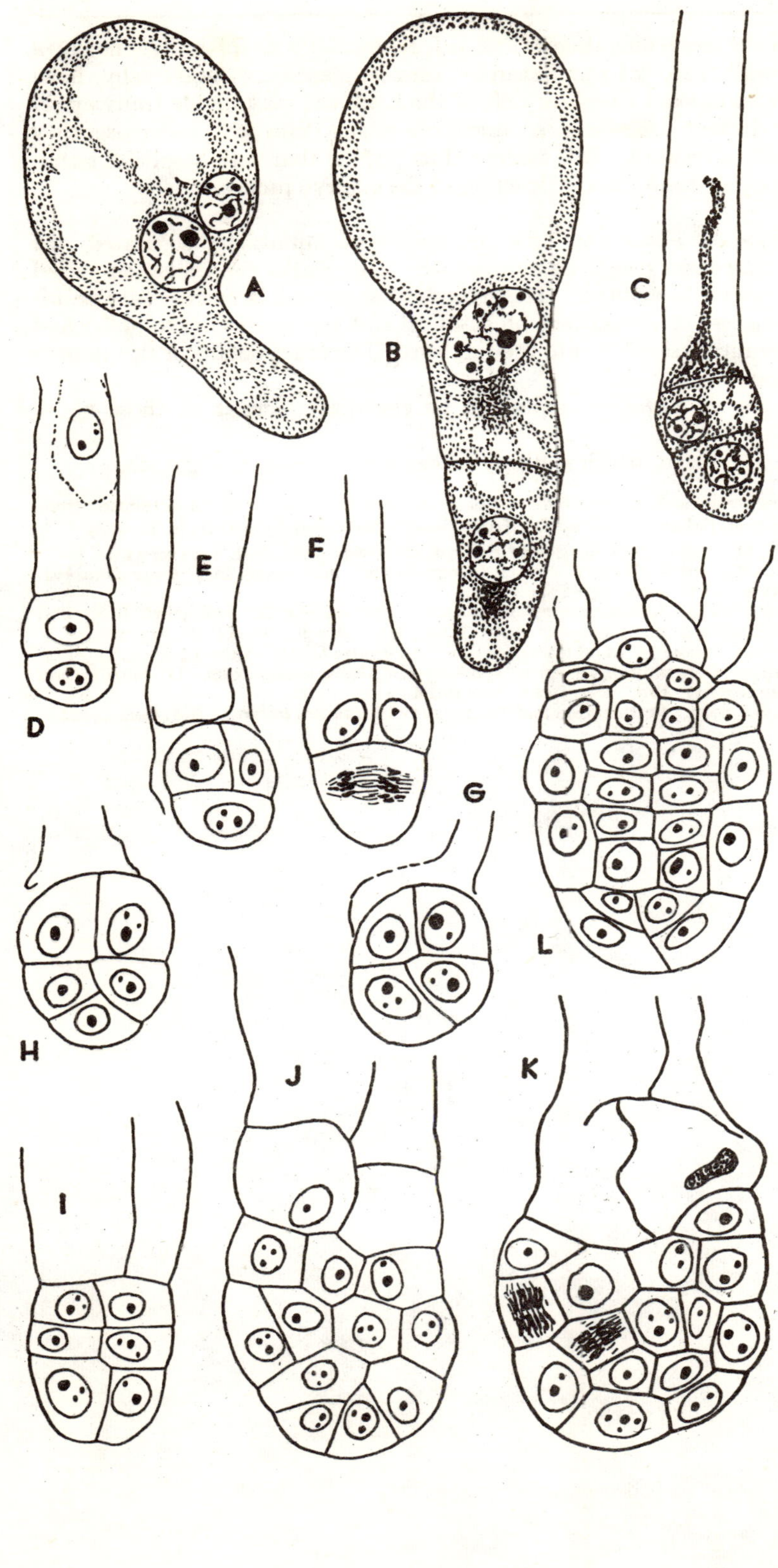
A
B
C
D
E
F
G
H
I
J
K
L

sections of the succeeding stages (FIG. 30J, K) until it is finally partitioned by a periclinal wall (FIG. 30L) and thereby loses its identity. Occasionally, however, both juxtaposed daughter cells of the basal cell may divide transversely before the division delimiting the apical cell occurs (FIG. 30I); otherwise, they occur shortly afterward. It is evident (FIG. 30J–L) that both basal and terminal cells contribute to the construction of the embryo proper.

Embryo:- At about the time the epidermal initials are separated, the secondary suspensor makes its appearance. This marks the transition period from proembryo to embryo. The secondary suspensor is erected by the addition of successive elongations of basal cells of the embryo (FIG. 30L) and finally it merges insensibly into the root cap. The exact details of this growth are not clear.

Events in *E. distachya* appear to be essential similar to those in *E. trifurca*.

Only one embryo, which is dicotyledonous, matures in a single seed.

The observations on the embryogeny of *E. trifurca* are original; slides were made from the identical material employed by Dr. LAND (kindly furnished by Dr. H. S. JURICA).

BERRIDGE, E. M. 1909. Fertilization in *Ephedra altissima*. Ann. Bot. 23: 509–512.
BERRIDGE, E. M., and E. SANDAY. 1907. Oogenesis and embryogeny in *Ephedra distachya*. New Phytol. 6: 127–134, 167–174.
HERZFELD, S. 1922. *Ephedra campylopoda* Mey. Morphologie der weiblichen Blüte und Befruchtungsvorgang. Denkschr. Acad. Wiss., Wien, Math.-nat. Kl. 98: 243–268.
KHAN, R. 1940. A note on "double fertilization" in *Ephedra foliata*. Current Sci. 9: 323–324.
KHAN, R. 1943. Contributions to the morphology of *Ephedra foliata* Boiss. II. Fertilization and embryogeny. Proc. Nat. Acad. Sci., India 13: 357–375.
LAND, W. J. G. 1907. Fertilization and embryogeny in *Ephedra trifurca*. Bot. Gaz. 44: 273–292.

Chapter XIV

GNETACEAE

The family includes one living genus: *Gnetum*. There is a certain degree of confusion between the accounts of the embryogeny, particularly in the older ones, unless it be that misinterpretations have been made in some respects. The species appear to differ considerably in details; this is so pronounced, in fact, that a thoroughgoing re-examination of as many species as possible in the light of modern knowledge of Gymnosperm morphology is highly desirable.

Fertilization:- In *Gnetum* two or three megaspores, each developed from a separate megasporocyte, usually grow into megagametophytes. The latter, when mature, have the shape of inverted flasks with a mass of cytoplasm and free nuclei at the lower end and a thin layer of cytoplasm containing a single row of nuclei along the sides. The total number of nuclei is 256, but in certain species the number may reach 512. After the microgametophyte comes into contact with the megagametophyte, a very significant development occurs. Up to this point, organization has been along the familiar gymnospermous lines but it now changes to a distinctly angiospermous condition. One, two or three of the nuclei at the upper end of the megagametophyte become differentiated from the others and definitely recognizable as egg nuclei (FIG. 31A). The position of these egg nuclei has no precise relation to that of the microgametophyte.

The microgametophyte usually contacts the megagametophyte at one side of the expanded portion (FIG. 31A). Both microgametes, the tube nucleus, and a certain amount of cytoplasm are discharged into the megagametophyte. The microgametes move to one of the eggs, leaving the tube nucleus behind. Subsequent events vary according to the species.

In an unidentified species the egg and two microgametes are found together in a definite chamber wholly or partially surrounded by endosperm cells (FIG. 31B, C). In all species it appears to be clear that only one of the microgametes actually fuses with the egg nucleus. In no case during the subsequent division of the fusion nucleus could paternal and maternal chromosomes be observed, which would indicate that the division is angiospermic rather than typically gymnospermic in nature (FIG. 31D).

The fertilized egg of *G. gnemon* is surrounded by a mass of cytoplasm and free nuclei, while in the other species it is enveloped by endosperm cells.

Proembryo:- In an unidentified species the fertilized egg divides two or three times to produce a small group of cells (FIG. 31E, G). There is thus no free-nuclear stage as in the Coniferophyta. Walls are always formed but there appears to be no precise order in the divisions and no definite arrangement of the cells. Each cell next elongates to form a suspensor which penetrates toward the bottom of the endosperm (FIG. 31F), most of whose growth has taken place in the lower portion of the megagametophyte. There is usually only a single large nucleus in each suspensor, but two have been observed. Walls have never been seen and there is little branching.

In *G. moluccense* the suspensors often grow outside the endosperm, usually between it and the nucellus, but sometimes even beyond the endosperm into the cellular tissue at the base of the nucellus. However, embryos are developed only within the endosperm. The manner of development of the embryo was

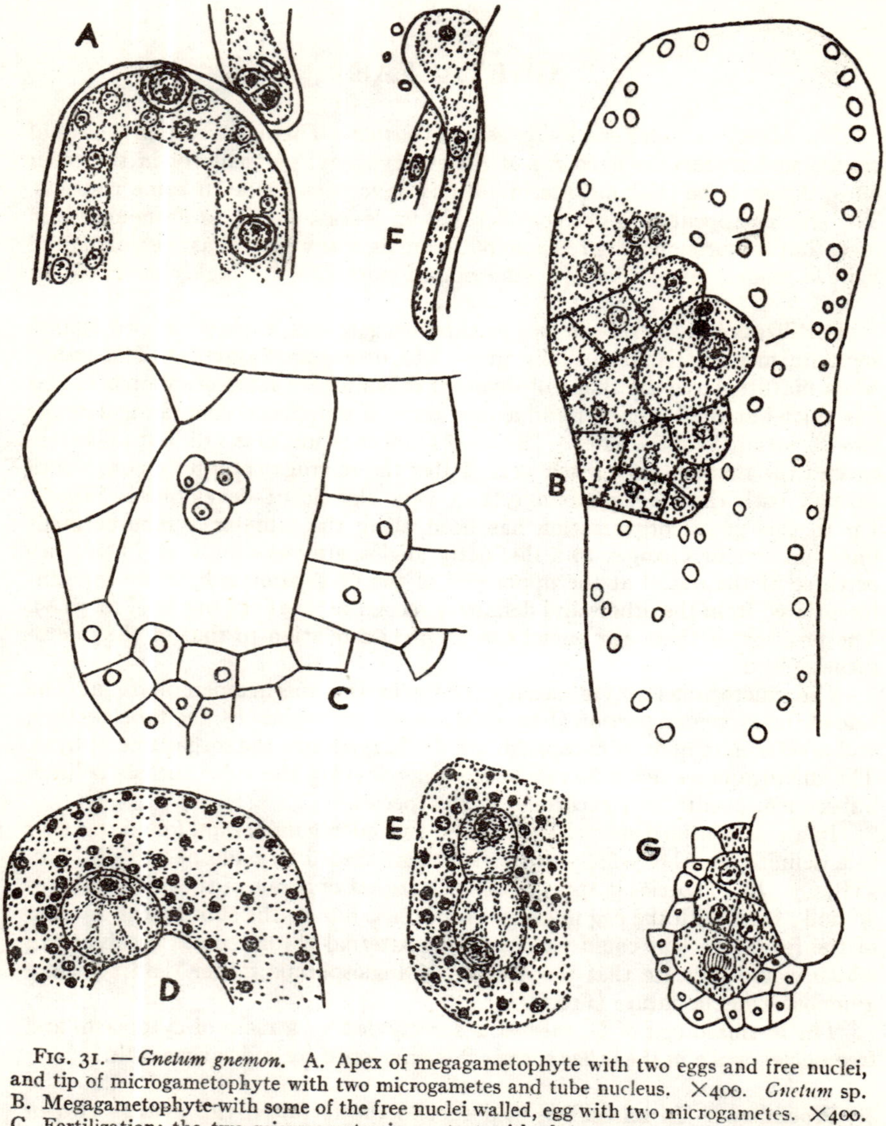

FIG. 31. — *Gnetum gnemon*. A. Apex of megagametophyte with two eggs and free nuclei, and tip of microgametophyte with two microgametes and tube nucleus. ×400. *Gnetum* sp. B. Megagametophyte with some of the free nuclei walled, egg with two microgametes. ×400. C. Fertilization: the two microgametes in contact with the egg. ×400. *Gnetum gnemon*. D. Fertilized egg surrounded by cytoplasm and free nuclei. ×400. E. Two-celled proembryo, surrounded by cytoplasm and nuclei. ×300. F. Suspensors developing from proembryonal cells. ×300. *Gnetum* sp. G. Several-celled proembryo. ×400. (*Redrawn from* THOMPSON 1916).

not followed out; the young embryo consists of a small group of irregularly arranged cells.

Only two cells are formed previous to the development of the suspensors in *G. gnemon*. The two suspensors grow down to penetrate the endosperm.

In certain unidentified species a coiled bundle of tubular suspensor-like structures is developed, with the one nucleus of each tube located near the tip. Such bundles do not occur in *G. funiculare* and *G. gnemon*; here the suspensor tubes soon separate from each other and ramify widely through the gametophytic tissues (FIG. 32A). In the unidentified species the cytoplasm at the tip of a suspensor tube becomes denser and surrounds a peculiar, elongated,

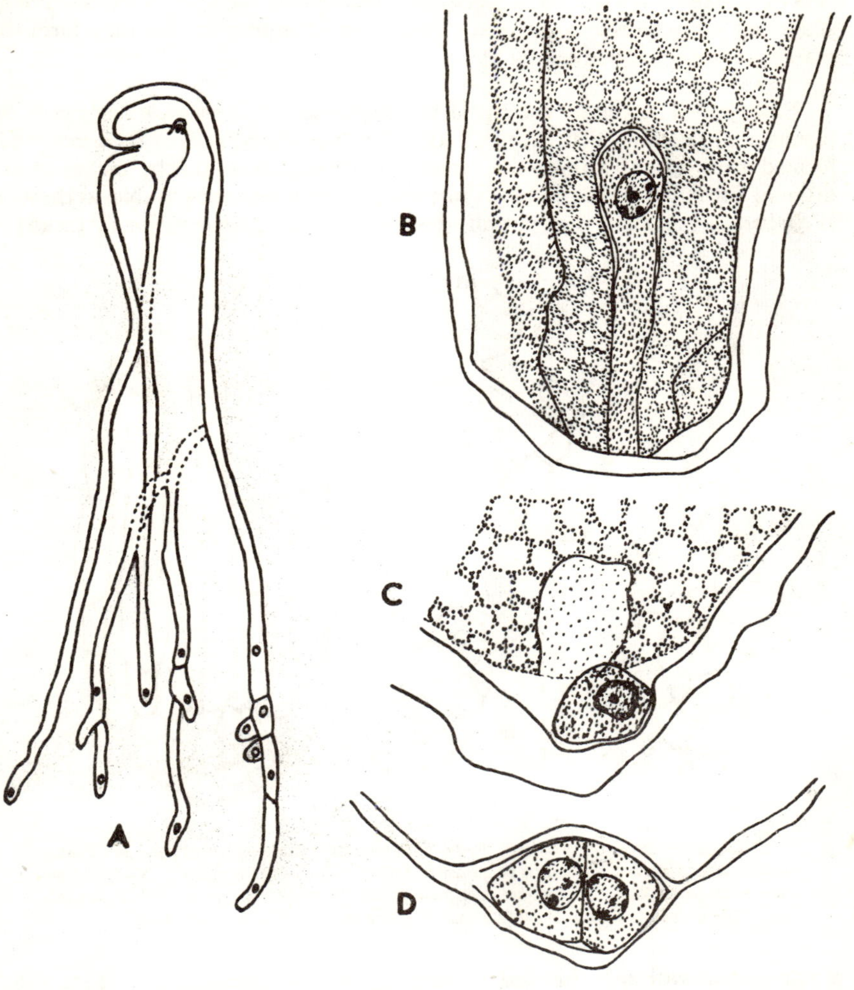

FIG. 32. — ***Gnetum funiculare.*** A. Branching of suspensor tubes. *Gnetum* sp. B. Top of primary suspensor with peculiar cell. ×425. C. Shortened end cell moving out of surrounding cytoplasm. ×425. D. Two-celled stage of secondary suspensor. ×425. (*Redrawn from* HAINING 1920).

pyriform cell containing a single large nucleus (FIG. 32B). The apex of this cell is closely applied to the end wall of the suspensor cell. The fruit is ripe at this time and further development proceeds after the abscission of the fruit. The suspensors swell to about five times their original size. The end cell becomes enlarged and moves out to crowd into a protrusion of the gelatinous wall of the tube (FIG. 32C). The cell is partitioned into two adjoining daughter cells (FIG. 32D), then each of these divides again in the perpendicular plane to form four cells. After this the cells continue to divide in an irregular manner to form an ovoid mass. The basal cells then elongate to produce the secondary suspensor. This structure is ribbon-like and folded in the gametophytic cavity. It is not found in *G. gnemon* and is unlike the secondary embryonal tubes of certain Coniferophyta. At the basal end of this secondary suspensor a chain of smaller cells with denser nuclei is differentiated, which may function in the proliferation of embryos. The cells at the apex of the secondary suspensor are actively meristematic and by rapid multiplication they form the massive embryonic body.

Embryo:- The cell at the tip of the suspensor of *G. gnemon* adds cells by means of anticlinal walls (FIG. 33A–D). The peripheral cells of the group thus formed grow laterally along the surface of the suspensor and by means of additional anticlinal walls produce short embryonic tubes comparable to those in *Welwitschia*. A cell is next cut off above the end of the suspensor by means of

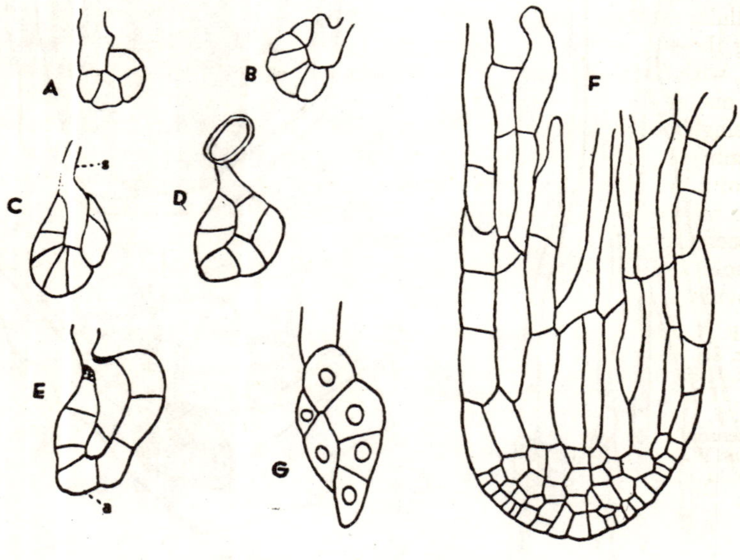

FIG. 33.—*Gnetum gnemon*. A–D. Developing embryos. ×175. E. Transverse wall formed near end of suspensor. ×175. F. Older embryo. ×175. *Gnetum moluccense*. G. Several-celled proembryo. Symbols: *a*, apical cell; *s*, suspensor. (A–F *redrawn from* BOWER 1882; G *redrawn from* THOMPSON 1916).

a transverse wall and this cell is then segmented longitudinally (FIG. 33E.) Periclinal as well as anticlinal walls are now erected in the lateral cells of the peripheral layer. What has all the appearance of being an apical cell is created at the apex when a cell centrally located divides by walls perpendicular to the outer surface of the embryo and inclined to one another (FIG. 33C–E). A large embryonic body is presently formed. A definite epidermis covers the cotyledons and hypocotyl but does not extend over the root region.

Growth of the developing embryo is very rapid (FIG. 33F). Shortly after the cotyledons commence growth a lateral protuberance appears on one side of the hypocotyl. The position of this protuberance is constant: it always lies on the under side of the embryo as the seed lies during germination, but has no constant relation to the position of the cotyledons. The protuberance during growth soon assumes a nearly cylindrical form. It arises through lateral extension of the tissues of the hypocotyl, the epidermis, cortex, vascular bundles and pith all participating in its formation. It is morphologically equivalent to the same structure found in *Welwitschia*, known as the "feeder," but in the latter genus the cortex and epidermis form the bulk of the body. The growth soon overtops the hypocotyl; after the embryo has protruded itself from the seed, the growth remains within the seed, expands and fills the whole of the space previously occupied by the embryo.

In some unidentified species the embryonic body is differentiated as in *G. gnemon*. At the apex a conical projection is produced, surrounded by a thick ridge from which the cotyledons develop. A "foot" projects from the side of the hypocotyl and may exceed the cotyledons in length; it remains in the seed as an absorbing organ.

Polyembryony seems to prevail in *Gnetum*. There is considerable variation in this respect, both between different species and within the same species. For instance, three of the four primary suspensors may develop extensive multicellular secondary suspensors and no one of the three shows any indication of being the successful one. Again, two of the four may be equally well developed, with the third grown to about one-third of the others before it aborts; or only one of the four may develop. Finally, several embryos may develop from a single primary suspensor, or a secondary suspensor may branch to form extra embryos. Sometimes chains of smaller cells may be observed in the tissues of a secondary suspensor; from these, outgrowths resembling normal embryos may develop. These are bud embryos. No cases were observed in which secondary suspensors fused to form a single embryo. No matter how many potential embryos may be produced, only one attains maturity in a given ovule.

Bower, F. O. 1882. The germination and embryogeny of *Gnetum gnemon*. Quart. Jour. Micr. Sci. 22: 278–298.
Coulter, J. M. 1908. The embryo sac and embryo of *Gnetum gnemon*. Bot. Gaz. 46: 43–49.
Haining, H. L. 1920. Development of embryo of *Gnetum*. Bot. Gaz. 70: 436–445.
Lotsy, J. P. 1899. Contribution to the life-history of the genus *Gnetum*. Ann. Jard. Bot. Buitenzorg II, 1: 46–114.
Thompson, W. P. 1916. Morphology and affinities of *Gnetum*. Amer. Jour. Bot. 3: 135–184.

Chapter XV

WELWITSCHIACEAE

The family is monogeneric and with only one species, *Welwitschia mirabilis.*

Fertilization:- Archegonia are not formed. In place of these, the megagametophyte contains between 1015 and 1360 free nuclei equally distributed throughout its mass. The megagametophyte next becomes clearly differentiated into an upper fertile and a lower mostly sterile region. The first comprises approximately the micropylar fourth of the megagametophyte; the nuclei are fewer in number and are functionally sexual. All, or at least the majority, of the more crowded nuclei contained in the lower three-fourths of the megagametophyte presently give rise to a sterile tissue which has been called the endosperm. However, a few of the nuclei in this region may remain functionally sexual and behave precisely like those of the upper region. In both regions a cleavage process, somewhat resembling that in *Ephedra* which differentiates the proembryonal eggs, separates the nuclei into compartments. The nuclei in each compartment in the micropylar region generally number two, rarely more than six (Fig. 34A), and the cells are also relatively larger than in the lower region. Here the number of nuclei may be as low as two, but is usually twelve or more. From this stage onward the two regions develop differently. The lower region will not be discussed further since it is not directly concerned with the development of the embryo.

The nuclei in the cells in the micropylar region increase in size and the cytoplasm becomes vacuolate. As the vacuoles enlarge, the bulk of the cytoplasm with the contained nuclei comes to lie against one wall of the compartment (FIG. 34B). As soon as this condition is reached that part of the wall begins to become extended in the form of a tube into which the cytoplasm and nuclei migrate (FIG. 34C). The tubes penetrate the megasporangial wall and grow into the axial core of the nucellar cone. The lower part of each tube is highly vacuolated, soon loses its cytoplasm and appears as an empty cell. The bulk of the cytoplasm, together with the nuclei, advances with the tip of the tube.

The morphological nature of these tubes poses a difficult problem. They were first designated as "secondary embryo-sacs," then as "prothallial tubes" and finally as "embryo-sac tubes." If an analogy with conditions in *Gnetum* is accepted, the cells appear to be multinucleate eggs which became tubular.

Each tube grows independently of the others. The direction of growth is upward and slightly outward for most of the tubes, but some may grow downward, or turn slightly outward and proceed more or less horizontally toward the integument. Most of those pointed upwards continue growth until a microgametophyte is encountered or if this does not happen development continues until the surface of the nucellar cone is reached, and they may even project into the micropyle. At first the nuclei in each tube lie close together but later generally become separated and arranged in single file. When contact is established with a microgametophyte one of the nuclei is commonly located considerably ahead of the others.

Rarely all of the nuclei within a given egg may fuse to form a single nucleus. As many as six nuclei may become fused into one.

Fertilization may occur in any part of the nucellus above the apex of the original megagametophyte, but generally takes place near the center of the

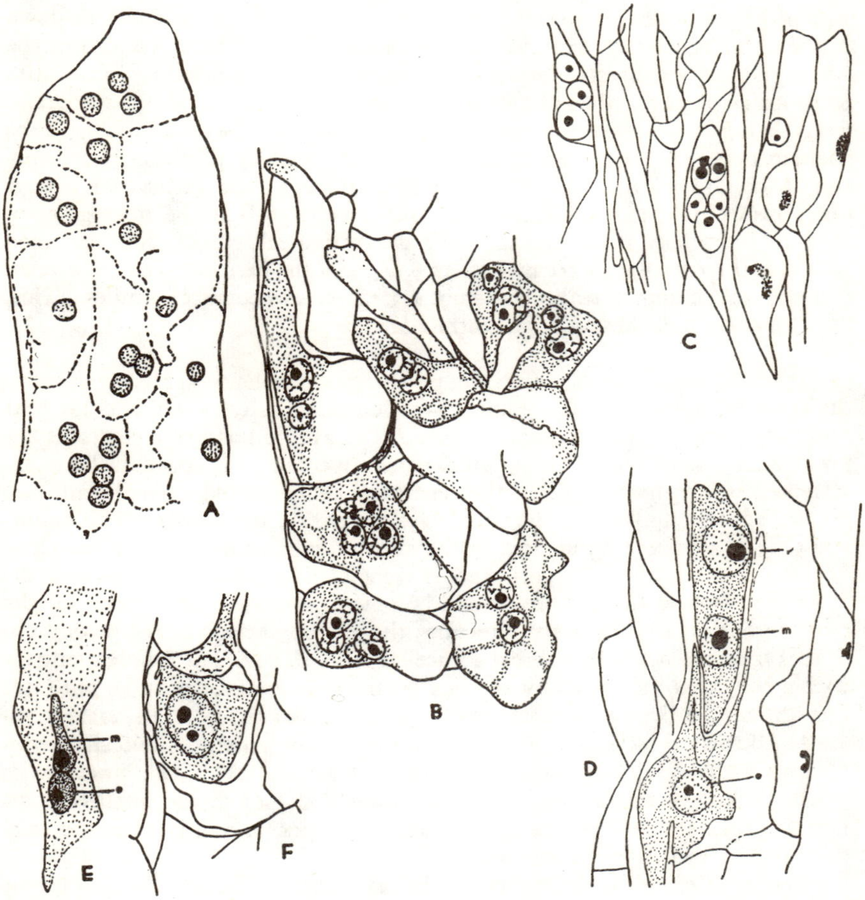

FIG. 34. — *Welw.tschia mirabilis*. A. Median longitudinal section through micropylar end of megagametophyte revealing the irregular planes of septation. Each compartment will form egg tubes into which their nuclei will pass. B. Commencement of tube growth. C. Median longitudinal section through tissue of nucellar cone lying immediately above the megagametophyte and into which the young egg tubes have penetrated. One tube contains five free sexual nuclei lying close together near the tip. D. Egg tube and microgametophyte (*p*) in contact. Egg nucleus (*e*) has entered cytoplasm of latter. Male nuclei (*m*) in generative cell are in normal condition. E. Generative cell (*m*) and egg tube (*e*) nuclei in contact. F. Fusion complete; respective nucleoli distinct. All × 525. (*Redrawn from* PEARSON 1909).

nucellar cone. The nucleus of the generative cell of the microgametophyte divides soon after the latter penetrates the nucellar cone. The two resulting nuclei are both potentially functional, though one nucleus is larger than the other and so situated that it is in advance. Frequently, however, one of the nuclei becomes disorganized and disappears. The tube nucleus precedes the generative cell but presently disappears. There is no stalk cell. At the time the egg tube and microgametophyte meet, therefore, there is only a generative cell present in the latter.

The two tubes may come into contact at their tips, or the egg tube becomes attached to the microgametophyte laterally in the neighborhood of the position occupied by the generative cell (FIG. 34D). The respective walls break down at the point of contact and that one of the egg nuclei which is ahead of the others then passes directly into the generative cell. This behavior, obviously, is the exact contrary of what happens during fertilization in all other Spermatophyta, but there appears to be no question regarding the authenticity of the phenomenon. No instances could be found in which fertilization clearly occurred within the egg tube. It is therefore probable that fertilization always takes place while the generative cell still remains within the microgametophyte. The nuclei of the egg tube and generative cell quickly fuse and then pass into the resting condition (FIG. 34E, F). In the first mitosis the paternal and maternal chromosomal sets are not segregated on separate spindles. This, as in *Gnetum*, indicates an angiospermic tendency.

Proembryo:- While the number of proembryos developed in a given nucellus probably do not exceed the number of microgametophytes which enter that nucellus, there are usually about twelve. The zygote boundary coincides approximately with that of the original generative cell. However, in those instances in which both nuclei of the generative cell persisted, only the anterior portion of the cytoplasm of the latter becomes part of the zygote. This doubtless accounts for the differences in the sizes of zygotes in the same nucellus. When formation of the zygote wall is completed the cell elongates and the cytoplasm toward the micropylar end becomes vacuolate. Elongation for the most part is in the downward direction; the micropylar end of the proembryo is always in contact with unbroken nucellar cells. The cell at first lies entirely within the canal excavated by the microgametophyte and egg tube, but later growth may be through hitherto undamaged nucellar tissue. The nucleus migrates downward as the cell elongates and presently undergoes mitosis to produce two cells separated by a transverse wall (FIG. 35A). The lower cell is smaller than the micropylar one. The latter does not divide further; it attains a considerable length as the "primary" suspensor and later as the axial part of the "secondary" suspensor.

The terminal cell is the embryonal initial proper. The first division in this cell is longitudinal (FIG. 35B) and the two resulting cells segment again in a plane perpendicular to that of the first division, whereupon each of the four daughter cells is divided transversely (FIG. 35C, D). The four resultant cells adjoining the primary suspensor cell are smaller than the outer tier of four cells (FIG. 35D). The next divisions occur in these smaller cells; the plane is radial and parallel to the long axis of the proembryo (FIG. 35E). The upper tier at this time possesses its full complement of eight cells; subsequently this tier gives rise to the inner cortical ring of the secondary suspensor and may therefore be designated as the "inner cortical tier" (*ict*, FIG. 35E).

The four terminal cells segment by means of radial longitudinal walls meeting in the axis of the proembryo. This tier hereafter consists of eight equal cells. In each of these cells an oblique wall is erected; the cells so separated form a ring lying above and outside the initial cells and beneath the inner cortical tier (FIG. 35F). The eight cells of this external ring divide again by longitudinal radial walls to produce the sixteen initial cells of the outer cortical ring. At the next stage the eight terminal initials divide by walls oriented transversely to the axis of the proembryo (FIG. 35G). Thus there is formed a plateau of eight cells which lie within the outer cortical ring and form what might be called the "embryonic plate."

The proembryo by this time has reached the upper part of the endosperm and the development of the secondary suspensor now commences. The first indication of this process is the elongation of the eight cells comprising the inner cortical tier (*ict*, FIG. 35E). These eight cells, together with the distal end of the primary suspensor, develop together as a unit; they are not isolated embryonal tubes which grow independently. The same is true of the cells of

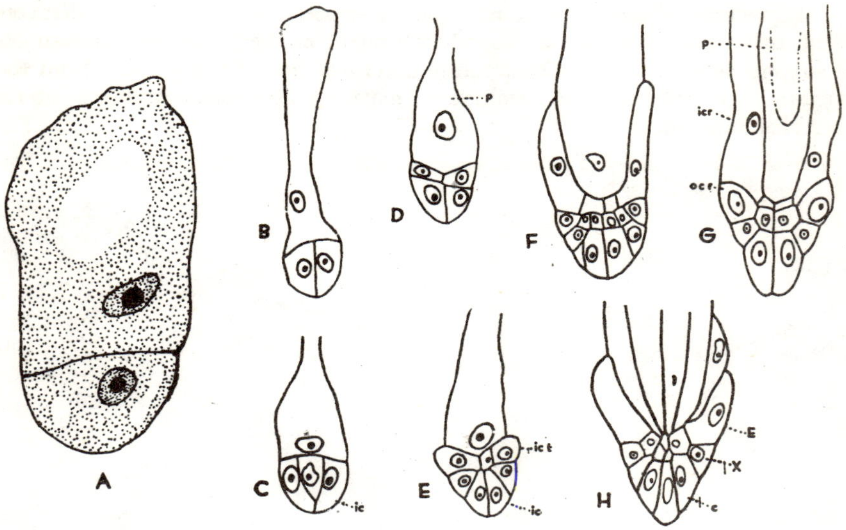

FIG. 35. — *Welwitschia mirabilis*. A. Two-celled proembryo. B. Three-celled proembryo. C. Proembryo with terminal group of four initial cells (*ic*). D. Nine-celled proembryo with four initial cells, with plateau of four cells above (*p*, pollen tube). E. Later stage, with two initial cells (*ic*) shown, and above them two of a ring of eight cells resulting from division of initial cells shown in D. F. Longitudinal section of proembryo, with two cells of inner cortical tier considerably elongated, and inner ends of two others sectioned transversely; also two cells of outer cortical ring (*ocr*) and, below them, two cells of the lower ring (*X*) which have been separated by oblique walls from the terminal initial cells; the section passes through three of the cap cells. G. A later stage than the preceding, in which the cells of the inner cortical ring (four shown) are now much elongated; those of the outer cortical ring (two visible) have commenced growth. The embryonic plate, lower ring and cap are each represented by two cells. H. A slightly older stage than the preceding; in oblique section, showing five cells of the embryonic plate and four of the cap. Symbols: *p*, microgametophyte; *ic*, initial cells; *ict*, inner cortical ring; *ocr*, outer cortical ring; *X*, lower ring; *E*, embryonal initials; *c*, cap cells. A × 525; all others ×230. (*Redrawn from* PEARSON 1909).

the outer cortical ring. The nucleus of the primary suspensor cell disintegrates and disappears during the early stages of elongation of the inner cortical ring. In the meantime the eight terminal initial cells segment in a manner similar to a preceding division by which the outer cortical ring was erected. A ring of eight cells, later increased to sixteen by radial longitudinal segmentation, is thus formed lying above and outside the initial cells and beneath the outer cortical ring. These sixteen cells were observed unchanged in the most advanced embryos encountered, but it is possible that they may give rise to a third ring of cortical suspensor cells. They may collectively be designated as the lower ring (X, FIG. 35H). The sixteen cells of the outer cortical ring elongate in precisely the same fashion as do those of the inner ring.

It will be observed that the advanced proembryo bears a remarkable resemblance to similar stages in *Araucaria angustifolia* (Coniferophyta: *Arau-*

cariaceae). The terminal group of eight cells formed when the lower ring-cells were separated from the initial cells is probably homologous with the group of cap cells in *A. angustifolia*, and the group of embryonic initials clearly have the same function.

When only one proembryo is present, it follows a fairly direct course toward the center of the endosperm. When several proembryos are developed, as is the usual case, the core of the upper third of the endosperm is completely demolished. As a rule, one of the several proembryos outgrows the others and thrusts its terminal group of cells deep into the endosperm and becomes the one to attain maturity. Only one mature embryo has ever been found in a single seed.

Dissections of embryo systems have never been made in *W. mirabilis*; consequently, many features of later development of the proembryo remain obscure. It is difficult to make accurate determination of these points from sections. Development of the embryo to maturity apparently occurs after the seed has been shed.

Pearson, H. H. W. 1909. Further observations in *Welwitschia*. Phil. Trans. Roy. Soc., London, B, 200: 331–402.

Chapter XVI

GENERAL CONSIDERATIONS

Scope and Definition of Angiosperm Embryology:- Angiosperm embryology may be divided into two divisions: General Embryology and Special and Comparative Embryology.

The first division may be subdivided into three groups of more or less related subjects:

1. *Embryonic Morphology.*— This subdivision embraces the static phases of embryology, and these depend upon whether the external aspects alone are considered, in which case it may be described as Embryography, or whether the internal aspects are under consideration, in which instance either Cytology, Histology or Anatomy proper are the terms to be employed as each occasion requires.

2. *Embryonic Physiology.*—This subdivision embraces the dynamic phases of embryology. These aspects are very rarely dealt with by plant embryologists and are generally left to plant physiologists. One familiar example is the growth *in vitro* of more or less immature embryos removed from seeds.

3. *Embryogeny.*—These phases, with which the great majority of embryologists are concerned, embrace the kinematic aspects of embryology. They may be separated as individual phases, with appropriate designations, as follows:

a. *Embryogenesis* proper: the origin of the embryo;
b. *Embryotectonics*: the architecture of the embryo;
c. *Embryogenergy*: the destination or functions of the embryo;
d. *Embryonomy*: the laws of embryonic development.

Special and Comparative Embryology as a division requires no subdivision since these topics relate to individual species, to well defined genera, occasionally even to entire families, and therefore follow phylogenetic lines more or less but there exists a certain degree of overlapping since species in very diverse and totally unrelated taxonomic families may follow the identical embryonomy.

Early Attempts at Embryonomic Classification:- Up to the present time no one has seriously attempted to formulate a definite scheme including all the embryonomic Types and Variations and with these either named or otherwise distinguished so as to be clearly recognizable. SOUÈGES, in one or two of his papers, mentions "Megaarchtypes" but the author has been unable to find a scheme employing such terms in any of this investigator's publications that could be located. Apparently SOUÈGES has done no more than to express the hope that a precise classification could some day be prepared. The beginning, therefore, is the one made by SCHNARF in his "Embryologie der Angiospermen"; he took the name of the family to which belonged the species whose embryonomy was most completely known and designated it as the Type. Thus, he employed the terms "Cruciferen-Typus," "Astereen-Typus," "Solanaceen-Typus," "Chenopodiaceen-Typus" and "Caryophyllaceen-Typus." These designations obviously are in German. The use of the word "Type" appears to have become generally accepted and the precedent established by SCHNARF is well worthy of perpetuation; the only alternative is to number the Types in some fashion, but this does not permit facile recognition of the individual Types and there are other objections. It seems to be best, all matters considered, to take the root of the family name rather than the latter itself and to attach to this root the Latin suffix -ad. SCHNARF's designations, with one exception, are retained. His use of the *Cruciferae* (*Brassicaceae*) must be abandoned in favor of the *Onagraceae* since species belonging to the latter family display a simpler, more diagrammatic and doubtless more fundamental type of embryonomy than does *Capsella bursa-pastoris*, the basis of the "Cruciferen-Typus."

Variations have never been specifically designated as such; they are being recognized for the first time in the present connection. A Type is fundamental; a Variation is based upon modifications of the fundamental type. These modifications are more or less precocious differentiations, differences in the number of cells which participate in the construction of the suspensor, in the form or very variable dimensions of this structure, or differences in the rapidity of the segmentations which may occur earlier or later in favor of the one or the other of the cells produced by the same division. It was once presumed that Variations could be separated into those derived by descendance and those derived by ascendance from the fundamental type, but this conception has been abandoned. Variations are most conveniently designated by the name of the genus whose embryonomy departs in a minor fashion from that of the Type to which it is attached. It is, of course, recognized that there may be variations in embryonomy among the species of a given genus but these are rare in occurrence. Such departures from the normal are often the result of the introduction of such factors as hybridization. When these species differences do occur, it is usually possible to distinguish the species which most closely conforms to the embryonomy of the Type, then to take this species as most typical and to consider the other species as variations (but without giving them separate recognition as Variations) of the typical species. Examples of this sort may be found in the *Ranunculaceae*, in which species of *Adonis*, *Actaea*, *Ranunculus* and other genera depart in various ways from the typical embryonomy of *Myosurus minimus*, which is considered to represent the standard Variation for the family.

The six Types which will be described in the next chapter have been tacitly accepted by those who have considered Angiosperm embryogeny in a broad sense and, merely to state a prophecy, it is believed that they will survive the test of time. Variations, on the other hand, are an entirely different matter. Not all are capable of as clear definition as is desirable and some should perhaps be dropped and the species concerned be assigned to some other Variation. As future research on uninvestigated species proceeds, it is inevitable that new Variations will need to be erected under the appropriate Types.

In the present connection the author has undertaken a task never before attempted by any embryologist: that of surveying all of the vast embryological literature that could be consulted in the original, in an endeavor to determine the assignment of each species to one or another of the Types and wherever possible to Variations. A very large number of published descriptions do not present sufficient detail to allow a precise assignment; occasionally recourse was had to the illustrations accompanying the original papers, but these are not always presented either in chronological order, in such a way that the sequence of segmentations could be perceived, or they represented nonmedian longitudinal sections. Altogether too many articles have included only isolated, usually rather late, stages, on which it is obviously impossible to base conclusions; such articles have been arbitrarily disregarded and the species concerned judged to be still uninvestigated.

Fundamental Laws of Embryonomy

Embryonomic Laws:- There are certain characteristics pertaining to embryonomy which partake of the nature of laws. Each of these laws is truly specific when applied to a given species. Those which succeed them during the ontogenesis of the embryo and which prevail in the development of definite tissues derived from the proembryonic initials or are already roughly delimited in the embryo are not specific but are merely characteristic of the tissues in those individual species whose construction they govern. Since the embryonomic laws are essentially specific, they provide the explanation for the fundamental organization of the embryo of a given species. To state the matter in other words, when the laws of embryonomy for a given species have been determined, they serve to define this species from the embryological standpoint. Certain species from very diverse families may follow the identical embryonomic laws in a general manner and may therefore be grouped together as a Type.

Embryonomic Formulae:- In order to elucidate the laws of embryonomy with the utmost precision and clarity and in as complete a manner as possible, the employment of formulae has become necessary. These formulae indicate precisely the relationships which the individual cells present to one another during the course of their segmentation, with respect to their origin, to their number, to their disposition and to their eventual destination or their histogenic functions. Only those cells engendered during the first four cell generations are considered in the present connection. These formulae may further be combined into tables of recapitulation which at a glance reveal both the resemblances and the differences.

In order to facilitate comparisons, the same cells and the embryonic regions to which they give rise may be given the same letter designations. The letters employed are those originally used by Souèges and are obviously abbreviations of French words, except for a few which were arbitrarily chosen and are not actually abbreviations. These letters have been so extensively employed that it is now very undesirable to change them into others based on another language. The terminal cell of the two-celled proembryo is given the designation *ca*, the basal cell that of *cb*. The two juxtaposed cells, daughters of *ca*, are designated as *q*; *m* is the upper daughter cell of *cb* and *ci* is the lower one; *n*, *o* and *p* are derivatives of *ci*; *l* and *l′* are the upper and lower groups respectively in a two-tiered octant proembryo and are together equivalent to *q*. The regions and their corresponding abbreviations are: *pco*, the cotyledonary region proper, from whose derivatives the cotyledons develop; *pvt*, the stem apex, in which the initials of the stem epidermis may have already been delimited (*iep*), or the initials of the stem cortex are recognizable (*iec′*); *phy*, the hypocotyledonary region, from which is derived the hypocotyl and the root; *co*, the root cap, and particularly its central portion (*co* has sometimes been used to indicate the cotyledons; *cot* is the proper designation for these structures); *s*, the suspensor *sensu stricto*; *icc*, the initials of the central cylinder of the stem; *iec*, the initials of the central cylinder of the root; finally, in certain cases, *ico*, the initials of the root cap.

When formulae are combined into a table of recapitulation, such a table is applicable to many species, either with exact superposition of all parts, or in a less rigorous manner (*i.e.*, as Variations) with allowance being made for one or more secondary differences. These tables may hold good: (*1*) in the number of the cells; (*2*) in the disposition of the cells; or (*3*) in the role played by the cells, especially those which give rise to the lower portion of the embryonic body and have no essential function in the construction of the embryo proper. Practically all other secondary differences arise at a later stage of development, beyond the fourth cell generation.

Certain species are thought to represent types derived by superposition; that is to say, by the imposition of the characteristics of one fundamental Type upon those of another. In precisely what light this peculiarity should be considered is a matter which can not be settled at the present time in view of the current scanty state of knowledge; it is a problem which only the future can solve.

Examples of the Application of Formulae:- *Senecio vulgaris*, the basis of the Senecio Variation, Asterad Type, may be taken to represent an example of the application of the laws of development in recapitulatory form. At the first cell generation, the bicellular proembryo consists of two cells disposed in two tiers, and their destination may be summarized as follows:

$$ca = pco + pvt$$
$$cb = phy + icc + iec + co + s$$

The formula is interpreted thus: *ca*, the terminal cell, engenders the cotyledonary region (*pco*) and the stem apex (*pvt*), while the basal cell (*cb*) gives rise to the hypocotyledonary region (*phy, icc*), to all parts of the root (*iec, co*) and to the suspensor (*s*). This obviously means that the basal cell contributes to the construction of the embryo proper.

At the second cell generation, the proembryo consists of four cells disposed in three tiers. The terminal cell has divided longitudinally and the basal cell transversely. The formula for this generation thus becomes:

$$q = pco + pvt$$
$$m = phy + icc$$
$$ci = iec + co + s$$

The interpretation is that the terminal cell region still remains one-tiered with the cells having the same destinations but is given a new designation (that of *q*), while the original basal cell has become divided into two tiers of one cell each and each tier has acquired new destinations. That is, the upper daughter cell of *cb*, now designated as *m*, gives rise to the hypocotyledonary region (*phy, icc*), while the lower daughter cell, now designated as *ci*, gives rise to all of the root (*iec, co*) and the suspensor (*s*).

At the third cell generation the proembryo consists of eight cells disposed in four tiers. At this time the original terminal cell has become resolved into quadrants, but still remains one-tiered; the second one-celled tier represented by *m* has divided longitudinally and hence still remains one-tiered with the same destinations; but the third one-celled tier of the preceding generation has divided transversely to give rise to two new tiers, each of which has but one cell, and are now designated respectively as *n* and *n'*. There has been a diversion of destinations brought about by the transverse division of cell *ci*: the upper daughter derivative, *n*, now gives rise to the initials of the root cortex, while lower daughter cell, *n'*, engenders the initials of the root cap and the suspensor.

Each of the eight cells of the third generation becomes bipartitioned to give rise to the fourth cell generation, which thus consists of sixteen cells that are arranged in five tiers:

$$\begin{aligned} q &= pco + pvt \\ m &= phy + icc \\ n &= iec \\ o &= co \\ p &= s \end{aligned}$$

The original terminal cell has become transformed into octants (in place of the quadrants of the preceding generation); tier *m* now includes four circum-axial cells in one tier and with the same destinations; region *n* remains unchanged except for the addition of a juxtaposed cell; while region *n'* has divided transversely into two new tiers of one cell each and with new destinations. The upper daughter cell of *n'*, now designated as *o*, has acquired the specific duty of delimiting the initials of the root cap, while the lower daughter cell, now designated as *p*, gives rise to the suspensor alone.

These formulae may now be brought together into a single table of recapitulation, which describes the complete proembryonic history of *Senecio vulgaris* at a glance:

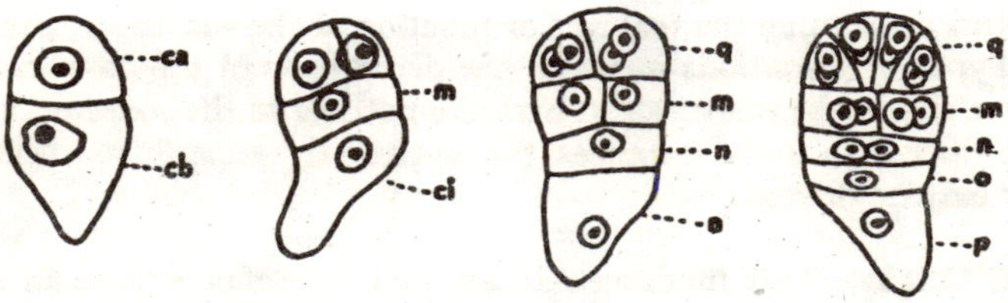

Recapitulatory Table for Senecio vulgaris
(from left to right)

I. First Cell Generation
Proembryo of two cells disposed in two tiers:

$$\begin{aligned} ca &= pco + pvt \\ cb &= phy + icc + iec + co + s \end{aligned}$$

II. Second Cell Generation
Proembryo of four cells disposed in three tiers:

$$\begin{aligned} q &= pco + pvt \\ m &= phy + icc \\ ci &= iec + co + s \end{aligned}$$

III. Third Cell Generation
Proembryo of eight cells disposed in four tiers:

$$\begin{aligned} q &= pco + pvt \\ m &= phy + icc \\ n &= iec \\ n' &= co + s \end{aligned}$$

IV. Fourth Cell Generation
Proembryo of sixteen cells disposed in five tiers:

$$\begin{aligned} q &= pco + pvt \\ m &= phy + icc \\ n &= iec \\ o &= co \\ p &= s \end{aligned}$$

As will be described in the chapter on Types and Variations, the identical recapitulatory scheme as presented above applies to such species as *Urtica pilulifera* and *Malva rotundifolia* among dicotyledonous plants and to *Muscari comosum* among monocotyledonous plants. These three species represent three large and diverse families, and the scheme further demonstrates that there is no fundamental difference between the Dicotyledons and Monocotyledons of the taxonomist insofar as proembryonic development is concerned.

Law of Parsimony:- The pertinency of this law has apparently not occurred to Souèges, who has formulated all of the other laws, but it is most decidedly applicable in the field of plant embryonomy. In fact, it may be the fundamental law and all the other laws are derived from it.

As originally expounded by WILLIAM OF OCCAM (ca. 1270 to ca. 1350), and known as "Occam's Razor," the law stated that "Entia non sunt multiplicandi praeter necessitatem." Later the same principle was discussed at greater length by Sir WILLIAM HAMILTON, a Scottish philosopher (1788–1856) and was called by him the "Law of Parsimony."

A very literal translation of the "Razor" is: "Entities are not to be multiplied beyond necessity." However, freer translations are possible and have been made. In so far as its application to plant embryology is concerned, this law may be stated thus: No more cells are produced by the embryo than are absolutely necessary.

The implications of this statement are obvious to any person who has ever carefully followed out the development of the proembryo of almost any plant. Each and every cell has a reason for its existence; its origin can be demonstrated; its destination can be determined; and its position is invariably the same. A superfluous cell would seriously upset the harmonious balance of developmental factors present in the proembryo, and would probably lead to the death of the latter.

The one flaw in the rigid application of the law is that we are not yet reasonably clear concerning the real role or function of the suspensor, particularly in those Types or Variations in which the derivatives of the basal cell, which usually give rise to the suspensor, contribute nothing to the construction of the embryo proper. In such instances the apparently superfluous cells of the suspensor may be disregarded.

Law of Origin:- This fundamental law may be defined thus: In any particular species, the sequences of cell formation may be established in such a manner that the origin of the cells may be defined in exact terms by referring to the one or to the other of the terms of the sequence.

It is hardly necessary to state that embryonic cells arise by simple successive bipartitionings of the zygote and its derivative cells. Their theoretical process of generation may be simply described in homely language, since there are no precise scientific terms that can be employed: the first two cells are sisters and, at the same time, daughters of the zygote. In the second cell generation the cells number four, two of which are sisters and first cousins of the other two; all are granddaughters of the zygote. At the third cell generation there are eight cells, all of which are great-granddaughters of the zygote; they form two groups of first cousins once removed and, in each of these groups, two sisters, as in the preceding cell generation, are first cousins of two other sisters.

One great difficulty is that segmentations are not necessarily synchronous in the proembryos of many plants; in the discussion in the preceding paragraph, synchronous division was taken for granted. Divisions may be more or less early or more or less late in the same cell generation. For example, the cells derived from the basal cell (*cb*) remain undivided during the second cell generation, while those derived from the terminal cell (*ca*) have completed their segmentation. The six cells subsequently appearing in the third cell generation offer complex relationships, yet it is nevertheless not difficult to determine their relationships. In other instances, only three cells appear in the second cell generation: two are sisters, nieces of the third cell and granddaughters of the zygote; the third is the aunt of the two others and daughter of the zygote.

To avoid the acute difficulties encountered in determining the origin and the parentage of the cells already present at the third cell generation, the procedure may be changed: in place of taking the zygote as the point of departure, one may select a cell of the youngest cell generation whose origin and parentage appears to have been probably determined with reference to the primordial embryonic cell (*ca*). Two practical rules may be employed: (1) at the third cell generation, the origin of the cells has been determined with reference to the primordial embryonic cell; these cells are simply daughters, granddaughters or great-granddaughters of this cell (*ca*); (2) beyond the third cell generation, the origin of the cells may be established by taking for the point of departure either one of the first two cells, one of the four elements of the proembryonic tetrad, one of the quadrants ordinarily differentiated at the third cell generation, one of the eight octants — upper or lower — which are generally individualized during the fourth cell generation, or finally one of the cells which serve certain specialized functions after the third cell generation and in which the origin from derivatives of cells nearer to the zygote may be clearly defined.

In the first case, one may take as the point of departure either the terminal cell (*ca*) or the basal cell (*cb*) of the two-celled proembryo. In by far the great majority of the Angiosperms the zygote divides transversely at the first segmentation, so that as a general rule one may take such a division for granted. One need not proceed further in such cases where all the cells of the embryo are derived from the terminal cell, the basal cell remaining undivided, as in *Sagina* and *Sagittaria*. It is again convenient to proceed in the same manner when a difference in the rapidity of the segmentations begins to appear during the first cell generation, either in the terminal cell as in the *Onagraceae* or *Euphorbiaceae* or in the basal cell as in *Erodium* or *Sherardia*.

Again, it is from the terminal cell of the two-celled proembryo that one can always determine the origin of the epiphysis of certain species, such as *Papaver rhoeas*, *Geum urbanum*, *Erodium circutarium* or *Myosotis hispida*. In the first species the epiphysis initial is the daughter of the terminal cell; in the next two, it is the granddaughter, and in the fourth species it is the great-granddaughter.

In the second case, one may proceed with one or the other of the four elements of the proembryonic tetrad differentiated during the second cell generation. This tetrad may theoretically be presented under four different aspects, to be described presently under the Law of Disposition. Generally, among the Angiosperms, one encounters either the so-called "T-form," composed of two upper juxtaposed cells and two lower superposed cells, or the "linear-form," consisting of a row of four cells. In the "T-form" the two upper elements, which are the mother cells of the two quadrants, are homodynamous and provide for identical lines during the entire proembryonic life; in the "linear-form" the four superposed elements may give different lines. Beginning at the top, these four cells are designated by the respective letters *l*, *l'* *m* and *ci*; the elements *m* and *ci* are called the intermediate cell and the lower cell, respectively, of the tetrad and are also encountered in the "T-form." One may easily establish the order of these elements and, by the aid of this order, determine with great precision the origin of certain definitely specialized cells.

The origin of the epiphysis initial has already been mentioned. It is also possible to determine the exact origin of the hypophysis initial from the derivatives of the intermediate cell, *m*, of the tetrad. The hypophysis is decidedly characteristic of the Onagrad Type. In the *Onagraceae*, for example, cell *m* is converted directly into the hypophysis initial; while in *Mentha viridis* the hypophysis cell is the upper daughter cell of *m*; in *Veronica arvensis* it is a granddaughter derivative of *m*; and in *Capsella bursa-pastoris* it may sometimes be a cell of an earlier generation.

In the third case, one may, when considering a form derived from a

"T-form" tetrad, commence the order with one of the quadrants ordinarily differentiated during the third cell generation. These quadrants are regularly arranged around the axis and are homodynamous; that is to say, they are endowed with the same constructive power during proembryonic life and obey the same rules during their process of segmentation.

In the fourth case, one may arrive at the closest decision by consideration of one of the octants, which are daughters of the quadrants and appear during the fourth cell generation. The octants are distributed in two tiers: there are four upper octants and four lower octants; the four octants of a given tier are again homodynamous and conform to the same rules in their divisions.

Finally, in the fifth case, one may, to begin a line of descent, commence with a cell already individualized and itself very precisely defined, as in the preceding cases, by its origin, which is most often very different according to the species. For example, one may take the epiphysis cell or the hypophysis cell and indicate by what filiation, from the respective derivatives of these cells, the stem apex and the root apex arise.

The law of origin is the most important of the embryogenic laws. The principal branch of embryogeny, embryogenesis, rests on this law. This branch may be defined as the science of the origin of the embryo and its component parts.

Law of Numbers:- The number of cells produced by different cell generations varies with the species and depends on the rapidity of the segmentation in the cells of the same generation.

The rapidity of the segmentation of a cell is calculated in determining the number of elements which this cell engenders in terms of successive generations. The number of elements engendered conditions the rapidity of the segmentation; it is the essential factor of determination. The law of numbers is identical with the law of rapidity and may be described in the same manner.

This law, like the other embryogenic laws, becomes evident in fundamentally different characters. It permits, among others, the establishment of particularly instructive resemblances between certain species.

The terminal cell of the two-celled proembryo generally segments more rapidly than does the basal cell and usually gives rise to all of the embryo proper. The basal cell usually contributes no more than the suspensor and portions of the root apex and divisions occur less rapidly. There are instances, however, in which the basal cell segments more rapidly to form an extensive structure, and it may also contribute a greater portion of the embryo proper.

1. *Rules of the first degree.*— Three examples of this first category of rules are available: (*a*) the first corresponds to those in which the segmentation proceeds at an equal rate in both the terminal and the basal cells; (*b*) the second to those in which the terminal cell segments more rapidly than the basal cell; and (*c*) the third to those in which the basal cell segments more rapidly than the terminal cell.

(*a*) In the first case, segmentation proceeds at an equal rate in both cells, so that a total of four elements appears in the second cell generation, eight in the third, sixteen in the fourth, and so on. Half of these cells at each generation are derived from the terminal cell and half from the basal cell. Examples of this type of segmentation occur in *Polygonum persicaria*, *Myosurus minimus* and in those species conforming to the Senecio Variation, Asterad Type.

(*b*) In the second case, the terminal cell segments more rapidly than does the basal cell. In the Onagrad Type and in the Juncus Variation thereof, for example, the difference becomes evident during the second cell generation; at this stage the proembryo remains tricellular for a long time, presenting two upper juxtaposed cells and a lower cell representing the undivided basal cell. The differences become even more pronounced during the succeeding stages. In *Nicotiana* and *Euphorbia* the variation in rapidity appears during the third cell generation: the proembryo, in lieu of being octocellular, is composed of only six elements, of which four originated from the terminal cell and two from the basal cell. Then, at the fourth cell generation, the proembryo consists of twelve elements instead of sixteen.

(*c*) In the third case, the basal cell segments more rapidly than does the terminal cell. This peculiarity frequently appears in the first stages. In *Capsella bursa-pastoris*, for example, it appears most often during the second cell generation in the tricellular embryo, in which the two lower cells arose from the basal cell and the terminal cell remains undivided. Again, in the following generation there is a six-celled proembryo consisting of two upper juxtaposed cells derived from the terminal cell and a row of four lower elements produced by the basal cell. However, after the formation of the quadrants, these differences become effaced: segmentation in the lower elements slackens, while the divisions increase in the portion derived from the terminal cell. Among other species, particularly in *Erodium cicutarium*, the difference in rapidity of segmentation is maintained for a much longer time in favor of the basal cell. At the third cell generation, the six-celled proembryo includes two upper cells engendered by the terminal cell and four lower cells produced by the basal cell; at the fourth generation the proembryo has become twelve-celled, generally with four cells issuing from the terminal cell and eight derived from the basal cell. The same proportions are maintained during the succeeding cell generations.

2. *Rules of the second degree.*— If, instead of the first two proembryonic cells, two elements of the second cell generation are considered (*e.g.*, the two daughter cells of the terminal cell, at the tetrad stage), it is possible to establish other laws of numbers which might be called rules of the second degree.

(*a*) In a "T-form" tetrad these two cells are juxtaposed, symmetrically oriented at the right and left of the proembryonic axis. These represent the mother cells of the quadrants; they are homodynamous and, although independent, proceed with the same degree of rapidity in their segmentations. It is possible to follow their divisions and enumerate very precisely the elements to which they give rise during the differentiation of the three histogens in their interior, particularly throughout all the Variations of the Onagrad Type in which the separation of the primordial cells of these three histogens is produced directly by tangential partitionings. Each of these two cells engenders, at exactly the same time, two quadrants in the first cell generation, four octants (two upper and two lower) at the second cell generation, eight elements at the third cell generation (of which four cells are epidermal initials and four are subepidermally internal) and sixteen cells at the fourth cell generation, of which eight are epidermal initials, four are periblem initials and four are plerome initials.

(*b*) In the "linear tetrad" the two daughter cells of the terminal cell are superposed; they maintain, as in the preceding case, an equal rate of division until their third or fourth cell generation; but most often, starting from their separation, segmentation proceeds more rapidly in one than in the other and the number of cells produced by each is therefore very different. These two upper cells of the linear tetrad are generally designated by the letters *l* and *l'*.

When these two upper cells are segmented by vertical walls they divide almost synchronously until the third, sometimes even until the fourth, cell generation and produce an equal number of cells.

When, of the two upper cells of the tetrad, cell *l* divides vertically and cell *l'* transversely, a difference in the rapidity of subsequent segmentations may appear either in favor of *l* or in favor of *l'*. Thus, in *Samolus valerandi*, *l* gives two juxtaposed cells in the first cell generation and four quadrants disposed in a horizontal plane in the second cell generation, then eight octants in the third generation, and in the fourth, by tangential divisions of the octants, sixteen elements of which eight are epidermal initials and eight are subepidermal internal cells. At the same time, cell *l'* engenders two superposed cells during the first cell generation, four cells disposed in a "T-shape" during the second cell generation, six elements only during the third, of which four are circumaxial and two are lower and superposed, and finally during the fourth cell generation twelve elements of which four are epidermal initials, four are subepidermal and internal and four are inferior and newly disposed in a "T-shape." In this instance, the difference in rapidity of segmentation is in favor of cell *l*. The reverse occurs in *Campanula patula*. The subterminal cell, *l'*, of the linear tetrad in this species produces more cells than does the upper cell, *l*.

3. *Rules of the third degree.*— The third cell generation is the one to which rules of the third degree apply and these are considered mainly in relation to the epiphysis and hypophysis initials, particularly the latter. These are both cells which have become specialized to serve clearly determined constructive functions. Both ordinarily divide by transverse walls at the first segmentation and engender a variable number of elements to construct the epiphyseal and hypophyseal tissues respectively.

The hypophysis initial of *Saxifraga granulata* may be taken as an example. This cell divides by a curved transverse wall shaped like a watch glass into two superposed elements which are designated by the letters *ec* and *co*; the upper one in general produces the initials of the root cortex while the lower one forms the primordium of the root cap. In the first cell generation *ec* engenders two cells oriented one under the other; at the second generation four cells in two dyads with one superior and the other inferior; and at the third generation eight elements in two tiers of four circumaxial cells. During this period, the element *co* gives, at the first generation, two juxtaposed cells; four circumaxial cells at the second generation; and at the third generation these four cells remain undivided. Segmentation proceeds more rapidly in *ec* than in *co*; the former produces eight cells, the latter four.

Law of Disposition:- In the course of normal embryonic development, the cells are constituted by divisions in clearly determined directions and appear to occupy positions in accordance with the role which they must play.

The laws thus far considered concern the relation of cells *to time*, while the law of disposition permits the definition of their relations *to space*. All of the elements which contribute to the construction of the embryo might be considered as the stones of which an edifice is built and with the aid of the language of geometry and architecture they may be used to determine clearly the form of these elements and to give the clue to their mode of groupment.

All the dispositions which might be envisaged are fundamentally governed by the axial symmetry which is presented, in a general manner among the higher plants, by the zygote and by the proembryo derived therefrom. The primordial embryonal cell possesses, in effect, by reason both of its external form and by its internal organization, a very distinct polarity and an axis of symmetry which will also be the axis of growth of the plant. At its upper germinative and cauline pole, negative geotropism is exhibited; at its lower vegetative and radicular pole, positive geotropism finds its point of application.

The direction of the walls of segmentation may be defined by taking account of their axial symmetry and by referring to the geometric figures which obey this symmetry. Cells separated by walls are thus similar to geometric solids; they possess the same faces, edges and angles which allows the precise statement of their forms and the comprehension of their mode of assemblage. A proembryonic stage may be unicellular or pluricellular; in a pluricellular stage, the elements of which it is composed offer diverse modes of juxtaposition which the law of disposition endeavors to describe. This branch of embryology which seeks to establish the laws governing the grouping of cells, the relations which they offer to each other in space, which are applied to describe all the architectural dispositions of the embryonic body, constitutes embryotectonics.

We may now proceed to discuss the application of the law of disposition during the course of the first four cell generations. In order to do so, the number of tiers constituted in terms of these four stages and the cellular composition of each stage is determined; otherwise, the distribution of the cells in the vertical sense, according to the axis of symmetry, and in the horizontal sense, in the different planes perpendicular to the axis.

At the first cell generation the proembryo presents two cells; the wall of segmentation is longitudinal, more or less oblique, or transverse, and the two cells do not present more than three dispositions: they are juxtaposed, more or less inclined one on the other, or superposed; and their form is more or less hemispherical. The first disposition is definitely known for one species, *Balanophora dioica*, and may be formed in certain species in which oblique walls so closely approach the vertical that they may be described as such (as in the Scabiosa Variation, Piperad Type). The third disposition occurs most commonly.

At the second cell generation, four dispositions are possible for the proembryonic tetrad derived from a bicellular proembryo composed of two superposed elements. In the first case, which is also the simplest one, it depends on the theoretical direction of the kinetic centers: the four cells are disposed in a tetrahedron, the directions of the divisions having been established according to two straight lines perpendicular to the preceding division and perpendicular among themselves, one parallel and the other perpendicular to the plane of the figure. In the second case the directions of the divisions are again perpendicular to the direction of the preceding division, but parallel among themselves, and the four new cells are disposed as before in two tiers of two elements each but in the same vertical plane. In the third case, one of the divisions is effected perpendicularly to the preceding division, the other parallel (*i.e.*, transversely); there is thus constituted two upper juxtaposed cells and two lower superposed cells — the tetrad presents three tiers, the so-called "T-shaped" tetrad. In the fourth case, the directions of the two divisions are parallel to the preceding one, oriented transversely, and the four elements thus engendered are superposed; there are four unicellular tiers; the tetrad thus has been described as "linear."

All these four dispositions have been observed among the Angiosperms; the last two are far more common than the first two. The "T-shaped" tetrad has been observed among almost all monocotyledons and often among dicotyledons; the "linear" tetrad thus far is known definitely among dicotyledons only. A reverse "T-shaped" tetrad, that is ⊥, composed of two upper superposed cells and two lower juxtaposed cells, has sometimes been produced by certain dicotyledonous genera (*Sedum*, *Chelidonium*), but it is capable of diverse interpretations and is of no particular significance. The first two dispositions occur normally among the Pteridophyta.

In the first two types of tetrads described above, the directions of the mitoses are clearly perpendicular to those that preceded them. In the first case, the walls are disposed vertically, parallel to the plane of the figure in the two upper elements and perpendicularly to the same plane in the two lower segments; in the second case, all the new walls are likewise vertical, but included in the plane of the figure. It results in two similar morulae, composed of eight elements corresponding to true octants. When a tetrad of either of these two types occurs among the Angiosperms, the vertical walls which are directed toward the lower radicular pole never exactly reach this pole; there is always a single lower cell and the axial symmetry is altered for the moment; the directions of the subsequent walls appear variable and the cells assume less regular dispositions. These processes of segmentation result in the construction of forms of geometrically spiral or irregular types and in this case it is difficult to invoke the law of disposition since the mode of assemblage offers neither constancy nor symmetry. The same remarks apply when, after the formation of a "T-shaped" or "linear" tetrad, the succeeding walls become inclined in all directions and the internal symmetry disappears.

Nevertheless, among the Angiosperms, there are usually developed from the derivatives of the "T-shaped" and "linear" tetrads forms attached to the geometric types called orthoradial, parallel or mixed, among which it is possible to distinguish a division in superposed tiers and an internal symmetry in the grouping of the constituent cells. In the case of the "T-shaped" tetrad, the eight cells generally engendered during the third cell generation may assume one of three dispositions. To these dispositions there may be attached a frequent modification (*Erodium*, *Geum*) resulting from the oblique direction of the upper wall, which is succeeded by two other walls perpendicular between themselves and perpendicular to the preceding wall; this gives rise to a tetrahedral grouping of the four new elements, which in the three other groupings are disposed regularly around the axis in a horizontal plane. These elements are described as quadrants, but they are not the true quadrants of the terminal cell and not those of the primordial embryonic cell as in the Pteridophyta. The four other cells of the eight-celled proembryo are disposed sometimes in three tiers or in a T-shape, sometimes in four unicellular tiers, or sometimes in two bicellular tiers. In the case of the "linear" tetrad the eight cells may present five dispositions. The four upper cells, corresponding to the quadrants, are always disposed in a vertical plane, of two tiers of two quadrants in the first two types; in three tiers, of which one is bicellular and the others unicellular, in the next two types; and in four unicellular tiers in the fifth type. The four lower cells are disposed in three tiers or in a T-shape in the first type and in four unicellular tiers in all the others. When a difference in the rapidity of the segmentations appears in the tetrad, sometimes in favor of the two upper cells or sometimes in favor of the two lower ones, hexacellular forms arise which may affect diverse dispositions. Three such forms occur most frequently: the first two are in the form of a "T-shaped" tetrad, the third in that of a "linear" tetrad. In the first and third of these types there are two lower cells which have remained undivided, while the two upper cells engender quadrants; in the second type the two upper cells have, on the contrary, remained undivided while the four elements constituted by the derivatives of the two lower cells are disposed in three tiers, of which one is bicellular and the others are unicellular. Among certain species, such as those belonging to the tribe *Chelidoneae* of the *Papaveraceae*, the hexacellular proembryo may assume any one of eight different dispositions; of these four are derived from a reversed "T-shaped" tetrad and the other four from a linear tetrad; between the forms consisting of three bicellular tiers and those of six unicellular tiers, there are all sorts of intermediate types.

At the fourth cell generation, numerous dispositions may be produced; only the principles will be discussed.

1. When the segmentations are synchronous, the proembryo consists of sixteen cells. The dispositions which the proembryos assume are dependent upon the different types of eight-celled proembryos differentiated during the third cell generation. Thus, in the first type of eight-celled proembryos derived from a "T-shaped" tetrad, two dispositions are possible. In one, transverse walls divide the quadrants into two tiers of octants; in the second, the walls are oblique and separate eight new cells which remain sensibly in the same plane. The other elements are arranged in the same manner in both instances. It is to this second disposition that the general law is applied which regulates the mode of grouping of cells among the forms which appear to assume the character of the most primitive Angiosperms. This law may be stated thus: The number of cells in a given tier is equal to the sum of the cells comprising all the tiers located beneath it.

2.· When a difference in the rapidity of segmentations occurs, either twelve-celled or ten-celled proembryos may appear during the fourth cell generation; the first includes four unicellular tiers below the octants, and the second includes two similarly unicellular tiers. If the basal cell of the two-celled proembryo remains wholly undivided, then the laws of development apply solely to the terminal cell. Four interesting modes of disposition which occur as a consequence may be described.

In *Sagittaria sagittaefolia* the basal cell becomes differentiated into a large haustorial vesicle; the terminal cell alone segments and replaces the primordial embryonic cell in its functions. It behaves exactly as the latter ordinarily would in the case of forms assigned to the Myosurus and Polygonum Variations; by successive bipartitionings it gives rise to a two-celled proembryo consisting of two superposed elements, then a "T-shaped" tetrad, an eight-celled proembryo of four tiers and finally a sixteen-celled proembryo of six tiers. — In *Sagina procumbens* the basal cell is likewise converted into a haustorial vesicle and the terminal cell alone segments. It possesses a linear tetrad, an eight-celled form in five tiers, then a proembryo of sixteen cells distributed in six tiers. In all these forms engendered by the terminal cell, the cells are disposed in a manner which may be very closely compared to the mode of groupment of the cells, derived directly from the primordial embryonic cell, in the forms corresponding to the Chenopodiad Type. — In *Saxifraga granulata* the basal cell becomes swollen but does not remain undivided; it produces four to eight large cells which to not enter into the construction of the embryo proper. The latter, as in the preceding example, takes its origin directly and entirely from the terminal cell, which forms a "T-shaped" tetrad, then at the quadrant stage a six-celled proembryo in three tiers, and at the octant stage, a twelve-celled proembryo of five tiers. In these forms, again, when examined anew under the reverse of the disposition at the fourth cell generation, it will be found that the position of the four lower cells conforms to those which are differentiated directly, as in *Euphorbia exigua*, from the derivatives of the primordial embryonic cell. — In *Androsaemum officinale* the basal cell does not become enlarged but segments transversely several times and gives rise to the lower portion of the filamentous suspensor. The terminal cell alone contributes to the construction of the embryo proper; it engenders a "T-shaped" tetrad, then six- and twelve-celled forms analogous to those of *Saxifraga granulata* in so far as the disposition of the cells is concerned. It may be added that all these types of construction, considered with the basal cell ignored, assume, from the point of view of comparative embryology, a very particular significance in the sense that it permits the understanding of all transitions between the forms in which the terminal cell gives rise to the embryo proper (*Capsella*, *Myosurus*) and those in which the embryo proper takes its origin from a terminal cell descended in the *n*th degree from the original terminal cell.

Finally, it is necessary to remark that in the multiple dispositions which embryonic forms may present, it is desirable to make allowance for two aspects: on the one side are ranged those which remain constant for a given species, and on the other side those which merely represent diverse manifestations of an interspecific polymorphism. To determine the differential specific characters, the first group is evidently of more value than the second; the second group, which contributes proof of a fault in the purity of the species concerned, serves only to inform us, according to the extent of their variations, of the role played by the intrinsic factors of hybridization.

Law of Destination:- The cells of the proembryo of a given species, when the development is normal, give rise to clearly determined parts, and always to the same parts, of the embryonic body.

This law appears to admit, in the fashion of a postulate, the specificity of the cell, besides which it affirms that each of these has a determined role to play, peculiar to itself, at the moment when it becomes separate, and reveals the constructive potentialities which are proper to it. Among plants the cells are not predetermined; they are isotropes or toti-potentials and their specificity is not evolved in proportion as the segmentation proceeds. Besides their true, actual potentialities, they conserve a more or less considerable total of virtual potentialities, thanks to which they may produce other parts of the body than those to which they normally give rise, provided the ordinary conditions of development become modified.

The law of destination is related very closely to that of disposition. It states, in effect, that the embryo and the plant represent a vertical edifice, that the fundamental regions of the body are situated in this edifice, at different levels corresponding to the uni- or pluricellular embryonic stages; while the law of disposition determines the mode of separation and the internal composition. It suffices generally to indicate the connection of these stages with the fundamental regions of the embryonic body or the order of origin of these organs to know the destination of the cell.

The law of destination, like the three preceding laws, reveals the really essential differences in embryogenic characters; it permits the establishment of the constant connections which, directly compared, constitute a source of precise indications of specific relationships. The determination of these relationships, which is the object of this branch of embryology, has been designated by the name embryogenergy and is in effect the science of work performed by the embryo, that of the constructive role of its parts and in sum the destination of the cells.

The relationships of the cells and the fundamental regions of the embryonic body, those which have previously been defined, have been established at different stages of development. Thus, the law of destination embraces a system of laws of several degrees, which allows one to arrive at more and more precise results and to the separation of a greater and greater number of groups of species. The laws of the first degree make known the destination of the two cells engendered at the first cell generation; the laws of the second degree the destination of the cells produced at the second cell generation, the laws of the third degree envisage the destination of the cells of the third cell generation, and so on. From the practical standpoint, and being given the actual state of our knowledge, it is scarcely possible to go beyond the third cell generation. No examples have been taken from among the Monocotyledons; by the lateral origin of their stem apices, they introduce complications in the tables of destination of the elements. Only the Dicotyledons are considered, and again only those of such plants as possess a geometric type of orthoradial, linear or mixed development, thus permitting the erection of recapitulative tables which are clearly distinct and easy to describe. Those dicotyledonous species attached to a spiral or irregular geometric type are not considered; to establish the laws with which these are concerned it would be necessary to enter into overly long discussions and to interpose interpretations which might be open to question.

1. *Laws of the first degree.* — The abbreviations which will be employed hereafter have already been described. The laws of the first degree concern the two cells resulting from the first division of the zygote; namely, the terminal cell (*ca*) and the basal cell (*cb*). Five specific groups are known from among the Dicotyledons.

The first group is defined by the formula:

$$ca = pco + pvt$$
$$cb = phy + icc + iec + co + s$$

This formula signifies that the terminal cell engenders the cotyledonary portion and the stem apex; that the basal cell gives rise to the hypocotyledonary portion, to the initials of the central cylinder of the root, to the root cap and to the suspensor. To this group belong such species as *Senecio vulgaris*, *Urtica pilulifera*, *Malva rotundifolia*, *Muscari comosum* and the tribe *Potentilleae* of the *Rosaceae*. *Lamium* and *Valerianella olitoria*, which have an irregular type of development, also belong in this group. All these species are assigned to the Asterad Type.

The second group presents the following regularity at the first generation:

$$ca = pco + pvt + \tfrac{1}{2}phy$$
$$cb = \tfrac{1}{2}phy + icc + iec + co + s$$

Here the terminal cell produces the cotyledonary portion, the stem apex and a portion of the hypocotyledonary axis, the second part of which is engendered by the basal cell. Belonging to this group are *Polygonum persicaria*, *Chenopodium bonus-henricus*, *Salix triandra* and *Myosotis hispida*. Some of these species are assigned to the Asterad Type, but most belong in the Chenopodiad Type.

The third group possesses the following formula:

$$ca = pco + pvt + phy + icc$$
$$cb = iec + co + s$$

In this case the terminal cell engenders the entire hypocotyledonary portion together with the initials of the central cylinder of the root. All species which conform to this formula are assigned to the Onagrad Type.

The fourth group is defined by the following formula:

$$ca = pco + pvt + phy + icc + iec$$
$$cb = co + s$$

Among the species included in this group, the basal cell produces only the central portion of the root cap and the suspensor; all the other portions of the embryo take their origin from the terminal cell. Among the genera may be mentioned *Nicotiana*, *Hyoscyamus*, most of the species in *Solanum* which develop regularly, *Petunia*, *Schizanthus* and *Papaver*. Also attached to this group are members of the tribe *Chelidoneae* of the *Papaveraceae*, the *Umbelliferae*, *Rubiaceae* and many species of *Solanum* which possess irregular and complex embryonic forms. Practically all such forms are assigned to the Solanad Type.

The fifth group may be characterized by the following formula:

$$ca = pco + pvt + phy + icc + iec + co + s \text{ (in part)}$$
$$cb = s \text{ (in part)}$$

In this group the terminal cell engenders all of the embryo proper as well as a portion of the suspensor, while the basal cell contributes a suspensor *sensu stricto* or only the lower portion of this suspensor which is often differentiated as a large haustorial vesicle. Belonging to this group are all species placed under the Caryophyllad Type.

2. *Laws of the second degree.* — These concern the destinations of the cells of the second cell generation. In order to establish these, each of the five groups separated according to the laws of the first degree will be taken up in order, and, in each of these groups, progress will be to new groups based on the differences which are revealed by the new formulae. The laws of the third and fourth degree may be established in the same manner and are based on the divisions which go along with the laws of the second and third degrees.

The first group consists, for the moment, of the species which all present at the second cell generation a "T-shaped" tetrad in three tiers, *q*, *m* and *ci*, whose destinations are represented in the following formula:

$$\begin{aligned} q &= pco + pvt \\ m &= phy + icc \\ ci &= iec + co + s \end{aligned}$$

These formulae are the same for all species in the group; they do not allow the erection of subgroups. At the third cell generation the proembryo is eight-celled and divided into four tiers; the destinations of these layers offer the differences which permit the separation of *Senecio vulgaris* and all the other species which conform to the same embryonic type on one side (A) and the tribe *Potentilleae* of the *Rosaceae* on the other (B):

$$\left.\begin{aligned} q &= pco + pvt \\ m &= phy + icc \\ n &= iec \\ n' &= co + s \end{aligned}\right\}\text{A} \qquad \left.\begin{aligned} q &= pco + pvt \\ m &= phy + icc \\ n &= iec + co + s \text{ (in part)} \\ n' &= s \text{ (in part)} \end{aligned}\right\}\text{B}$$

It will be observed from these formulae that the differences which cause the separation reside in the destinations of the two lower elements, *n* and *n'*, separated at the third cell generation, at the expense of the lower cell of the tetrad, *ci*, differentiated during the second cell generation.

In the second group, the tribe *Persicarieae* of the *Polygonaceae* and *Salix triandra* possess a "T-shaped" tetrad in three tiers at the second cell generation, while *Chenopodium bonus-henricus* and *Myosotis hispida* have a "linear" tetrad of four tiers. These formulae are clearly different and allow the disposition in the same subgroup of the first two species (A), which remain united during the following generations; then in the second subgroup of *Chenopodium* (B) and a third subgroup including *Myosotis* (C):

$$\left.\begin{aligned} q &= pco + pvt + \tfrac{1}{2}phy \\ m &= \tfrac{1}{2}phy + icc + iec \\ ci &= co + s \end{aligned}\right\}\text{A} \quad \left.\begin{aligned} l &= pco + pvt \\ l' &= \tfrac{1}{2}phy \\ m &= \tfrac{1}{2}phy + icc + iec \\ ci &= co + s \end{aligned}\right\}\text{B} \quad \left.\begin{aligned} l &= pco + pvt \\ l' &= \tfrac{1}{2}phy \\ m &= \tfrac{1}{2}phy + icc \\ ci &= iec + co + s \end{aligned}\right\}\text{C}$$

The differences between B and C reside in the destinations of the two lower elements *m* and *ci* of the "linear" tetrad. During the succeeding generations the differences between these two subgroups become accentuated and lead to the formation of new separations.

In the third group, all the species possess at the second cell generation a "T-shaped" tetrad in three tiers; but the destinations of these tiers are different and the erection of two subgroups is thus permitted:

$$\left.\begin{aligned} q &= pco + pvt + phy + icc \\ m &= iec + co \\ ci &= s \end{aligned}\right\}\text{A} \qquad \left.\begin{aligned} q &= pco + pvt + phy + icc \\ m &= iec + co + s \text{ (in part)} \\ ci &= s \text{ (in part)} \end{aligned}\right\}\text{B}$$

To subgroup A belong *Myosurus minimus, Euphorbia exigua,* the *Onagraceae, Reseda luteola,* which the law of destination does not allow to be separated during the following generations; to the subgroup B are assigned *Ruta graveolens, Capsella bursa-pastoris,* and all species with a more or less elongated filamentous suspensor which conform to the same embryonic type (*Scrophulariaceae, Mentha, Glechoma, Plantago, Lythrum, Verbena, Hippurus*). In this last subgroup they constitute at the third cell generation an eight-celled proembryo arranged in six tiers, whose destinations offer new differences and again permit two divisions:

$$\left.\begin{aligned} l &= pco + pvt \\ l' &= phy + icc \\ d &= iec + co \\ f + n + n' &= s \end{aligned}\right\} \text{a} \qquad \left.\begin{aligned} l &= pco + pvt \\ l' &= phy + icc \\ d &= iec + co + s \text{ (in part)} \\ f + n + n' &= s \text{ (in part)} \end{aligned}\right\} \text{b}$$

In division *a* are ranged *Ruta graveolens, Lythrum salicaria,* and *Mentha viridis,* species in which the cell *d* which occupies the apex of the suspensor is converted into the hypophysis cell; in division *b* are placed those species in which the hypophysis cell originates from a daughter cell, granddaughter cell or even a later daughter cell of *d,* while the other descendants of *d* enter into the construction of the suspensor. However, these differences are of secondary value and, moreover, are subject to variation. All the species of this group are closely related.

The fourth group consists of those species which, at the second cell generation, possess a "linear" tetrad of four tiers, *l, l', m* and *ci.* The law of destination separates these into two groups. To the first (A) belong the *Solanaceae*; to the second (B) is assigned *Papaver rhoeas*:

$$\left.\begin{aligned} l &= pco + pvt \\ l' &= phy + icc + iec \\ m &= co + s \text{ (in part)} \\ ci &= s \text{ (in part)} \end{aligned}\right\} \text{A} \qquad \left.\begin{aligned} l &= pvt \\ l' &= pco + phy + icc + iec \\ m &= co + s \text{ (in part)} \\ ci &= s \text{ (in part)} \end{aligned}\right\} \text{B}$$

The first is again split up at the third cell generation. In *Nicotiana,* for example, the proembryo produced at this stage presents the same four tiers differentiated during the second cell generation. In *Hyoscyamus, Atropha, Petunia,* etc., the eight-celled proembryo engendered is separated into eight tiers, the two cells *m* and *ci* having been segmented transversely to give a total of four superposed elements; the upper daughter cell of *m* produces the primordium of the root cap, *co,* and the three other elements form a short suspensor. It is true that during the fourth cell generation these differences between the various species of the *Solanaceae* are effaced or become less and less important, the modifications on which they depend consisting in the variable number of cells of which the lower portion of the suspensor is made up. This, after all, does not essentially change the law of destination. The differences appearing during the second and third cell generations conserve all their value; they are related to the law of numbers and give evidence of precocious differentiation.

The fifth group consists of those species in which the basal cell of the two-celled proembryo does not participate in any manner in the construction of the embryo proper. It produces a large haustorial vesicle, sometimes the lowermost elements of the suspensor or finally sometimes all of a suspensor *sensu stricto.* It is therefore entirely unnecessary to envisage their destinations in the construction of the fundamental regions of the embryonic body or of the plant. The laws of development, in this case, are applicable to the terminal cell alone, which replaces the primordial embryonic cell (the zygote) in its constructive functions.

At the second cell generation in the zygote, otherwise the first division of the terminal cell, there is produced at the expense of this cell two superposed cells, *cc* and *cd*, on which the laws of destination may be based. This law leads to the separation of five well marked divisions.

I. In the first case, the destinations of the two cells, *cc* and *cd*, give the following formula:

$$\left.\begin{aligned} cc &= pco + pvt + phy + icc \\ cd &= iec + co + s \end{aligned}\right\} \text{A}$$

Sedum acre, Saxifraga granulata and the *Hypericaceae* may be ranged in this division. Among all these species the destinations of the cells are the same at the first four cell generations; the differences which have been observed reside solely in the disposition, juxtaposed or superposed, of the lowermost elements connecting the embryo proper with the cells produced by the basal cell.

II. In the second case, the destinations of the cells *cc* and *cd* lead to the following expressions:

$$\left.\begin{aligned} cc &= pco + pvt \\ cd &= phy + icc + iec + co + s \end{aligned}\right\} \text{B}$$

Sagina procumbens and *Campanula patula* respond to this formula. The latter becomes separated from the former at the second cell generation; it gives a "T-shaped" tetrad stage in three tiers (a), while the other gives a "linear" tetrad in four tiers (b); the destinations of the elements of these two tetrads are evidently different:

$$\left.\begin{aligned} cc &= pco + pvt \\ ph &= phy + icc \\ h &= iec + co + s \text{ (upper portion)} \end{aligned}\right\} \text{a}$$

$$\left.\begin{aligned} l &= pvt \\ l' &= pco \\ m &= phy + icc + iec \\ ci &= co + s \end{aligned}\right\} \text{b}$$

From other points of view, the differential characters between these two species are very numerous.

III. In the third case, the formulae relative to the law of destination of the two cells *cc* and *cd* may be written as follows:

$$\left.\begin{aligned} cc &= pco + pvt + phy \\ cd &= icc + iec + co + s \text{ (upper part)} \end{aligned}\right\} \text{c}$$

Only *Samolus valerandi* thus far is known to conform to this formula.

IV. In the fourth case the formula is:

$$\left.\begin{aligned} cc &= pco + pvt + phy + icc \\ cd &= iec + co \end{aligned}\right\} \text{d}$$

At the moment *Drosera rotundifolia* is the only species which responds to this formula.

V. In the fifth case, the formula of the law is thus modified:

$$\left.\begin{aligned} cc &= pco + pvt + phy + icc + iec + co \\ cd &= s \text{ (for the greater part)} \end{aligned}\right\} \text{e}$$

Only *Medicago lupulina* so far is known to belong to this final group.

ANTHOPHYTA

Chapter XVII

TYPES AND VARIATIONS

In by far the great majority of species in the Anthophyta the first wall formed in the zygote is transverse, but there are authentic instances of the more or less vertical partitioning of the zygote, as in *Balanophora dioica* and *Peperomia pellucida*. All cases of longitudinal segmentation of the zygote are assigned to the Piperad Type.

Six principal types of embryonomy are recognized in the Anthophyta and each is based upon the operation of each and every one of the embryogenic laws discussed in the preceding chapter. Specifically, outside of the Piperad Type, the separation of one Type from another is based primarily on the behavior of the terminal cell of the two-celled proembryo during the second and subsequent cell generations, and less directly upon the behavior of the basal cell but decidedly upon the fact whether this cell contributes something or nothing to the construction of the embryo proper.

Description of the Types:- The six types are perhaps best described according to the following key:

A. The zygote divides by a wall which is either vertical or so obliquely oriented that it may be assumed to be essentially longitudinal .... **Piperad Type**

B. The zygote divides by a transverse wall.

- I. The terminal cell divides by a longitudinal wall during the second cell generation.
  - a. The basal cell has no or only a minor part in the construction of the embryo proper .... **Onagrad Type**
  - b. Both basal and terminal cells contribute to the construction of the embryo proper .... **Asterad Type**
- II. The terminal cell divides by a transverse wall during the second cell generation.
  - a. The basal cell has no essential part in the formation of the embryo proper.
    - 1. The basal cell undergoes no further division and becomes a large suspensor cell (when a suspensor of two or more cells is formed, these are always derivatives of the terminal cell) .... **Caryophyllad Type**
    - 2. The basal cell usually forms a suspensor of two or more cells .**Solanad Type**
  - b. The basal cell contributes more or less to the construction of the embryo proper .... **Chenopodiad Type**

The majority of accounts which have been described in the literature in sufficient detail can be fitted into one or another of the Types. Most of the reports, moreover, can be fitted into the Variations which have been created under each of the Types. During the long course of evolution of individual species it is inevitable that minor but significant departures have been made from each of the generalized Types. The erection of new and additional Variations is to be expected as more and more species are thoroughly investigated. The point which should be borne in mind is that the six Types as described are incontrovertibly fundamental since they are logically based upon the behavior of the two cells resulting from the first division of the zygote and their respective contributions to the sum total of the mature embryo. There is no other logical interpretation of the sequence of events which culminate in a viable embryo.

The major Types, together with the important Variations assigned to each, may now be described in detail. In the following chapter all families in which sufficient work has been published will be taken up in sequence and the nature of the embryonomic Type or Types prevalent in each family will be described.

I. Piperad Type

This Type includes all known instances in which the zygote is segmented by a strictly longitudinal wall or by one that is so inclined to the vertical that it may be considered as being longitudinal in the light of subsequent behavior of the two daughter cells. The species assigned to this Type are relatively few in number and all belong to dicotyledonous families.

It is with some misgiving that the root of the family *Piperaceae* is used to designate this Type inasmuch as the genus *Piper* is practically unknown embryologically. *Peperomia pellucida* is the species upon which the diagnosis of the Type rests, since the other genera (*Balanophora* and *Scabiosa*) upon which the Variations are based belong to families whose names for valid reasons can not be used. The *Balanophoraceae* consist principally of species which reproduce apomictically, while *Scabiosa* is the only investigated species in a large family (*Dipsacaceae*) and there is too much danger that other species in that family may be found not to agree embryonomically with *Scabiosa*.

The Variations so far recognizable differ sharply, as is indicated by the following descriptive key:

Second walls vertical in each daughter cell of the two-celled proembryo.
  Second walls not curved .... **Piperad Type**
  Second walls curved and attached at both ends to the primary longitudinal wall ... . **Balanophora Variation**
Second walls transverse (*i.e.*, perpendicular to the primary wall) in each daughter cell.
  Suspensor and cotyledons rudimentary; multiple embryos not formed.................................... **Scabiosa Variation**
  Suspensor greatly elongated, often tortuous; multiple proembryos often formed; cotyledons presumably lacking .... **Dendrophthoe Variation**

The principal feature of embryos conforming to the Piperad Type is that histogens, with the general exception of epidermal initials or epidermal cells, are ordinarily unrecognizable. A suspensor and cotyledons are also not formed, except for a rudimentary suspensor and cotyledons in *Scabiosa*. In *Balanophora* the reduction in the number of cells in the proembryo reaches an extreme.

II. Onagrad Type

The embryogeny of the family *Onagraceae* is the most simplified and the construction of the embryo is decidedly schematic. There are, of course, exceptions to this rule, particularly in the genera *Zauschneria* and *Epilobium* and in certain Variations. The Type is based upon the embryonomy of *Godetia amoena*, *Oenothera biennis* and other species, in all of which the development is absolutely identical.

At the first cell generation the destination of the two cells is:

$$ca = pco + pvt + phy + icc$$
$$cb = iec + co + s$$

At the second cell generation the terminal cell has divided longitudinally and the basal cell transversely:

$$q = pco + pvt + phy + icc$$
$$m = iec + co$$
$$ci = s$$

The feature of this generation is the individualization of the middle cell (m)

which immediately becomes the hypophysis initial and engenders the initials of the root and the root cap. At the third cell generation the two upper juxtaposed cells are each segmented vertically to engender quadrants, while the other two cells have remained undivided. The destinations are still the same. At the fourth cell generation all cells have divided, more or less in the transverse plane:

$$l = pco + pvt$$
$$l' = phy + icc$$
$$m = iec + co = \text{(hypophysis)}$$
$$n + n' = s$$

The quadrants have divided transversely to produce the octant stage and *ci* has divided transversely to produce two cells (n and n') which form the short suspensor. The peculiarity of this generation is the form of the wall produced in cell *m*. In the Type this wall is curved and at both ends is attached to the transverse wall originally separating *ca* from *cb* (or, which is the same thing, *q* from *m*). The result is an upper daughter cell more or less lenticular in shape and a larger lower daughter cell with the general shape of a watch glass. The initials of the root originate from the former and those of the root cap from the latter. In certain of the Variations, as well as occasionally in certain species of the *Onagraceae* (such as *Eulobus sceptrostigma*), the wall in *m* is diagonal or oblique, with only one end attached to the original wall segmenting the zygote. A second diagonal wall is later produced; since a vertical wall is the first to be erected in the lower derivative of *m*, the end result is essentially the same in both cases. Again, as in the Lotus Variation, a hypophysis may be entirely lacking; or cell *m* may segment once or twice transversely and the hypophysis cell is either a daughter or a granddaughter or later derivative of *m*. In still other instances cell *m* or a derivative which functions as a hypophysis cell is segmented vertically. No matter what the plane of the first division in the hypophysis cell, those parts of the embryo which this cell or initial engenders are always the same.

Quadrants are not always formed, as in the Catalpa and Ruta Variations, but eventually octants are produced which give the appearance of being exactly like those in the Type and which have the same destinations.

An epiphysis is present in one Variation only, that of the Trifolium Variation. This particular Variation may possibly more properly be assigned to the Asterad Type since the embryonomy is much like that of the two Variations (*Geum* and *Erodium*) in this Type which also possess an epiphysis.

Variations:- The Variations which have been assigned to the Onagrad Type are best characterized by means of the following descriptive key. In this, as in the keys accompanying the descriptions of the other types, those characters most readily recognized have been chosen wherever possible. When the distinction between two or more Variations is necessarily based upon somewhat obscure ~~details~~, these have been simplified as much as possible.

Mature embryo dicotyledonous or essentially so.
  Quadrants are regularly formed (not always in Lotus Variation).
    Epiphysis absent.
      Hypophysis present.
        End cell or cells of suspensor swollen and vesicular.
          Suspensor elongated with solitary end cell .... **Capsella Variation**
          Suspensor consisting of one tier of eight cells .... **Sonerila Variation**
        End cell of suspensor neither swollen nor vesicular.
          Hypophysis initial is cell *m* of the three-celled proembryo.

First division in hypophysis initial is typically curved. . . . **Onagrad Type**
First division in hypophysis initial is vertical or somewhat oblique (such divisions may abnormally occur in the Onagrad Type).
At third cell generation: six cells disposed in three tiers. . . . . **Euphorbia Variation**
At third cell generation: eight cells disposed in four tiers. . . . . **Myosurus Variation**
Hypophysis cell is a derivative of cell *m* of the three-celled proembryo.
Hypophysis cell is the upper daughter cell of cell *m* . . . **Lythrum Variation**
Hypophysis cell is a granddaughter or later derivative of cell *m*.
First division in hypophysis cell is vertical . . . . **Alyssum Variation**
First division in hypophysis cell is curved (as in Onagrad Type).
At fourth cell generation (12 cells disposed in 6 tiers): eight are derived from *ca* and four from *cb* . . . . **Mentha Variation**
At fourth cell generation (12 cells disposed in 6 tiers): four cells are derived from *ca* and eight from *cb* . . . . **Veronica Variation**
Hypophysis absent . . . . **Lotus Variation**
Epiphysis present . . . . **Trifolium Variation**
Quadrants are not regularly formed.
Daughter juxtaposed cells of terminal cell are both segmented transversely . . . . **Catalpa Variation**
Daughter juxtaposed cells of terminal cell are segmented one transversely and the other longitudinally . . . . **Ruta Variation**
Mature embryo not dicotyledonous.
Basal cell either degenerates quickly or forms a massive structure . . **Lilium Variation**
Basal cell behaves otherwise.
Basal cell becomes two-celled or two-nucleate . . . . **Heloniopsis Variation**
Basal cell contributes the initials of the root cortex and root cap . . . . **Juncus Variation**

It should be noted that the Onagrad Type serves both as a Type and as a Variation simultaneously, as do most of the other Types.

The Variations may now be characterized in fuller detail, but for complete descriptions, reference should be made to the genus concerned under the appropriate family in the succeeding chapter.

Capsella Variation. — This Variation is based upon *Capsella bursa-pastoris*, one of the most comprehensively studied of plants from the embryological viewpoint. There has been considerable misinterpretation by many of the earlier workers, several of whom contended that the basal cell of the two-celled proembryo did not divide further, but who did not actually observe mitoses in this cell. The supreme importance of actually observing mitoses and the orientation of their spindles is thus vividly emphasized.

This Variation differs from the Type in that the difference in rapidity of the sequence of divisions in the products of the terminal and basal cells respectively does not become apparent until the fourth cell generation; in the elongated suspensor; and in that the hypophysis is developed at the expense of an upper daughter derivative of cell *m*.

Euphorbia Variation. — This Variation bears a very close resemblance to that of the Myosurus Variation. The main difference is that in the latter there are eight cells disposed in four tiers at the third cell generation, whereas in the Euphorbia Variation there are six cells disposed in three tiers.

Myosurus Variation. — Species belonging to the family *Ranunculaceae* display a complicated type of embryogeny, whose interpretation is rendered doubly difficult by the presence of curved walls (as seen in transverse sections) during the later stages of development, yet the embryonomy is essentially like that of the Onagrad Type. The first division in the hypophysis cell is vertical.

Lythrum Variation. — In *Lythrum salicaria* the development up to the octant stage corresponds to that in the Type. Cell *m* divides transversely;

the upper daughter derivative becomes the hypophysis cell, while the lower daughter cell together with derivatives of the basal cell form a small suspensor.

Alyssum Variation. — *Alyssum* is closely related to *Capsella* and in general has the same type of embryonomy, but appears to differ at two points. The first divergence is that it is claimed that the basal cell of the three-celled proembryo does not divide further, all of the very long suspensor being derived from the intermediate cell (*m*). The suspicion immediately arises that this is a misinterpretation of exactly the same nature as happened in earlier accounts of the embryogeny of *Capsella*, but against this is the fact that the basal cell at the three-celled stage is extremely long and hence probably incapable of dividing. If the latter assumption is correct, then all the cells of the suspensor, totaling from eight to fifteen, are derived from cell *m*, as is the hypophysis cell. The second divergence is that the first division in the hypophysis cell is vertical. All the other histogens are delimited as in *Capsella*.

Mentha Variation. — This and the following Variations are distinguished by the fact that the hypophysis cell is a granddaughter or even a later derivative of the intermediate cell, *m*.

Veronica Variation. — At the fourth cell generation the Mentha and Veronica Variations are both characterized by having twelve cells disposed in six tiers, but in the Mentha Variation eight of these cells are derived from the terminal cell and four from the basal cell, whereas in the Veronica Variation four are derived from the terminal cell and eight from the basal cell (that is, an exact reversal of the respective derivations).

Lotus Variation. — As with the following Variation, assignment to the Onagrad Type is on the basis of the longitudinal division of the terminal cell of the three-celled proembryo (the other two cells being derived from the basal cell) and the noncontribution of the basal cell to the construction of the embryo proper. This Variation is to the Onagrad Type what the Datura Variation is to the Solanad Type. The Variation is further characterized by the absence of a hypophysis cell.

Trifolium Variation. — This Variation is the equivalent under the Onagrad Type of the Geum or Erodium Variations in the Asterad Type. All essential portions of the embryo are derived from the terminal cell of the two-celled proembryo, while the basal cell forms a short suspensor.

Catalpa Variation. — The difference from the Onagrad Type resides in the mode of formation of the quadrants. There is also less constancy in the processes of development of the proembryonic tetrad, in the succession of divisions in the construction of the cotyledonary portion and in the hypophysis. This Variation is probably closest to the Veronica Variation.

Ruta Variation. — The terminal cell usually divides by an oblique wall and subsequently behaves as in the Geum Variation, Asterad Type. When the terminal cell divides vertically in the typical fashion, the behavior of the daughter cells is peculiar: one daughter divides by a longitudinal wall and the other by a transverse wall. The formation of the octant stage is then brought about by the formation of longitudinal walls in the transversely divided derivatives and by transverse walls in the others.

Lilium Variation. — It is extremely difficult to determine the precise assignment of the monocotyledonous genus *Lilium*. The terminal cell of the two-celled proembryo segments in the longitudinal plane and it is highly probable that the root cap is contributed by the basal cell. Assignment of the genus as a Variation under the Onagrad Type is more tentative than definite.

Heloniopsis Variation. — This Variation is to the Onagrad Type what the Muscari Variation is to the Asterad Type. The basal cell becomes two-celled or two-nucleate and takes no part in the construction of the embryo proper, all of which is derived from the terminal cell.

Juncus Variation. — The embryogeny of the *Juncaceae* provides characteristics comparable to those found among the *Ranunculaceae* (especially to *Myosurus minimus*) and among the *Brassicaceae* (*Alyssum* and *Capsella* particularly). The extreme precocity of the differentiation of the epidermal initials especially characterizes this Variation; this feature occurs in no other Angiosperm (except possibly in *Orobanche hederacea*), and may be considered as marking this Variation as the most highly evolved one under the Onagrad Type. Octants are not formed.

III. Asterad Type

The Asterad Type characterizes practically all investigated species included in the *Asteraceae* (*Compositae, sensu latu*), as well as many species in other families.

Briefly, the four-celled stage consists, as in the Onagrad Type, of three tiers: two adjoining cells derived from the terminal cell of the two-celled proembryo, and two superposed cells (the upper of which is designated as the middle cell, *m*), derived from the original basal cell. Each upper cell divides by a wall oriented perpendicularly to the first vertical wall to form the quadrant stage. These quadrant cells behave like the four upper octant cells of *Capsella* or *Ranunculus*, in the sense that they develop into the cotyledonary region. The middle cell divides similarly into four quadrant cells which lie in a plane and later give rise to the hypocotyl and the plerome initials of the root. Through further divisions of the basal cell, whose sequence differs according to the species, a short filament of cells is formed; from these are derived the initials of the root periblem and the root cap as well as a presumable hypophysis initial. The differentiation of the histogens commences in the quadrant cells derived from the terminal cell by means of oblique walls.

Proof that the basal cell contributes a considerable proportion of the sum total of an embryo developed according to the Asterad Type has been furnished in a most ingenious manner.* *Crepis capillaris* served as the subject; the normal embryo develops exactly according to the Asterad Type. Pollinated capitula were subjected to X-ray irradiation after fertilization had occurred but before the zygotes had divided. The irradiation presumably induced a chromatid aberration (an "A" chromosome deficient for a portion of its long arm). Five or six days later the ovules were fixed and processed into slides. One embryo was obtained which had numerous division figures in which the chromosomes could be clearly observed. In the basal portion of the embryo the complements appeared normal; in the region in or near the cotyledons each figure contained a deficient "A" chromosome. The presence of the deficient chromosome in the upper part of the embryo is most plausibly explained by assuming that the chromatid deficiency was induced at the time of irradiation and that the terminal cell of the two-celled proembryo received the deficient daughter chromosome during the first zygotic mitosis while the normal chromosome went into the basal cell. The cells derived from the terminal cell would then possess chromosome complements which are deficient while the derivative cells of the basal cell would contain only normal chromosomes. On the basis of this explanation of the situation, it is possible to calculate the amount of embryonic tissue in the six-day embryo which was derived from the basal cell

* Personal communication from Dr. Ladley Husted. (A brief abstract of his unpublished results appeared in: Virginia Jour. Sci. 2: 184. 1941.)

by determining the location of the boundary between cells containing a deficient "A" chromosome and those containing normal "A" chromosomes. The result reveals that the terminal cell contributed 64 to 68% of the embryo while the basal cell contributed 32 to 36%. This investigation should remove all doubt whether the basal cell can contribute a substantial proportion of the body of the embryo.

The Variations which are assigned to the Asterad Type may be characterized as follows:

Suspensor present.
- Mature embryo dicotyledonous.
  - Epiphysis present.
    - Tier *n* (upper daughter cell of *ci*) contributes to the construction of the suspensor .... **Geum Variation**
    - Tier *n* does not contribute to the construction of the suspensor .... **Erodium Variation**
  - Epiphysis absent.
    - Tier *m* gives rise to the hypocotyledonary region alone.
      - Suspensor consists of one large vesicular cell and one or two additional cells.
        - Tier *l′* forms the short upper portion of the hypocotyl, tier *m* the remainder .... **Polygonum Variation**
        - Tier *m* alone forms the hypocotyl in its entirety .... **Urtica Variation**
      - Suspensor consists of more than two cells, none being vesicular .... **Lamium Variation**
    - Tier *m* gives rise to certain initials of the root as well as to all of the hypocotyledonary region.
      - Tier *m* gives rise to the plerome initials of the root only .... **Asterad Type (or Senecio Variation)**
      - Tier *m* gives rise to all of the root initials (except for root cap) .... **Oxalis Variation**
- Mature embryo monocotyledonous.
  - Tier *n* engenders the initials of the root proper only.......... **Muscari Variation**
  - Tier *n* engenders the root cap, coleorhiza and epiblast.......... **Poa Variation**

Suspensor absent; basal and terminal cells contribute about equally to bulk of embryo .... **Penaea Variation**

The Variations may now be described in greater detail.

Geum Variation. — In this Variation the peculiarity consists of the formation of oblique walls in the terminal cell which eventually result in the appearance of an epiphysis (primordium of the stem apex) in the developing embryo. The first longitudinal wall in the terminal cell is pronouncedly oblique and one daughter cell (*a*) is placed in a higher position than the sister cell (*b*). The latter is then partitioned by a longitudinal wall and the former by an oblique wall which is at right angles to the wall in *b*. A wedge-shaped upper cell results from the division in *a* and this is the epiphysis initial. The cells subsequently developed from the sister of this cell and from *b* go into the formation of the upper half of the embryo and the cotyledons. In the meantime the middle cell divides longitudinally and the basal cell transversely, so that there are four cell tiers beneath the epiphysis initial, designated *q*, *m*, *n* and *n′* respectively. From *q* arises the cotyledonary region, as was just mentioned, from *m* comes the hypocotyl; from *n* or a daughter cell arises a typical hypophysis which behaves as in *Capsella*, and often a portion of the suspensor, the remainder of which is derived from the basal cell.

Erodium Variation. — An epiphysis initial arises in this Variation precisely as in the Geum Variation and the middle cell goes into the formation of the hypocotyl. The upper daughter cell resulting from the transsection of the basal cell eventually participates in the formation of the embryo but the hypophysis does not develop in the customary manner. The other daughter cell of the basal cell develops into a massive suspensor composed of several rows of cells. The main difference between the Erodium and Geum Variations is that region *n* does not contribute to the suspensor in the former.

Polygonum Variation. — Several different genera in the *Polygonaceae* exhibit a modified Asterad Type embryogeny. A four-celled stage is composed of two adjoining sister cells derived from the terminal cell, a middle cell and a basal cell derived from the original basal cell of the two-celled proembryo. The middle cell participates in the formation of the hypocotyl while the basal cell contributes the hypophysis. The octant stage occurs next; the upper tier of four cells develops into the cotyledonary region, and the lower four-celled tier forms the upper part of the hypocotyl. Two longitudinal walls are formed at right angles to each other in the middle cell; these cells divide further and produce the periblem initials of the root tip as well as the lower portion of the hypocotyl. The daughter cells derived from the basal cell which adjoin the derivatives of the middle cell produce the hypophysis; the other daughter cells develop a suspensor of variable dimensions.

Urtica Variation. — *Urtica pilulifera* follows the Asterad Type only in that both cells of the two-celled proembryo contribute to the sum total of the embryo. The earlier stages of development up to the quadrant stage agree with those species that follow the Asterad Type rigidly, but thereafter the sequence of divisions follows another pattern. Octant cells are not formed; instead, in each quadrant cell a diagonal wall is formed in such a manner that it extends from near where the internal walls meet at the bases to the center of the external wall, the resulting daughter cells thus appearing wedge-shaped in longitudinal section. Derivatives of these cells contribute only the cotyledonary region of the mature embryo; all the remainder of the latter is developed from the middle and basal cells. The middle cell contributes the hypocotyl and the plerome initials of the root tip. The initials of the root cortex, the root cap and the short suspensor are all derived from the basal cell.

Lamium Variation. — The embryogeny of *Lamium purpureum* greatly resembles that of *Urtica pilulifera.* The much elongated zygote is divided by a transversely oblique wall; from the terminal cell a quadrant group is developed as in *U. pilulifera* and this group also divides further in the same way by means of oblique walls. The middle cell becomes transformed into a quadrant by the formation of two longitudinal walls oriented at right angles to each other. These cells, together with the upper daughter cell originating from the transverse division of the basal cell, go into the formation of the hypocotyl.

Oxalis Variation. — The differences which separate this Variation from the Asterad Type reside in the origin and mode of individualization of the initials of the root cortex, as well as, and in consequence thereof, in the delimitation of the regions *m*, *n* and *o* of the sixteen-celled proembryo. That is, in the Type tier *m* gives rise only to the plerome initials of the root, whereas in the Oxalis Variation tier *m* gives rise to all of the root initials except for those of the root cap.

Muscari and Poa Variations. — These are the only Variations assigned to the Asterad Type in which the mature embryos are monocotyledonous. The first three cell generations are much alike, but differences become pronounced during the third generation. In both Variations the derivatives of the terminal cell give rise only to the cotyledon (or including associated structures), but the derivatives of the basal cell (*m*, *n* and n^1) do not have the same destinations. In the Muscari Variation tier *m* gives rise to the hypocotyledonary region and stem apex; in the Poa Variation, this region, in addition to these two structures, also engenders the lower lip of the coleoptile, the initials of the central cylinder and the root cortex. The last two structures are provided by tier *n* in the Muscari Variation; in the other Variation this tier engenders the root cap, coleorhiza and epiblast. Tier *o* furnishes the root cap in the Muscari Variation but in the Poa Variation it combines with tier *p* to develop that part corresponding to a suspensor but generally called the hypoblast.

Penaea Variation. — All the products of both basal and terminal cells go into the construction of the embryo. There is no suspensor.

IV. Solanad Type

The Solanad Type is based upon a generalized scheme involving the development of a linear proembryonic tetrad. At the first cell generation the destinations of the two cells proves that the basal cell has no function in the organization of the embryo proper:

$$ca = pcc + pvt + phy + icc + iec$$
$$cb = co + s$$

There is a division of destinations at the second cell generation:

$$l = pco + pvt$$
$$l' = phy + icc + iec$$
$$m = co + s \text{ (in part)}$$
$$ci = s \text{ (in part)}$$

The third cell generation typically includes six cells disposed in four tiers, or more commonly eight cells in either five, six, or eight tiers, with the destination of each tier remaining the same as in the second cell generation.

Variations:- The Variations assigned to the Solanad Type may be distinguished by means of the following key:

Embryo proper terminal; linear proembryo a tetrad.
- Hypophysis initial absent (initials of root cortex derived from lowermost cells of *phy*).
  - Suspensor short, consisting of few cells; filamentous.
    - Destinations at third cell generation as in the schematic type (tier $l = pco + pvt$).
      - Proembryo normally consisting of eight cells (rarely six) at the third cell generation.
        - Cells arranged in five tiers .... **Hyoscyamus Variation**
        - Cells arranged in six tiers .... **Nicotiana Variation**
        - Cells arranged in eight tiers .... **Physalis I Variation**
      - Proembryo consisting of a variable number of cells at the third cell generation .... **Datura Variation**
    - Destinations at third cell generation otherwise than in the schematic type.
      - Tier $l = pvt$ .... **Papaver Variation**
      - Tier $l = pco + pvt + phy$ .... **Physalis II Variation**
      - Tier $l = pco + pvt + phy + icc + iec$ .... **Physalis III Variation**
  - Suspensor elongated, cells variously swollen and becoming haustorial .... **Sherardia Variation**
- Hypophysis initial present .... **Linum Variation**

Embryo proper intercalary; linear proembryo becoming as much as fifteen cells long .... **Hydnora Variation**

A more complete discussion of those Variations which bear the names of solanaceous genera is presented under the *Solanaceae* in the next chapter. The nonsolanaceous genera are discussed below.

Papaver Variation. — This Variation closely resembles the Nicotiana Variation in most respects, save that at the second and succeeding cell generations tier *l* engenders only the stem apex (through the precocious erection of an epiphysis).

Sherardia Variation. — The elements *l* and *l'* of the four-celled proembryo may divide by either horizontal or vertical walls, while *m* and *ci* always divide by transverse walls. As a result, the eight-celled stage may consist of either six, seven or eight tiers of cells, as in many of the solanaceous genera. In general, tier *l* goes into the formation of the cotyledonary region and *l'* into the hypocotyl, while *m* and *ci* form a thick suspensor which is filamentous in the proximal region and composed of a number of bladder-like haustorial cells in the distal region. The first portion is derived from *m*, the other from *ci*. The cell of the suspensor adjoining the embryo proper may possibly be a presumptive hypophysis and contributes to the middle portion of the root cap.

Linum Variation. — Tier *l* divides by a longitudinal wall; both *l'* and *m* may divide by either a longitudinal or a transverse wall. Tier *l*, in any case, develops into the cotyledonary region. If *l'* divides longitudinally, it gives rise to the hypocotyl; if it divides transversely, then the hypocotyl arises only from the upper daughter cell, while the lower one goes into the development of the suspensor Cell *m* may likewise behave differently; either *m* itself or a daughter cell becomes a typical hypophysis cell.

Hydnora Variation. — A linear proembryo of as many as fifteen cells is produced; the embryo proper arises from about the fifth to the ninth cells inclusive, having a linear suspensor below and a row of three to five cells above the embryonic body. The embryo proper is thus intercalary in origin. The mature embryo is undifferentiated.

V. Chenopodiad Type

✓The Type is based primarily on the embryonomy of *Chenopodium bonus-henricus* (*Chenopodiaceae*). At the first cell generation the destination of the two cells is:

$$ca = pco + pvt + \tfrac{1}{2}phy$$
$$cb = \tfrac{1}{2}phy + icc + iec + co + s$$

It is therefore evident that the basal cell contributes a substantial portion of the embryo proper. At the second cell generation the proembryo consists of four cells arranged in four tiers, with the destinations:

$$l = pco + pvt$$
$$l' = \tfrac{1}{2}phy$$
$$m = \tfrac{1}{2}phy + icc + iec$$
$$ci = co + s$$

At the third cell generation tiers *l*, *l'* and *m* have divided vertically and cell *ci* transversely into *n* and *n'*. There are thus eight cells disposed in five tiers and the destinations are:

$$l = pco + pvt$$
$$l' = \tfrac{1}{2}phy$$
$$m = \tfrac{1}{2}phy + icc + iec$$
$$n = co + s \text{ (part)}$$
$$n' = s \text{ (part)}$$

In the next cell generation quadrants are produced in tiers *l*, *l'* and *m*, while *n* and *n'* both are segmented transversely. The destinations of the first three tiers remain the same as in the third cell generation, but

$$h = co$$
$$k + o + p = s$$

The Variations may be distinguished by the following key:

Epiphysis not present .... Chenopodiad Type
Epiphysis present
Cell m of the tetrad divides transversely into d and f ($d = co; f = s$ (part)) .... Polemonium Variation
Cell m divides vertically ($m = \frac{1}{2}phy + icc$) .... Myosotis Variation

The Type has already been discussed; to describe the Variations further:

Polemonium Variation. — This embryonomic Variation is markedly similar to the Myosotis Variation in the structure of the six- and twelve-celled proembryos, by the mode of division of the element l of the tetrad and by the differentiation of an epiphysis. On the other hand, it approaches the Solanad Type through the destination of the four elements of the proembryonic tetrad, and by the origin and mode of separation of the initials of the root cortex. In effect, this Variation may be considered as a transition form between the Chenopodiad Type and the Solanad Type.

Myosotis Variation. — The terminal cell, l, of the four-celled proembryo divides by an obliquely longitudinal wall into two unequal cells and subsequently develops in the same manner as in the Geum Variation, Asterad Type, to form an epiphysis initial. The other cells which have arisen from l form the cotyledonary regions while l' and m behave as in *Chenopodium bonus-henricus* to produce the hypocotyl. From ci are derived two daughter cells, of which the upper functions as a typical hypophysis and the other develops the root cap and the periblem initials of the root tip. The lower of the two daughter cells of ci develops into a short suspensor.

At the second cell generation, the destinations of the tetrad elements differ from the corresponding ones in *C. bonus-henricus*:

$$\begin{aligned} l &= pco + pvt \\ l' &= \tfrac{1}{2}phy \\ m &= \tfrac{1}{2}phy + icc \\ ci &= iec + co + s \end{aligned}$$

Again, at the fourth cell generation there are twelve cells disposed in five tiers, in contrast to the sixteen cells disposed in seven tiers found in *C. bonus-henricus*:

$$\begin{aligned} l &= pco + pvt \\ l' &= \tfrac{1}{2}phy \\ m &= \tfrac{1}{2}phy + icc \\ n &= iec + co \\ n' &= s \end{aligned}$$

VI. Caryophyllad Type

The Caryophyllad Type was originally based upon the embryonomy of *Sagina procumbens* (*Caryophyllaceae*), but since then it has become necessary to broaden the definition to include many other genera from both dicotyledonous and monocotyledonous families. The feature characterizing most of the species included is the failure of the basal cell of the two-celled proembryo to segment; in other species it may divide once or twice. In either case the cell or the two cells become much enlarged.

Variations:- The following key serves to characterize the Variations assigned to the Caryophyllad Type:

Mature embryo dicotyledonous.
Epiphysis present.
Hypophyseal region absent.

Suspensor short with large end cell .... **Sagina Variation**
Suspensor elongated; all cells much extended............... **Vaccaria Variation**
Hypophyseal region present.
Suspensor absent (there are two large vesicular cells which soon become detached). .... **Hypecoum Variation**
Suspensor present; consisting of three vesicular cells, the lower two with free nuclei. .... **Corydalis Variation**
Epiphysis absent.
Basal cell (*cb*) never or very rarely divides again.
Hypophysis present, behaving as in the typical Onagrad Type .... **Sedum Variation**
Hypophysis absent .... **Medicago Variation**
Basal cell forms a suspensor.
Basal cell divides longitudinally to form two juxtaposed inflated cells .... **Myriophyllum Variation**
Basal cell divides transversely; daughter cells are usually again segmented .... **Drosera Variation**
Basal cell divides quite irregularly to form a comparatively large suspensor. .... **Fumaria Variation**
Mature embryo not dicotyledonous.
Mature embryo with one cotyledon.
Hypocotyl hollow in upper portion .... **Ruppia Variation**
Hypocotyl solid throughout.
Cotyledon derived from tiers *l* and *l'* of the proembryonic tetrad; stem apex derived from tier *m* .... **Sagittaria Variation**
Cotyledon and stem apex derived from tier *l* ....**Zannichellia Variation**
Mature embryo acotyledonous .... **Aglaonema Variation**

To discuss the Variations more fully:

Sagina Variation. — As in the Chenopodiad Type, the terminal cell divides twice to form a linear row of four cells, to which are given the respective designations *l* (the terminal daughter cell), *l'*, *cm* (or *m*) and *ci*, in descending order. Of these four cells, three become divided by longitudinal walls and the fourth by a transverse wall; so that these divisions give the embryo the appearance of having five tiers (the original basal cell of the two-celled proembryo is not counted), namely, *l*, *l'*, *cm*, *n* and *n'*. The next divisions occur in the three upper layers *l*, *l'* and *m* and are longitudinal and at right angles to the first vertical walls: these result in a series of three quadrants. A longitudinal wall is thereupon erected in *n* and a transverse wall in *n'*. The six tiers of cells thus produced finally give rise to the various portions of the mature embryo as follows: from *l* is developed an epiphysis initial as in *Geum* and *Myosotis*; *l'* gives rise to the cotyledons; the hypocotyl develops from *m*; the initials of the root cap originate in *n*; *n'* divides transversely to form two cells, *o* and *p*, which form a kind of short suspensor to which the large original basal cell of the two-celled proembryo is attached.

Vaccaria Variation. — Most species whose embryogeny follows the Caryophyllad Type possess only one large suspensor cell. In *Vaccaria* and *Dianthus*, however, the suspensor is composed of two or more often three equally large cells. Their exact derivation is not clear, but it might indicate that the original basal cell is capable of dividing once or twice.

Hypecoum Variation. — The distinguishing characters of this Variation are that (*1*) the zygote segments obliquely into two unequal cells; (*2*) the basal cell never divides again, becoming an inflated vesicle; (*3*) the terminal cell divides diagonally into two cells, *cc* and *cd*; (*4*) *cd* does not divide again but becomes inflated, though not to such an extent as *cb*; (*5*) all of the embryo proper is derived from *cc*; (*6*) the embryo proper soon becomes detached from the inflated cells; and (*7*) massive epiphyseal and hypophyseal regions are later formed. There is no suspensor in the strict sense.

Corydalis Variation. — This Variation, insofar as the derivatives of the terminal cell are concerned, much resembles the Hypecoum Variation; the principal difference is that a suspensor of three large vesicular cells occurs. The basal cell never divides again, but eventually contains a large number of free nuclei; the middle vesicular cell, derived from the terminal cell, also contains several free nuclei.

Sedum Variation. — The basal cell of the two-celled proembryo does not divide again; all of the embryo proper, as well as the suspensor, is developed from the terminal cell. The octant embryo is developed as in the Onagrad Type; the origin and behavior of the hypophysis initial likewise follows that Type. An epiphysis region is not formed.

Medicago Variation. — The terminal cell of the two-celled proembryo divides transversely into two superposed cells, *cc* and *cd*, while the basal cell, *cb*, generally remains undivided. There are many points of resemblance with the Sagina Variation in other respects.

Fumaria Variation. — The terminal cell behaves almost exactly as in the Medicago Variation, but the basal cell divides in a rather irregular manner to produce a suspensor which, for a considerable period, is much larger than the embryo proper.

Myriophyllum Variation. — Four species, *Myriophyllum alternifolium*, *Portulaca oleracea*, *Pyrola rotundifolia*, *Samolus valerandi*, and perhaps others, form a closely related embryogenic group. The Variation is distinguished by the behavior of the middle cell (*m*). An epiphysis region is not formed.

Drosera Variation. — This Variation closely resembles the Sedum Variation in so far as the development of the embryo proper is concerned. The filamentous proembryo follows the Solanad Type in development and the disposition of the quadrants agrees likewise, while the development of the hypophysis region corresponds to that prevalent in the Asterad Type. There is thus a superposition of other Types upon the fundamental Caryophyllad Type.

Ruppia Variation. — Early proembryonic development is typically that of the Sagina Variation. Later there is developed a single more or less tubular cotyledon and a massive cylindrical hypocotyl which is hollow in the upper portion.

Sagittaria Variation. — The basal cell becomes transformed directly into a large vesicular cell. The terminal cell is segmented into three superposed cells, of which the top one is next partitioned vertically. The two upper juxtaposed cells give rise to the cotyledon; the intermediate cell engenders the upper portion of the hypocotyl and the "stem apex," while the derivatives of the lower cell differentiate the lower portion of the hypocotyl, the initials of the cortex, the root cap and the suspensor proper.

This Variation includes all species in the Alismataceae that have been investigated as well as many other monocotyledonous species.

Zannichellia Variation. — This Variation is similar to the Sagittaria Variation, save that the cotyledon and stem apex are both derived from the uppermost cell of the linear tetrad.

THE SOUÈGES SYSTEM OF CLASSIFICATION

It is important that the reader bear in mind that the present book was completed before the author became aware that the distinguished French embryologist SOUÈGES had formulated a classificatory scheme for the various modes of embryonal development. The Type-and-Variation scheme of classification was devised in complete ignorance of any other scheme save the provisional one of SCHNARF. If the SOUÈGES classification had been available at the beginning, a determined effort undoubtedly would have been made to adopt it, with minor changes and considerable amplification. However, the author's scheme was prepared in good faith and is being retained both on that ground and for the important reason that the SOUÈGES scheme embraces only those species which he himself had previously investigated. It takes no account whatever of embryological studies made by others and permits little or no flexibility.

In any event, it is necessary to outline the SOUÈGES system and to compare it with that of the author.

Basis of the Souèges System:- Four categories of embryonomic types are distinguished:

Definable by the laws of embryonomy	1. Fundamental types or archetypes	
	2. Secondary or derivative types	Adaptation by condensation or convergence
	3. Superposed types	Consequences of hybridization
Not definable by the laws of embryonomy	4. Irregular types	

For the most part, only those species conforming to the archetypes are further considered in the classificatory scheme. Two Grand Divisions, or first and second "periods" respectively, are recognized. In the First Grand Division or first "period" the zygote conforms to the laws of embryonomy; in the Second Grand Division or second "period," it is the terminal cell of the bicellular proembryo which is designated as conforming to the embryonomic laws and the basal cell is omitted from further consideration.

The First Grand Division is subdivided into three Series, designated as A, B and C, respectively. Series A is characterized by a proembryonic tetrad of the so-called "T-shape," in which the terminal cell of the two-celled proembryo is segmented by a vertical wall. This type of tetrad is considered to be the most primitive (with this conclusion the author agrees, as is evidenced by the designation of the Onagrad Type as most primitive). Series B is denoted by a "T-shaped" tetrad in which the longitudinal wall in the terminal cell is oriented obliquely (rather than strictly vertical as in Series A). In Series C the tetrad consists of four superposed proembryonic units (the "linear" tetrad). Series C is subdivided into three groups of Subseries, depending upon the type of segmentation of the two upper cells of the tetrad or of their derivative cells.

The three series in the Second Grand Division, together with the Subseries in Series C, are essentially similar to those in the First Grand Division, with the basal cell of the bicellular proembryo entirely left out of consideration (it plays no part whatever in the development of the embryo).

Six Megarchtypes constitute the units of periodic classification in each of the two Grand Divisions. Each is defined by a simple formula, based in the First Grand Division on the behavior of the basal cell (*cb*) of the two-celled proembryo, and in the Second Grand Division on the behavior of cell *cd*, which in this Division is the equivalent of cell *cb* of the First Grand Division. The formulae for the Megarchtypes are as follows (for the Second Grand Division, substitute *cd* for *cb*):

Megarchtype I: $cb = pvt + phy + icc + iec + co + s$
Megarchtype II: $cb = phy + icc + iec + co + s$
Megarchtype III: $cb = \frac{1}{2} phy + icc + iec + co + s$
Megarchtype IV: $cb = iec + co + s$
Megarchtype V: $cb = co + s$
Megarchtype VI: $cb = s$

The next classification is that of Groups, which are arranged according to the Megarchtype under each Series or Subseries.

In his earlier investigations, it is apparent that SOUÈGES laid greater emphasis on the behavior of the terminal cell of the two-celled proembryo and its derivatives than on that of the basal cell and its derivative cells. The author, on the other hand, considered that equal importance should be attached to both cells, although his Types depend mainly on the sequence of events in the terminal cell. It will be noted from the outline of the SOUÈGES classification as given above that emphasis has shifted to the basal cell and its derivatives. For certain embryonomic schemes this may be the logical procedure, but in many others it is clear that serious difficulties immediately arise.

Critique of the Souèges System:- Any proposed system of classification, whatever is included in such a system, is always open to criticism. (One might recall that never since the time of LINNAEUS have the plant taxonomists been able to agree on any system for the classification of plants.) The real test of a classification resides in its practical application — its utility to others who wish to make use of it. Both SOUÈGES' and the author's systems are tentative to a certain degree or in some respects, and they involve slightly different concepts, but it should be emphasized that both are based on the operation of the laws of embryonomy. There can be no criticism on the latter point, but in other respects disagreements are to be expected.

The basic requirement of the SOUÈGES system is an extremely high degree of accuracy in following out the cell-by-cell development of the proembryo of a particular species, together with a profound knowledge of the embryonomy of all other species hitherto investigated with equal carefulness. Few investigators, to judge from the voluminous literature, either possess the ability to make confirmed, detailed studies, or have the requisite knowledge. It is difficult, even after prolonged study, to fit a newly investigated species into the SOUÈGES classification. The author was cognizant of the difficulties, even before becoming acquainted with SOUÈGES' scheme, that would confront other investigators, and therefore allowed somewhat wider latitude in preparing his classification.

Theoretically, designation of a given species in the SOUÈGES classification is on a quintuple basis, although in practice only the Group and Megarchtype are indicated, with the reader expected to supply the rest of the classification from his acquaintance with the entire scheme. This becomes confusing; it is psychologically impossible for most persons to retain such a detailed classification in mind, so that its cumbersomeness defeats its purpose. The practice in other biological fields, as based upon experience, has been to give easily recognized names to classificatory entities. It was for these practical reasons that the Types and Variations in the author's scheme were given names which were readily recognized and each of which indicated adherence to some specific embryonomic entity that could be easily recalled.

Separation of Series A and B in the First Grand Division is a debatable matter. SOUÈGES attaches more significance to the oblique orientation of the longitudinal wall in the terminal cell of the bicellular proembryo than the facts appear to warrant. He includes in Series B only those species in which this wall is constantly oblique and in which an epiphysis initial is immediately delimited in the larger of the two daughter cells, but ignores those known instances in which the orientation of the wall may vary from the vertical to the oblique even though no epiphysis is developed. It is significant that no examples are cited for the corresponding Series B in the Second Grand Division. In the First Grand Division, Series A and B should perhaps best be consolidated into one Series, with two Subseries, precisely as has been done with the Subseries under Series C, since the differences are of the same degree. If this were done, the Subseries would be equivalent to the author's Variations, whatever the Type might be.

The main objection to the Souèges system, as already indicated, is that it concerns only those species which he himself had investigated, but even then not all the species which he had previously studied are included. It is unfortunate, though unquestionably justifiable, that Souèges distrusted embryological studies made by others; but the fact remains that an effort should be made include all published accounts in any classificatory scheme, however mediocre or inaccurate such investigations might be considered.

Furthermore, since the Souèges classification embraces only those species whose development conforms to rigid embryonomic laws, no provision is made for the host of species included under superposed and irregular types. One is left wondering what to do if such an instance were to be encountered. Nor is there any equivalent of the author's Piperad Type; certainly the constant vertical partitioning of the zygote is as significant as the more customary transverse segmentation.

Souèges appears, moreover, to believe that superposed and irregular types are the consequence of hybridization. It is questionable whether this is the case in every instance; it would first be necessary to prove that the species concerned actually was a hybrid. It has been postulated that two species, even if included in the same genus, are incapable of being successfully crossed if each follows a different set of embryonomic laws, the reasoning being that it was impossible to reconcile the differences in the laws. In the case of species whose embryonomy is described as being a superposition of two different sets of laws, it is evident that some other explanation of the presumed superposition must be sought. "Superposition" is fundamentally a contradiction when applied to embryonomic laws; it is inconceivable how two divergent sets of laws can operate simultaneously within one individual proembryo.

Johansen, D. A. 1945. A critical survey of the present status of plant embryology. Bot. Review 11: 87–107.
Souèges, R. 1936. La segmentation. Deuxième fascicule: III — Les phénomènes externes. IV — Les blastomères. Actualités Sci. et Indust. 318. Paris, Hermann et Cie.
Souèges, R. 1937. Les lois du développement. Actualités Sci. et Indust. 521. Paris, Hermann et Cie.
Souèges, R. 1939. Embryogénie et classification. Deuxième fascicule: Essai d'un système embryogénique. Actualités Sci. et Indust. 781. Paris, Hermann et Cie.
Souèges, R. 1941. Variantes dans les deux premiers groupes des périodes de la classification embryogénique. Bull. Soc. Bot. France 88: 602–608.

Chapter XVIII

SPECIAL AND COMPARATIVE EMBRYOLOGY

Magnoliales*

Extremely little information is available concerning the embryology of the Magnoliales and it all relates to the family *Magnoliaceae*. Nothing has been presented for the six remaining families: *Winteraceae*, *Schizandraceae*, *Himantandraceae*, *Lactoridaceae*, *Trochodendraceae* and *Cercidiphyllaceae*.

Magnoliaceae:- The basal cell of the two-celled proembryo of *Magnolia grandiflora* divides longitudinally and the terminal cell later becomes partitioned by a similar wall. The two daughter cells of the terminal cell are next segmented by longitudinal walls perpendicular to the first vertical wall. Later divisions proceed slowly and are described as not being uniform in direction. These divisions were not followed out in detail.

This species would appear to conform to the Onagrad Type, and possibly to the Myosurus Variation. The description for *Magnolia virginiana* agrees substantially with that for *M. grandiflora*.

Tricotyledonous embryos may occasionally occur in *M. grandiflora*; the three cotyledons are equidistant from one another as in the normal type with two cotyledons.

Earle, T. T. 1938. Embryology of certain Ranales. Bot. Gaz. 100: 257–275.
Maneval, W. E. 1914. The development of *Magnolia* and *Liriodendron*, including a discussion of the primitiveness of the *Magnoliaceae*. Bot. Gaz. 57: 1–31.

Anonales

Nothing has been described of the embryogeny of any species included in either of the two families, *Anonaceae* and *Eupomatiaceae*.

Laurales

The *Lauraceae* and *Monimiaceae* have been investigated; the *Gomortegaceae*, *Hernandiaceae* and *Myristicaceae* remain unstudied.

The first division in the zygote of *Sassafras variifolium* may be either transverse or vertical. The exact longitudinal partitioning of the zygote is a rare phenomenon. The next stage is that of two tiers of two cells each, followed

* Arrangement of the orders and families is according to Hutchinson's "The Families of Flowering Plants," vol. 1, Dicotyledons, and vol. 2, Monocotyledons. However, certain families merged with others by Hutchinson, or which were described after the appearance of either volume, are recognized. In some instances, synonyms are cited for orders or families when such are in common usage.

In all cases, generic and specific epithets employed in the original articles are used; when the article concerned is a very old one and important taxonomic changes were made in the meantime (whenever such were known to the author), the new names are cited in parentheses. This is emphatically not a taxonomical treatise, and no particular effort was made to check every name appearing in the morphological literature. This is a duty which belongs to the authors of original articles.

by one of three tiers of presumably two cells each. This sequence of events brings *S. variifolium* into very close agreement with the Scabiosa Variation, Piperad Type.

Persea americana conforms to the Penaea Variation, Asterad Type. *Cinnamomum camphora*, however, develops differently, as there is a distinct suspensor present.

COY, G. V. 1928. Morphology of *Sassafras* in relation to phylogeny of Angiosperms. Bot. Gaz. 86: 149–171.
GUILIANI, A. 1928. Contributo all'embriogenia del *Cinnamomum camphora*. Bull. Bot. Univ. Napoli 9: 33–40.
SCHROEDER, C. A. 1943. The ontogeny, floral anatomy and embryology of *Persea americana* Miller. Thesis, Univ. of California at Los Angeles.

Monimiaceae :- The genus *Siparuna* is completely sterile.

The description of the embryogeny of *Peumus boldus* is confusing and the original figures are arranged in anything but the correct sequence. The terminal cell of the two-celled proembryo apparently divides transversely but may divide longitudinally according to the description although the figures scarcely bear out this claim. The basal cell may divide longitudinally (which is unlikely, to judge from the figures), transversely or somewhat diagonally; most often, it appears to segment transversely to engender a row of about four superposed cells. Whether the basal cell contributes anything to the construction of the embryo proper is problematical. If the terminal cell divides transversely, then two obliquely vertical walls are erected in the upper daughter cell. Although it is highly probable that misinterpretations of the sequence of events were made, the figures do present certain resemblances to the Oxalis Variation, Asterad Type.

HEILBORN, O. 1931. Studies on the taxonomy, geographical distribution and embryology of the genus *Siparuna* Aubl. Svensk Bot. Tidskr. 25: 202–228.
MAURITZON, J. 1935. Zur Embryologie von *Peumus boldus*. Arch. Botanico 11: 317–327.

Ranales

All four families have been investigated. The *Cabombaceae* are a relatively new family separated from the *Nymphaceae* so that species properly placed in the first family are treated under the second one in nearly all older accounts.

Ranunculaceae :- Since *Myosurus minimus* may be considered as the species whose embryonomy is most representative of the *Ranunculaceae* as a whole, it may be taken up first and described in detail.

Myosurus. — In *M. minimus* the four-celled proembryo is erected as in the typical Onagrad Type; that is, it is composed of two juxtaposed cells derived from the terminal cell and two superposed lower cells derived from the basal cell (FIG. 36A). The two upper cells are next divided vertically perpendicular to the first wall: the daughter cells constitute the quadrant stage. The middle cell, which represents the hypophysis initial (*h*), is divided vertically and the lowest cell (*s*) is divided transversely (FIG. 36B). In the next stage, the quadrants are divided transversely to become octants, and the two hypophyseal cells are also segmented vertically to give four cells in the same plane, while the suspensor initials multiply in an irregular fashion to form a short suspensor (FIG. 36C).

The next divisions occur in the lower group of octants (FIG. 36C). The wall is vertical but as its orientation can not be observed clearly in longitudinal sections of the proembryo, it is best described in transverse sections through the regions concerned (FIG. 37E–H). The outer insertion of the wall is in the center of the peripheral wall of the octant cell and the other insertion is on one of the vertical walls separating the octant from its neighbor; the new wall is more or less curved (FIG. 37E). In the adjoining cell (proceeding in a clockwise direction) the inner insertion of the new wall in that cell is on the wall away from that on which the new wall was inserted in the first cell. Thus the in-

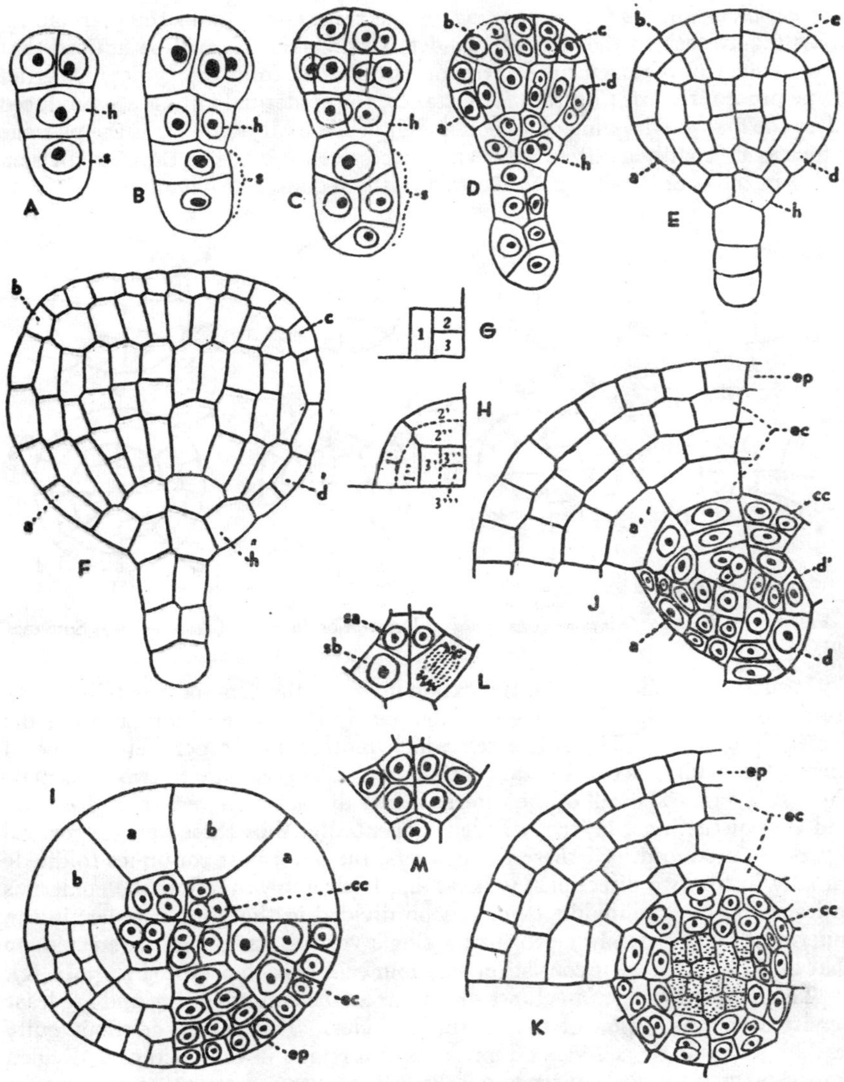

FIG. 36. — *Myosurus minimus*. Description in text. A, B × 650; C × 620; D–G × 600; L × 580; M, N × 560; Q–S × 790; T, U × 830. (*Redrawn from* SOUÈGES 1911*).

sertions proceed until all four cells in the layer are bipartitioned. One of the daughter cells (*2*, FIG. 37B) is more or less rectangular and four-sided in outline, while the other (*1*) is more or less triangular and three-sided. The following division occurs in each of the rectangular cells (*2*) and divides it nearly equally

* In these and in most of the subsequent figures credited to SOUÈGES, it should be observed that when two nuclei (or occasionally more nuclei) are depicted within the limits of a single cell, this does not denote a binucleate condition, but is merely a device to indicate that another cell (or even more cells, if more than two nuclei are shown) exists below the level of the outlined cell.

in the vertical plane (FIG. 37F); then similar walls are erected in each of the triangular cells (*1*), but here the resulting daughter cells are very unequal in size and shape (FIG. 37G). The outer daughter cell shortly after is divided by a tangential wall oriented perpendicularly both to the preceding wall in the triangular cell and to the peripheral wall. As a result of these divisions in the lower octants, the three histogens have been delimited: the epidermal initials, the periblem initials (corresponding to the stem cortex) and the plerome initials (equivalent to the central cylinder) (FIG. 37H). It will be noticed that this manner of histogen differentiation is different from the typical Onagrad Type procedure. In *Capsella*, for instance, the epidermal initials are produced after the first partitioning of an octant cell, while in *Myosurus* it is the plerome which is first individualized after two successive divisions — the first division being in the octant cell and the second in the rectangular daughter cell.

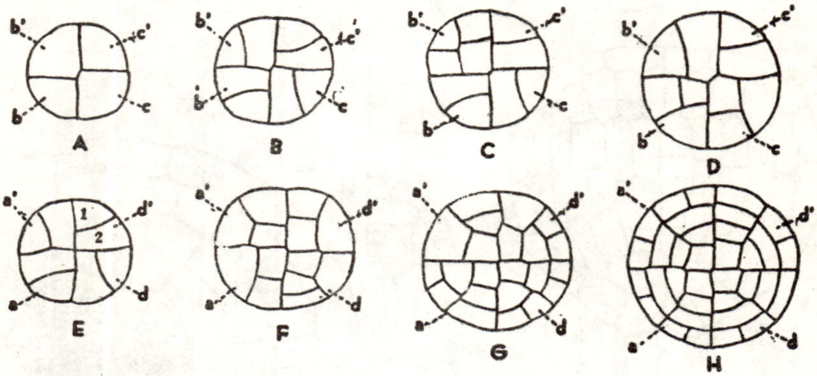

FIG. 37. — *Myosurus minimus* (*Continued*). Description in text. (*Redrawn from* SOUÈGES 1911).

It suffices to follow cell multiplication in one of the demi-octant cells arising from *a* or *d* in FIG. 37E to describe completely the mode of formation of the cortex (FIG. 37G). The single cell which represents the periblem in one of these demi-octants is divided tangentially and thus gives rise to two concentric cortical layers. The cell of the external layer divides by a vertical radial wall, and that of the inner layer is divided tangentially: thus three definite cortical layers are produced. Of these three layers, the outermost continues to divide radially in the two directions, vertical and horizontal, to follow the epidermis in its extension; the middle layer is soon divided in the same manner; in the innermost layer scarcely more than a single vertical radial wall is erected, so that at the adult stage it consists of only four cells in each octant (FIG. 36H–K).

The limits of the demi-octants disappear as the embryo grows and are least persistent in the region of the central cylinder. This region develops quite tardily. The single cell which represents this region in each octant is divided transversely into two superposed cells. Transverse sections of this region again are more useful than those microtomed in the longitudinal plane to demonstrate the development. A wall perpendicular to one of the meridional walls separating the octants divides the plerome cell into two cells which are rectangular in shape. The internal cell, by a wall perpendicular to the preceding one, is divided into two new cells of the same form but naturally much smaller (FIG. 36G). The innermost of these two cells then follows the same mode of division as the mother cell from which it was derived; that is, in it a transverse wall is formed, then two vertical walls oriented perpendicularly to one another

(Fig. 36H, *3′*, *3″* and *3‴*). The two external cells, however, segment differently. They first divide transversely, then tangentially, to engender two concentric plerome layers; the outermost of these two layers has every appearance of being the pericycle (Fig. 36H, *1′*, *1″*, *2′*, *2″*). The further development of the central cylinder proved to be too difficult to follow in detail.

In the upper layer of octant cells (Fig. 37A–D), each of these cells is first divided exactly as in the lower layer of octants (Fig. 37B). The second wall occurs in the equivalent of cell *2* of the lower octants; it is also curved and perpendicular to the preceding wall and separates a median and an external cell (Fig. 37C, D). The cell neighboring the embryonic axis is soon divided transversely to engender two superposed cells. During this time the cell which is triangular in section is divided by a tangential wall parallel to the contour of the embryo. Two layers are thus differentiated in the upper octant region of the embryo: the epidermis and a subepidermal layer which furnishes the initials of the cortex at the summit of the stem and plays the greatest role in the formation of the cotyledons. More strictly speaking, one may consider the inner cell as the initial of the cortex and the outer cell as the mother cell of a cotyledon.

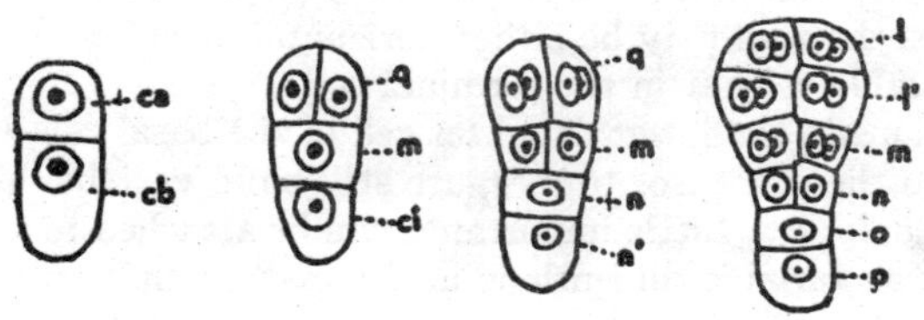

Recapitulatory Table for Myosurus minimus
(from left to right)

I. First Cell Generation
Proembryo of two cells disposed in two tiers:
ca=*pco*+*pvt*+*phy*+*icc*
cb=*iec*+*co*+*s*

II. Second Cell Generation
Proembryo of four cells disposed in three tiers:
q=*pco*+*pvt*+*phy*+*icc*
m=*iec*+*co*
ci=*s*

III. Third Cell Generation
Proembryo of eight cells disposed in four tiers:
q=*pco*+*pvt*+*phy*+*icc*
m=*iec*+*co*
n+*n′*+*s*

IV. Fourth Cell Generation
Proembryo of sixteen cells disposed in six tiers:
l =*pco*+*pvt*
l′=*phy*+*icc*
m=*iec*+*co*
n+*o*+*p*=*s*

Each cell of the quadrant hypophyseal region is in the form of a pyramid with a triangular base and whose summit is directed toward the mass of the embryo (Fig. 36D). The first wall is oriented parallel to the base of the pyramid and divides the cell into two segments — an apical or internal segment (*sa*) and a basal or external segment (*sb*) (Fig. 36L). The latter divides first by a radial wall which follows the external surface of the embryo in a parallel arc. This results in the formation of two cells of dissimilar shapes: one elongated in the embryonic axis and the other adjoining it at one side. The elongated cell is generally the one to divide next; the wall is radial but oriented perpendicular to the preceding one and bisects the cell about equally. Thus the basal segment of the primitive quadrant at this moment is divided into three cells: one central and two peripheral (Fig. 36M). The apical segment (*sa*) behaves, in the following stages, like the primitive quadrant cell: it generally divides by a tangential wall oriented parallel to the base. Consequently, a single cell is always found at the summit of the pyramid, and this, together with those of the three other hypophyseal quadrants, is included among the initials of the cortex. Presently there is developed a massive asymmetrical hypophysis (FIG. 36F).

The recapitulatory table illustrating the rules of development in *Myosurus minimus* will be found on the previous page.

Adonis. — The sequence of segmentations in *Adonis* does not follow any definite rules in contrast to those prevailing in *Myosurus minimus.*

The terminal cell of the two-celled proembryo is smaller than the basa cell. The second division occurs in the latter and separates an upper flatteneld cell and a larger lower cell. The former appears distinctly to be the hypophysis initial. The first wall formed in the terminal cell is extremely variable in its orientation; in fact, it may occupy every imaginable position. It may be vertical as in *Myosurus minimus*; or, on the contrary, it may be perfectly horizontal or more or less diagonal. When it is oblique, the end of one wall may be inserted on the wall which originally separated the terminal from the basal cell. Oblique walls always result in daughter cells of very different forms and dimensions.

The hypophysis initial divides oftenest by a wall close to the vertical, but at other times the wall may be either horizontal or oblique. Behavior in this cell is comparable to that in the terminal cell.

The divisions in the lower daughter cell of the basal cell, which is destined to give rise to the suspensor proper, are still more variable in orientation but are generally oblique. Little importance can be attached to this portion of the embryo; it is of variable dimensions in the mature embryo but is rarely over two cells in width.

When the hypophysis initial divides transversely, or more or less obliquely, it is the upper daughter cell which is the hypophysis proper; the lower cell later contributes an indefinite number of cells to the suspensor. This agrees with conditions in *Clematis* (*v. infra*). Again, when the terminal cell segments transversely, it is difficult to affirm whether the lower cell thus engendered gives rise to the hypophysis or whether it enters into the construction of the embryo proper. An examination of older embryos permits both interpretations. At this stage of development there are three zones or regions in the proembryo: (*1*) an upper zone or embryo proper characterized by large cells, limited exteriorly by thicker membranes, and containing dense cytoplasm and relatively large nuclei. Cellular multiplication is more active in this region; (*2*) a median zone or hypophyseal region composed of smaller cells that are flattened and with fairly dense cytoplasm; (*3*) a lower zone or suspensor, balloon-shaped, composed of large cells less rich in cytoplasm and with large vacuoles and much smaller nuclei.

The process of differentiation of the quadrants varies according to the position of the first wall in the terminal cell. Quadrant cells disposed in a horizontal plane and separated by two truly cruciate walls have never been observed in any species of *Adonis.* One of the four cells appears always so situated that it occupies a position at the summit of the embryo. In a general manner, the nucleus of the cell which offers the greatest surface of contact with the hypophysis cell or initial divides in a horizontal plane; while that of the cell nearest the summit, on the contrary, divides in a vertical plane.

It is not possible to interpret in a simple manner all the instances of embryos which have passed the quadrant stage. At this time, the four cells appear to form a tetrahedron, one of whose faces reposes on the hypophysis cell, but save for the latter feature, the disposition of the walls is extremely variable.

It is very rare that the disposition of the octants in the typical Onagrad Type occurs in *Adonis*. The most that can be said is that there are eight cells, two of which are contributed by each cell of the quadrant, and more or less disposed as four upper octants and four lower octants. Each octant cell undergoes two successive divisions to give rise to a group of four cells; each of these, in turn, gives rise to still another group of four cells each without definite orientation. The peripheral cells thus resulting are flattened and are disposed in a concentric layer of which the external cells constitute the epidermis and the others the first layers of the cortex. It is futile to attempt to describe with precision the sequence of partitionings in the central portion of the embryo.

The sequence of divisions in the hypophysis cell is very much like that in *Myosurus minimus*. The first wall is inclined in the most diverse positions, and the succeeding ones are very irregular.

Clematis. — The first transverse division in the zygote of *Clematis recta* separates the cell into two unequal portions, the basal cell being the larger one. The second division occurs in the basal cell and is transverse; the third is in the terminal cell and divides it obliquely into two unequal cells. Next the two derivatives of the basal cell divide simultaneously in the transverse plane. The smaller cell of the two derivatives divides obliquely by a wall oriented at right angles to the first wall, then a similar wall is erected in the sister cell. The uppermost cell of the row of four derived from the original basal cell now divides vertically. Hereafter, it is extremely difficult to assign a fixed rule to the sequence of divisions in the embryo proper. The position of the walls is particularly hard to determine since many of them are curved or oblique.

Each of the four cells of the quadrant stage divides once, then each daughter cell divides again. The resultant four cells in each original cell of the quadrant are very diversely arranged; in general, one is internal and the other three are external. When this happens, the external cells are undoubtedly epidermal initials. Additional walls, both transverse and vertical, are laid down in the interior of the developing embryo, which increases but little in size during this time. The young embryo at this stage is cylindrical in shape, then becomes pyriform just before the cotyledons begin to develop.

The cells of the suspensor now commence to divide, generally in the vertical and transverse planes but also obliquely.

The upper daughter cell of the row of four cells originally derived from the basal cell is the hypophysis cell. The first division, as has already been mentioned, is vertical. Presently the region commences to divide actively and forms a considerable mass of cells very difficult to distinguish from the adjoining suspensor.

Clematis flammula is much like *C. recta* embryonomically but the number of cells in the suspensor is larger and the nuclei of the embryo portion are more or less smaller. *C. campaniflora*, *C. viticella*, *C. integrifolia* and *C. vitalba* also appear to agree with *C. recta*.

Anemone. — *Anenone japonica* is embryonomically homologous, for the most part, with *Myosurus minimus*.

A. dichotoma and *A. pulsatilla* are similar in the embryonomy, but differ somewhat from *A. japonica*. They are also little different from *Clematis recta*, at least during the earlier proembryonic stages.

Cimicifuga. — In the two-celled proembryo of *Cimicifuga racemosa* the basal cell divides first, in the vertical plane, then the terminal cell divides by an oblique wall. Divisions thereafter are described as being irregular and comparatively slow. The cotyledons remain relatively undeveloped, appearing as two small knobs of tissue on the mature embryo.

It is not possible, from the extremely meager account, to place this species definitely, but theoretically it should conform to the Myosurus Variation.

Eranthis. — *Eranthis hyemalis* displays an extremely irregular development of the proembryo and embryo; it follows the Myosurus Variation. The zygote does not divide until about a month after fertilization has occurred. The earlier proembryonic divisions are typically similar to those in *Ranunculus*; a very much elongated and poorly differentiated embryo is presently formed, in which no histogens are clearly recognizable and even the epidermal initials are very tardily delimited. A single cotyledon is produced.

Actaea. — *Actaea alba* casually appears to conform in the proembryonic stages to the Myosurus Variation.

Ranunculus. — A considerable number of species in *Ranunculus* have been investigated and all conform substantially to the Myosurus Variation. There are minor divergencies which might warrant the erection of a separate Variation to accommodate the genus *Ranunculus*, but it appears to be best to place all the species together under the Myosurus Variation for the time being.

Ranunculus ficaria (*Ficaria verna*) has been the subject of numerous investigations. This species in certain regions has more or less lost the ability to reproduce sexually. However, fertilization may be effected. In some localities, the resultant embryos disintegrate before progressing much beyond the few-celled proembryonic stages, but in other localities the embryos attain maturity and the seeds are viable. The conclusions reached by most of the earlier investigators are mainly speculative and based on incomplete studies. The proembryonic stages are very irregular. When the mature achene dehisces, the embryo is still in the proembryonic stage; there is no clear differentiation of any of the usual histogens, although there is some evidence of epidermal initial delimitation. Further growth of the proembryo into the embryo does not commence until some time after the seed commences germination. When growth begins anew, the embryo develops in such a manner that it has only one longitudinal plane of symmetry. This means that a series of longitudinal sections of the embryo microtomed in one plane are very different from a similar series sectioned at right angles to the first plane. Therefore, unless the sections are very carefully inspected, it is easy to arrive at false conclusions. The upper region of the embryo is the first to commence development. This becomes slightly indented at the apex, concave on one side and convex on the other, thus appearing somewhat kidney-shaped when observed in transverse section. The apical concave end of the embryo becomes still further enlarged and gives rise to the two lobes of the cotyledonary lamina, which at first are folded together in such a manner that the tissues which subsequently become the upper surface of the two lobes are in contact with one another. The cotyledonary petiole next elongates rapidly; the petiole is grooved on the upper portion at maturity but is only slightly flattened or circular in transverse section throughout the greater part of its length. A second concavity at the basal end of the petiole represents the base of the original lateral concavity present at a younger stage. Certain cells with large nuclei and prominent nucleoli which early develop near the base of the lateral concavity mark the position where the stem apex subsequently arises, lateral to the cotyledon and surrounded by its sheathing base. Opposite the cotyledon there is developed a small hump of tissue, occupying the position where the second cotyledon would have arisen if it were actually present. Its position and mode of origin indicate that it is the rudiment of a second cotyledon which fails to develop. This hump, like the other cotyledon, is supplied with a provascular strand. As a matter of fact, seedlings with two cotyledons have been observed, hence the rudimentary cotyledon is actually capable of normal development.

Ficaria. — *Ficaria ranunculoides* has an embryo which is acotyledonous at maturity. The embryonomy is a combination of those of *Myosurus minimus* and *Adonis autumnalis.* It approaches that of *Myosurus:* (*1*) in the general constitution of the proembryo of three superposed cells, of which the upper cell gives rise to the embryo proper, the middle one to the hypophyseal group and the lower to the suspensor; (*2*) in the general disposition of the quadrants in a horizontal plane, separated by two cruciate walls; (*3*) in the processes of the formation of the octants by the division in a vertical direction of the four nuclei of the quadrants, and (*5*) in the dimensions of the cells and those of the embryo itself. On the other hand, the following characteristics are identical with those of *Adonis;* (*1*) a certain freedom in the formation of the quadrants and octants; (*2*) the complete irregularity of the first divisions in the interior of the octants; (*3*) the sudden appearance of cells of the second order; (*4*) the lack of, or very tardy, differentiation of the histogens — epidermal initials, periblem and plerome, and (*5*) the development of the hypophyseal group is sometimes predominant, sometimes quite tardy, which gives embryos of the same age very different aspects.

Aconitum. — The account of the embryonomy of *Aconitum napellus* is in such an unsatisfactory condition that no definite conclusions can be drawn. The basal cell of the two-celled proembryo is described as dividing vertically before the terminal cell segments.

Arzt, T. 1933. Über die Embryobildung von Pseudomonokotylen. Beih. Bot. Centralbl. 50: 671–696.
Coulter, J. M. 1898. Contribution to the life-history of *Ranunculus.* Bot. Gaz. 25: 73–88.
Earle, T. T. 1938. Embryology of certain Ranales. Bot. Gaz. 100: 257–275.
Loschnigg, F. 1926. *In:* Winkler, H., Bausteine zu einer Monographie von *Ficaria.* Beitr. Biol. Pflanzen 14: 335–358.
Metcalfe, C. R. 1936. An interpretation of the morphology of the single cotyledon of *Ranunculus ficaria* based on embryology and seedling anatomy. Ann. Bot. 50: 103–120.
Mottier, D. M. 1905. The embryology of some anomalous dicotyledons. Ann. Bot. 19: 447–463.
Osterwalder, A. 1898. Beiträge zur Embryologie von *Aconitum napellus.* Flora 85: 254–292.
Souèges, R. Recherches sur l'embryogénie des Renonculacées. Bull. Soc. Bot. France 57: 242–250; 266–275. 1910; 58: 128–135, 144–151, 542–549, 629–636, 718–725. 1911; 59: 23–31, 474–482, 545–550, 602–609. 1912; 60: 237–243, 283–289, 506–514, 542–549, 615–621. 1913.
Wolter, H. 1933. Bausteine zu einer Monographie von *Ficaria.* VIII. Über Bestäubung, Fruchtbildung und Keimung bei *Ficaria verna* Huds. Beitr. Biol. Pflanzen 21: 219–255.

Cabombaceae :- *Cabomba piauhiensis* and *Brasenia purpurea* develop similarly; in the first species, a proembryo of three or four cells in linear arrangement is first developed. Apparently the lower two cells are derived from the basal cell, which contributes nothing to the construction of the embryo proper. A cotyledonary ridge is produced in both species; it extends almost entirely around the periphery of the embryo.

It does not seem to be possible to make a definite assignment of these two species.

Cook, M. T. 1906. The embryogeny of some Cuban *Nymphaeaceae.* Bot. Gaz. 42: 376–392.

Ceratophyllaceae :- The first zygotic cytokinesis of *Ceratophyllum submersum* is transverse, after which vertical walls are erected in both daughter cells. Each daughter derived from the basal cell is next partitioned transversely. Thereafter there is no apparent order in the sequence of divisions and both basal and terminal cells appear to contribute about equally to the further growth of the embryo. No suspensor is developed. The cotyledonary lobes arise from a common primordium, as in the *Nymphaeaceae*. In fact, the embryonomy of these two families displays remarkable resemblances. Assignment is to the Penaea Variation, Asterad Type.

STRASBURGER, E. 1902. Ein Beitrag zur Kenntnis von *Ceratophyllum submersum* und phylogenetische Erörterungen. Jahrb. Wiss. Bot. 37: 477–526.

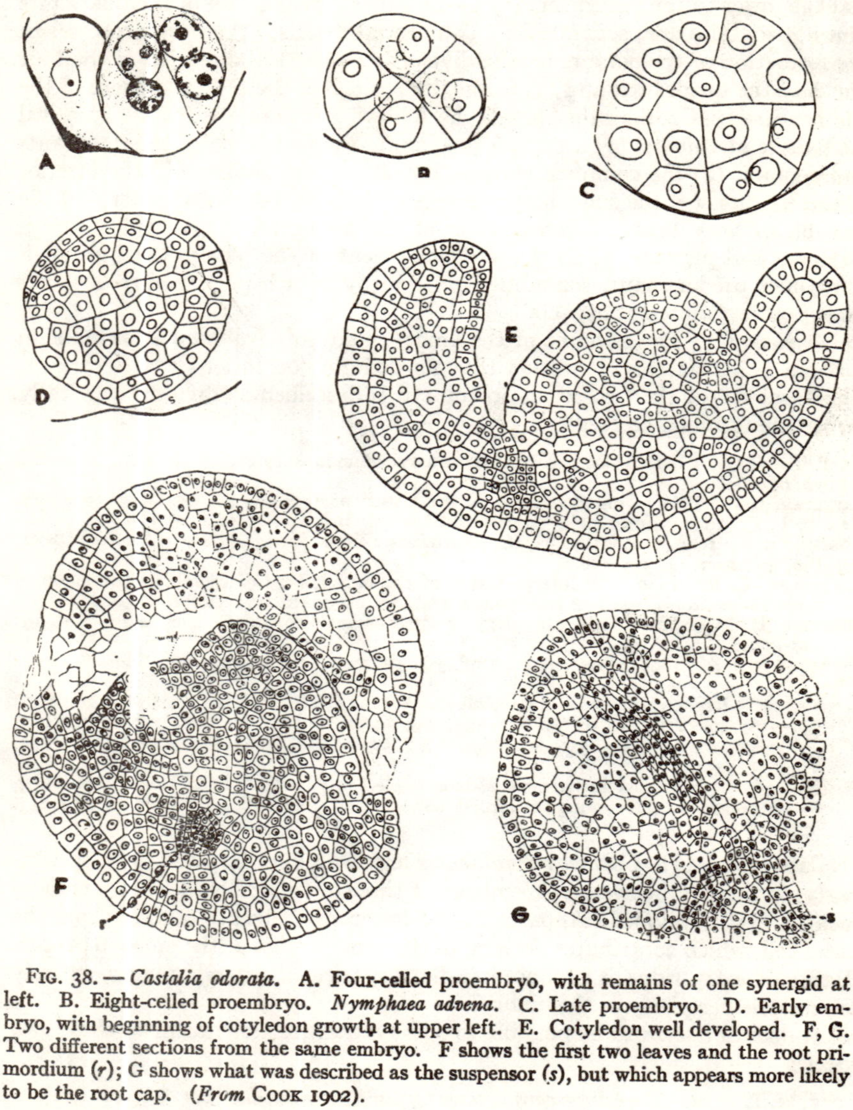

FIG. 38. — *Castalia odorata*. A. Four-celled proembryo, with remains of one synergid at left. B. Eight-celled proembryo. *Nymphaea advena*. C. Late proembryo. D. Early embryo, with beginning of cotyledon growth at upper left. E. Cotyledon well developed. F, G. Two different sections from the same embryo. F shows the first two leaves and the root primordium (*r*); G shows what was described as the suspensor (*s*), but which appears more likely to be the root cap. (*From* COOK 1902).

Nymphaeaceae:- *Nymphaea advena* and *Castalia odorata* are very similar in their embryonomy. A Cuban species, apparently identical with the northern form of *N. advena*, forms the basis of the most accurate account available. In this form, the zygote divides transversely, followed by two cruciate longitudinal walls in each cell to produce an eight-celled stage (FIG. 38A–B). This development permits assignment to the Penaea Variation, Asterad Type.

Later cell divisions are somewhat variable but the developing embryo retains its spherical shape (FIG. 38C–D). Presently there is an extensive growth on the side toward the micropyle to form a suspensor, and a flattening and broadening development on the other side. From the latter end a cotyledonary ridge is produced, extending almost completely around the periphery. The cotyledonary lobes are next developed from the cotyledonary ridge (FIG. 38E). There is a great deal of variation in the prominence of this dicotyledonous character. The cotyledonary ridge surrounds and overhangs the stem apex. The phenomenon has been claimed to be a monocotyledonous character, but this is scarcely accurate since the cotyledonary ridge is lateral in origin and not terminal. It merely represents the growth of cotyledonary protuberances almost entirely around the periphery of the proembryo, instead of being localized at two places as is customary among dicotyledonous plants, and the true dicotyledonous character does not become apparent until a later stage in the development of the embryo arrives.

A suspensor is absent in *Nelumbo lutea*, but a filamentous structure may be formed in *N. odorata* and *N. lotus*. *Nymphaea* is without a suspensor but later develops a latent suspensor-like structure.

Castalia ampla and *C. pubescens* differ in their embryonomy from *C. odorata*. In the first species, a proembryo of as many as six cells linearally arranged is first developed, but nothing is said as to which are derived from the terminal and which from the basal cell. The upper two cells are next partitioned by longitudinal walls, followed by similar walls in each daughter cell oriented perpendicular to the first wall. *C. pubescens* develops similarly save that the proembryo is usually only four cells long when the first longitudinal divisions occur. The cotyledonary ridge in this species extends about two-thirds of the way around the periphery. A reinvestigation is needed to determine the assignment of these two species.

CONARD, H. S. 1905. The Waterlilies. Carnegie Inst. Wash. Publ. 4.
COOK, M. T. 1902. Development of the embryo-sac and embryo of *Castalia odorata* and *Nymphaea advena*. Bull. Torrey Bot. Club 29: 211–220.
COOK, M. T. 1906. The embryogeny of some Cuban *Nymphaeaceae*. Bot. Gaz. 42: 376–392.
COOK, M. T. 1909. Notes on the embryology of the *Nymphaeaceae*. Bot. Gaz. 48: 56–60.

Berberidales

Nothing has been described of the embryogeny of the *Lardizabalaceae* *Sargentadoxaceae* or *Menispermaceae*, while that of the *Berberidaceae* and *Circaeasteraceae* is superficially known.

Berberidaceae:- The embryogeny of the family is in an unsatisfactory condition and only general conclusions are possible. *Berberis nepalensis*, *Caulophyllum thalictroides* and *Podophyllum emodi* inferentially conform to the Myosurus Variation, Onagrad Type.

An early investigation on *Podophyllum peltatum*, *Caulophyllum thalictroides* and *Jeffersonia diphylla* purported to show that a crescentic ridge was formed on the distal end of the embryo and that this then bifurcated to form the two cotyledons. Later researches, however, indicate that the cotyledons of *C. thalictroides* do not arise in this manner but appear as two moundlike protuberances occupying opposite sides of the margin of the distal end of the embryo. Again, other studies show that the two cotyledons of *P. peltatum* originate as two independent protuberances similarly located. A cotyledonary tube is then developed through intercalary growth.

In *Mahonia aquifolium* the proembryo forms a tier of six cells, of which the fourth and fifth cells become longitudinally segmented, followed by all except the lowermost cell. Subsequent stages are too crudely drawn to permit conclusions.

Podophyllum peltatum displays the same irregularities in the planes of the proembryonic divisions as occur in many of the Ranunculaceae; consequently, assignment is to the Myosurus Variation, Onagrad Type. A tricotyledonous embryo has been described.

ARZT, T. 1933. Über die Embryobildung von Pseudomonokotylen. Beih. Bot. Centralbl. 50: 671–696.

BUTTERS, F. K. 1909. The seeds and seedlings of *Caulophyllum thalictroides*. Minn. Bot. Studies 4: 11–33.

CLARK, L. 1923. The embryogeny of *Podophyllum peltatum*. Minnesota Studies in Pl. Science 1923: 111–126.

JOHRI, B. M. 1935. The gametophytes of *Berberis nepalensis* Spreng. Proc. Indian Acad. Sci. 1: 640–649.

MAURITZON, J. 1936. Zur Embryologie der *Berberidaceen*. Meddel. Göteborgs Bot. Trädg. 11: 1–18.

Circaeasteraceae:- Information concerning the embryonomy of *Circaeaster agrestis* is very scanty. Of the proembryonic stages, a two-celled and a five-celled stage are depicted. The latter consists of a linear row of cells. Assignment may possibly be to the Chenopodiad Type.

JUNELL, S. 1931. Die Entwicklungsgeschichte von *Circaeaster agrestis*. Svensk Bot. Tidskr. 25: 238–270.

Aristolochiales

The *Aristolochiaceae* and *Nepenthaceae* are terrae incognitae embryologically.

Cytinaceae (Rafflesiaceae):- The terminal cell of the two-celled proembryo of *Rafflesia patma* divides transversely, following which both daughter cells, as well as the basal cell, divide vertically. The wall in the middle cell may be oriented at right angles to those in the basal cell and in the upper daughter cell, or those in the derivatives of the terminal cell may be in nearly the same plane with that of the basal cell at a right angle. The basal region subsequently remains bicellular. Quadrants are next formed in the upper daughter cell derived from the terminal cell. The oldest stage depicted is that of a proembryo consisting of five tiers of cells, of which at least the upper two are of four cells each and the lowest tier (the basal or suspensor region) contains only two cells.

Brugmansia zippelii agrees substantially with *R. patma*.

Mitrastemon yamamotoi has a four-celled embryo in the mature seed. This consists of two superposed basal cells and two juxtaposed terminal cells.

The evidence is so scanty that assignment of the above species is difficult; there are resemblances to both the Solanad and Caryophyllad Types.

ERNST, A., and E. SCHMID. 1913. Über Blüte und Frucht von *Rafflesia*. Ann. Jard. Bot. Buitenzorg II, 12: 1–58.

WATANABE, K. 1933. Biologie von *Mitrastemon yamamotoi* Makino (*Rafflesiaceae*). I. Frücht und Samen. Bot. Mag. Tokyo 47: 798–805. (In Japanese.)

Hydnoraceae:- The embryonomy of *Hydnora africana* is peculiar. The two-celled proembryo continues transverse segmentation until a linear row of as many as fifteen cells is produced, but nothing is said as to the proportion derived from the basal and terminal cells respectively. Next longitudinal walls are erected beginning in about the fifth cell from the terminal end and extending through the intervening cells as far as the ninth. These cells alone engender the embryo proper. The cells at the basal end gradually disintegrate, but those above the embryonic body persisted in the oldest embryos observed. The latter are undifferentiated.

The disposition of this peculiar mode of embryonomy is doubtless best under the Solanad Type. There are certain resemblances to the Sherardia Variation, but the erection of a new Variation to accommodate the genus appears to be justified and may be designated as the Hydnora Variation.

DASTUR, R. H. 1922. Notes on the development of the ovule, embryosac and embryo of *Hydnora africana* Thunb. Trans. Roy. Soc. South Africa 10: 27–31.

Piperales

No accounts are available for the *Sauruaceae* and *Lacistemaceae*.

Piperaceae:- The zygote of *Peperoma pellucida* has been described and figured as being longitudinally divided, with subsequent vertical then transverse walls in each daughter cell. There is no differentiation in the embryos of ripe seeds. *P. reflexa* appears to develop similarly. These two species conform to the Piperad Type.

The first wall in the zygote of *Heckeria umbellata* is obliquely oriented, and the arrangement of the walls in older embryos makes it appear certain that this is the general rule; assignment therefore is the same as with the two preceding species. The mature embryo is an undifferentiated globular mass with a slight projection which may be a secondarily developed, rudimentary suspensor. The mature embryo of *P. pellucida* lacks a suspensor but is otherwise similar.

CAMPBELL, D. H. 1901. The embryo-sac of *Peperomia*. Ann. Bot. 15: 103–117.
FAGERLIND, F. 1940. Die Entwicklung des Embryosackes bei *Peperomia pellucida* Kunth. Arkiv f. Bot. 29A: No. 17 (15 p.).
FISHER, G. C. 1914. Seed development in the genus *Peperomia*. Bull. Torrey Bot. Club 41: 137–156.
JOHNSON, D. S. 1900. On the endosperm and embryo of *Peperomia pellucida*. Bot. Gaz. 30: 1–11.
JOHNSON, D. S. 1902. On the development of certain *Piperaceae*. Bot. Gaz. 34: 321–340.

Chloranthaceae:- The embryonomy of *Hedyosmum nutans* is partially illustrated but not described. The proembryo develops into a linear row of ten to eleven cells, when the outermost cell is partitioned longitudinally. The cells are very much flattened. The developing embryo has a massive suspensor and from the figure it is clear that the basal cell contributes to the construction of the embryo. The cotyledons have not been formed by the time the seed is ripe.

Assignment of this species would appear to be to the Chenopodiad Type.

EDWARDS, J. C. 1920. Flower and seed of *Hedyosmum nutans*. Bot. Gaz. 70: 409–424.

Rhoeadales

Papaveraceae:- The embryonomy of *Papaver rhoeas* conforms to that of of the Papaver Variation, Solanad Type. As in the typical Solanad Type of development a linear tetrad is erected. The destinations at the first cell generation are:

$$ca = pvt + pco + phy + icc + iec$$
$$cb = co + s$$

At the second cell generation, however, the destinations are radically changed:

$l = pvt$
$l' = pco + phy + icc + iec$
$m = co + s$ (in part)
$ci = s$ (in part)

and it is therefore apparent that tier *l* engenders only the stem apex, which comes about through the precocious differentiation of an epiphysis.

Chelidonium majus is very similar to *Papaver rhoeas* embryonomically. The basal cell does not divide more than once and the wall may be oriented in almost any direction. The two species agree in the formation of a linear tetrad (the basal cell is ignored), in the development of somewhat unusual six-, eight- and sixteen-celled proembryos; in the differentiation of an epiphysis which becomes elevated as a prominent cone at the upper end of the embryo and which is differentiated from cell *l* of the tetrad and furnishes the initials of the epidermis and the stem cortex at the apex of the stem; in the origin of the initials of the root cortex from the lowermost elements of the hypocotyledonary portion; and in the participation of the upper cell of the suspensor only in the construction of the central portion of the root cap.

The table of recapitulation which will serve for *Papaver rhoeas* and probably most of the *Papaveraceae* follows:

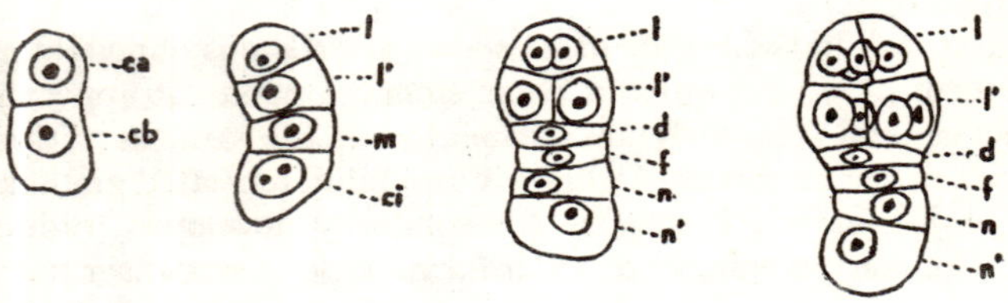

Recapitulatory Table for Papaver rhoeas
(from left to right)

I. First Cell Generation
Proembryo of two cells disposed in two tiers:
$ca = pvt + pco + phy + icc + iec$
$cb = co + s$

II. Second Cell Generation
Proembryo of four cells disposed in four tiers:
$l = pvt$
$l' = pco + phy + icc + iec$
$m = co + s$ (in part)
$ci = s$ (in part)

III. Third Cell Generation
Proembryo of eight cells disposed in six tiers:
Arrangement and destinations as in second generation.

IV. Fourth Cell Generation
Proembryo of twelve cells disposed in six tiers:
$l = pvt$
$l' = pco + phy + icc + iec$
$d = co$
$f + n + n' = s$

Although there is no description and the figures are inadequate, *Sanguinaria canadensis* probably conforms to the Solanad Type.

Souèges, R. 1926. Embryogénie des *Papavéracées*. Développement du proembryon chez le *Papaver rhoeas* L. C. R. Acad. Sci. Paris 183: 902–904.
Souèges, R. 1936. Embryogénie des *Papavéracées*. Les derniers stades du développement de l'embryon chez le *Papaver rhoeas* L. C. R. Acad. Sci. Paris 183: 1119–1120.
Souèges, R. 1936. Embryogénie des *Papavéracées*. Développement de l'embryon chez le *Chelidonium majus* L. C. R. Acad. Sci. Paris 203: 678–680.
Souèges, R. 1937. Développement de l'embryon chez le *Chelidonium majus* L. Ann. Sci. Nat., Bot. X, 19: 445–466.
Surface, F. M. 1905. Contribution to the history of *Sanguinaria canadensis*. Ohio Nat. 6: 379–385.

Fumariaceae:- *Hypecoum procumbens* behaves in a somewhat peculiar fashion. The zygote divides by a markedly oblique wall. The basal cell never divides again, but becomes a greatly swollen vesicle. The terminal cell, *ca*, is located at one side of the vesicle; strictly speaking, the wall erected following the first division in *ca* should be described as vertical, but subsequent events demonstrate that it is really transverse. It looks more as if it were oblique in most sections. Cell *ca*, like the zygote, is segmented into two unequal cells, *cc* and *cd*. The latter does not divide again but becomes inflated, though not to such an extent as *cb*. Cell *cc* segments transversely into *ce* and *cf*; *ce* in turn divides transversely into *cg* and *ch*; while *cf* generally segments vertically, occasionally transversely — there is no constant order. At about this time the embryo proper becomes separated from the two inflated cells; subsequently no suspensor proper is formed. Cell *cg* later divides transversely into *ci* and *cj*; the former of these two daughter cells becomes an epiphysis initial. The derivatives of *cf* presently develop into a massive hypophyseal region. The embryogeny of *H. procumbens* departs so radically from that of all other species that it can only constitute a variation, the Hypecoum Variation of the Caryophyllad Type.

Fumaria officinalis is the species on which the Fumaria Variation of the same Type is based. The terminal cell alone develops all of the embryo proper, almost exactly as in *Medicago lupulina* (Medicago Variation, of the same Type). The basal cell, however, by means of rather irregular divisions produces a huge suspensor which is considerably larger than the embryo proper until the latter is almost mature.

Corydalis lutea, together with *C. ochroleuca*, forms still another embryonomic type. The zygote segments transversely; the basal cell does not divide again, but its nucleus divides first into two free nuclei, then these divide into four and finally there is a large number of nuclei which are mostly peripherally located. The terminal cell, *ca*, segments transversely into *cc* and *cd*. The latter daughter cell does not divide again, but its nucleus divides repeatedly and the cell becomes a second vesicle of the suspensor. Cell *cc* segments transversely into *ce* and *cf*. The latter element behaves like cell *cd*, but its nucleus remains undivided for a long time; it constitutes a third vesicle of the suspensor. The three vesicles elongate into a tube, surmounted by the small terminal cell. Eventually it divides transversely into *cg* and *ch*. Cell *ch* becomes partitioned vertically; a little later *cg* segments transversely into *ci* and *cj*. Cell *ci* eventually forms an epiphyseal region, while cell *ch* finally develops a hypophyseal region. These two species of *Corydalis* together constitute the Corydalis Variation, Caryophyllad Type.

Corydalis cheilanthifolia, unlike the two preceding species, however, is considered to adhere more closely to the Trifolium than to the Corydalis Variation.

GUIGNARD, L. 1903. La formation et le développement de l'embryon chez l'*Hypecoum*. Jour. de Bot. 17: 33–44.

HEGELMAIER, F. 1878. Vergleichende Untersuchungen über die Entwicklung dicotyledoner Keime mit Berücksichtigung der Pseudo-monokotyledonen. Stuttgart.

SOUÈGES, R. 1941. Embryogénie des Fumariacées. L'origine du corps de l'embryon chez le *Fumaria officinalis* L. C. R. Acad. Scr. Paris 213: 528–530.

SOUÈGES, R. 1941. Embryogénie des *Fumariacées*. La différenciation des régions fondamentales du corps chez le *Fumaria officinalis* L. C. R. Acad. Sci. Paris 213: 699–701.

SOUÈGES, R. 1943. Embryogénie des *Fumariacées*. L'origine et les premières divisions de la cellule embryonnaire proprement dit chez l'*Hypecoum procumbens* L. C. R. Acad. Sci. Paris 216: 310–311.

SOUÈGES, R. 1943. Embryogénie des *Fumariacées*. La différenciation des régions fondamentales du corps chez l'*Hypecoum procumbens* L. C. R. Acad. Sci. Paris 216: 354–356.

SOUÈGES, R. 1946. Embryogénie des *Fumariacées*. Les premiers termes du développement de l'embryon chez le *Corydalis lutea* DC. C. R. Acad. Sci. Paris 222: 161–163.

SOUÈGES, R. 1946. Embryogénie des *Fumariacées*. Développement de l'embryon chez le *Corydalis cheilanthifolia* Hemsl. C. R. Acad. Sci. Paris 222: 523–524.

SOUÈGES, R. 1946. Embryogénie des *Fumariacées*. La différenciation des régions fondamentales du corps chez le *Corydalis lutea* DC. C. R. Acad. Sci. Paris 222: 699–701.

Loasales

The *Turneraceae* have not been investigated.

Loasaceae:- *Loasa lateritia*, on the whole, appears to agree with the Hyoscyamus Variation, Solanad Type, but there are some resemblances to other Variations under the same Type.

CRÉTÉ, P. 1946. Embryogénie des Loasacées. Développement de l'embryon chez le *Loasa lateritia* Gill. C. R. Acad. Sci. Paris 222: 920–921.

Capparidales

The *Tovariaceae* remain uninvestigated.

Capparidaceae:- The proembryo of *Cleome serrata*, *C. monophylla* and *Gynandropsis pentaphylla* consists of a linear row of three or four cells, while those of *Capparis frondosa* and *C. galeata* consist of from seven to nine similar cells. All of these species appear to follow the Alyssum Variation, Onagrad Type, very closely.

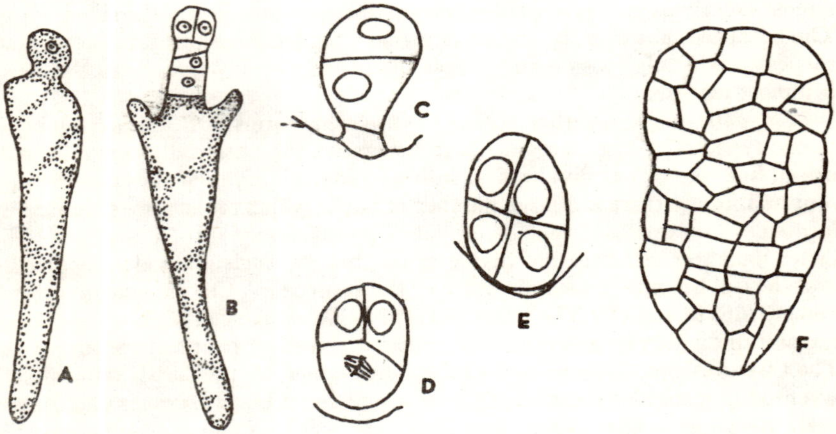

FIG. 39. — *Polanisia trachysperma*. A. Zygote. ×230. B. Proembryo with peculiar suspensor cell. ×230. *Moringa oleifera*. C–E. Development of proembryo. ×460. F. Young embryo. ×460. (A, B *redrawn from* MAURITZON 1935; C–F *redrawn from* PURI 1941).

The zygote and proembryo of *Polanisia trachysperma*, on the other hand have a peculiar shape. The zygote becomes very much elongated and nearly cylindrical in shape (FIG. 39A); then a protuberance is formed at the outer end and the nucleus is included within this rounded portion, while the cell becomes broader below the constriction and projections appear at each side. Details of the further development are meager, but it appears that the first transverse wall is erected in the protuberance above the constriction (FIG. 39B). The later embryogeny apparently also conforms to the Alyssum Variation.

A later and more accurate investigation of *Cleome chelidonii*, however, reveals that this species follows the Lythrum Variation, Onagrad Type, precisely.

Adventitious embryony is very prevalent in *Capparis frondosa*; as many as twenty nucellar embryos have been found attached to the same internal nucellar epidermis.

MAURITZON, J. 1935. Die Embryologie einiger *Capparidaceen* sowie von *Tovaria pendula*. Arkiv f. Bot. 26A, No. 15:1–14.

RAGHAVAN, T. S. 1937. The life-history of *Cleome chelidonii* Linn. fil. Jour. Linn. Soc. London, Bot. 51: 43–72.
RAO, V. S. 1938. Studies on *Capparidaceae*. III. Genus *Capparis*. Jour. Indian Bot. Soc. 17: 69–80.

Moringaceae:- The terminal cell of the bicellular proembryo of *Moringa oleifera* is segmented vertically, but occasionally the wall may be transverse (FIG. 39C–E). In the former case the basal cell divides vertically. The further divisions are rather irregular (FIG. 39F); the developing embryo is bulbous with what is claimed to be a massive suspensor. The two cotyledons are generally somewhat unequal in width and the larger one, although single near the hypocotyl, becomes bilobed downward. This tricotyledonous condition is apparently a case of schizocotyly. Assignment is not possible with certainty since there may be a misinterpretation of the nature of the suspensor. On the basis of the illustrations, the species might be referable to the Penaea Variation, Asterad Type.

PURI, V. 1941. The life-history of *Moringa oleifera* Lamk. Jour. Indian Bot. Soc. 20: 263–284

Cruciferales

Brassicaceae (Cruciferae):- The family is most conveniently discussed by separating it into the two Variations (*Capsella* and *Alyssum*) which are recognized. Both are under the Onagrad Type.

Capsella Variation. — This Variation is based upon *Capsella bursa-pastoris*, one of the most comprehensively studied of plants from the embryological standpoint. There has been considerable misinterpretation by many of the earlier workers, several of whom contended that the basal cell of the two-celled proembryo did not divide further, but who did not actually observe mitoses in this cell. The great importance of actually observing mitoses and the orientation of their spindles is thus vividly emphasized.

The second division is transverse in the basal cell (FIG. 40B). The third divides the terminal cell longitudinally into two adjoining cells (FIG. 40C). In *Lepidium sativum*, however, the third division occurs in the middle cell (*m*) and divides it transversely. There is thus some variation as to the cell in which the third division occurs, but in the long run it is of little significance. In *Lepidium* the next division bisects the terminal cell in the longitudinal plane; then, in both *Lepidium* and *Capsella*, each of the two terminal cells is divided by a wall oriented perpendicularly to the wall first formed. This constitutes the quadrant stage (FIG. 40E). Next each of the quadrant cells is, usually simultaneously, bisected by a transverse wall to give rise to the octant stage (FIG. 40F). There are two superposed tiers of four cells each: the upper tier gives rise to the stem apex and cotyledons, the lower tier to the hypocotyl. In each cell of the octant embryo a periclinal wall cuts off an inner and an outer cell. Each outer cell is an epidermal initial and the inner cells shortly divide by longitudinal walls to engender the periblem and plerome initials (FIG. 40G).

By the time the embryo has reached the quadrant stage, the suspensor cells have divided to form a row of five to seven cells in *Capsella* (FIG. 40F), with about five in *Lepidium*. The last cell of the suspensor rarely divides, but becomes swollen and vesicular. While the periclinal walls are forming in the octant cells the number of cells in the suspensor increases to as many as ten.

The cell at the top of the row of suspensor cells now assumes a significant role. To this cell HANSTEIN gave the name "hypophysis." It is very difficult to determine whether this cell originated from the middle or the basal cell of

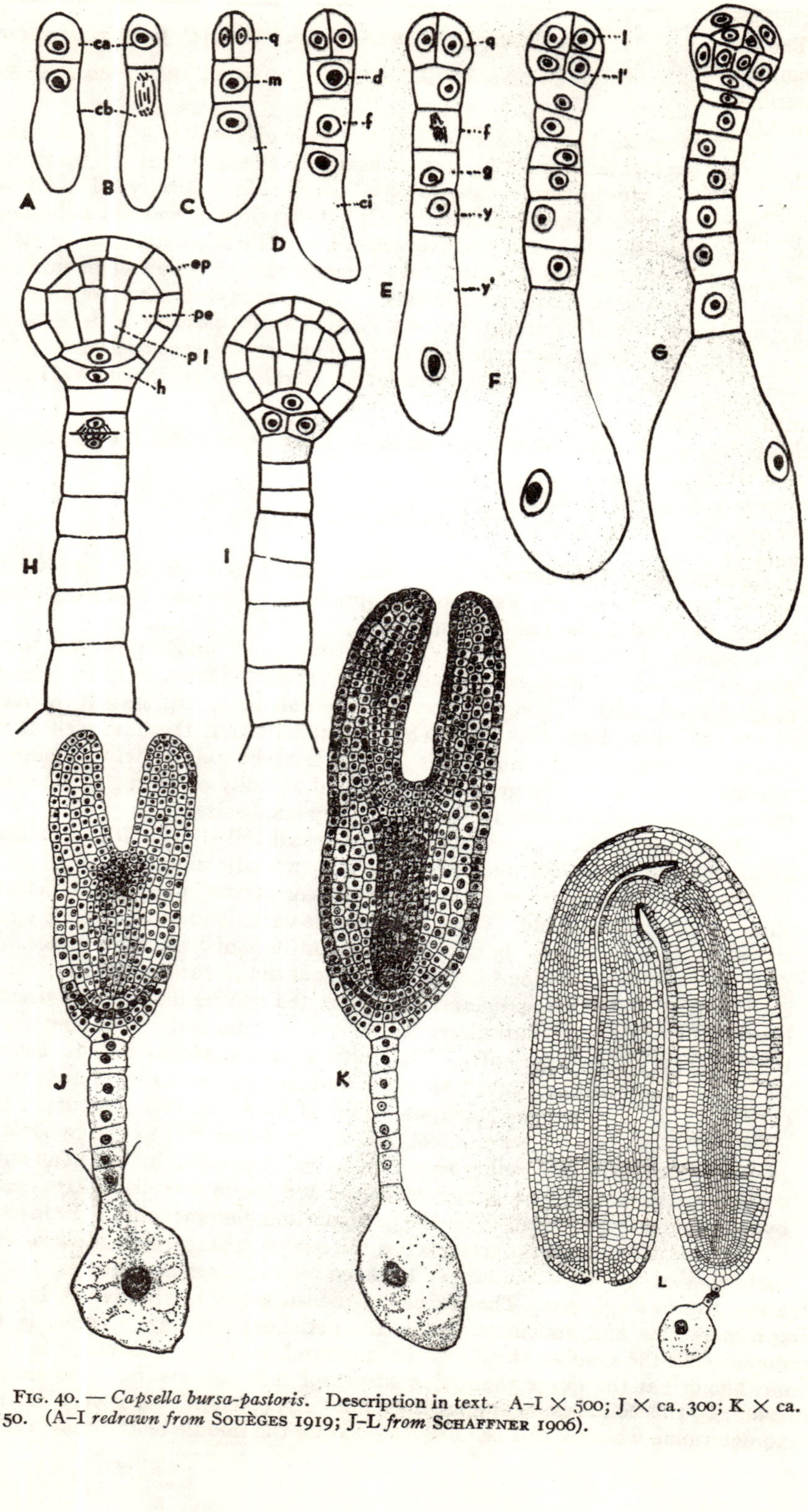

FIG. 40. — *Capsella bursa-pastoris*. Description in text. A–I × 500; J × ca. 300; K × ca. 150. (A–I *redrawn from* SOUÈGES 1919; J–L *from* SCHAFFNER 1906).

the three-celled proembryo, but the probability is that it arose from the former. The first division in this cell does not occur until the epidermal cells of the embryo proper have become delimited and four to six cells are present in the periblem-plerome region; the wall is exactly like the form developed in the middle cell of the typical Onagrad Type (FIG. 40H, I).

Presently longitudinal walls are erected in both upper and lower tiers of the embryo proper, together with anticlinal walls in the epidermal initials in both regions. Cell divisions soon become more numerous at opposite ends of the outer peripheral region of the upper tier and these initiate the cotyledons; the embryo becomes more or less cordate in longitudinal section (FIG. 40J). The periblem and plerome regions of the hypocotyl increase through the addition of new cells, which for the most part divide transversely. The former is rarely over three cell rows in thickness as seen in transverse section. A pericycle becomes differentiated in the outer row of plerome cells. The epidermal, periblem and plerome initials of the root tip become functional but it is not until the embryo has become nearly mature that their contributions can be clearly recognized (FIG. 40K). The epidermal cells contribute the initials from which the root cap is developed. The stem apex is not formed until the embryo is nearly mature. The ovule itself is curved in a horseshoe-shape and the growing cotyledons conform to the space imposed by the curvature of the ovule (FIG. 40L). The table of recapitulation for *Capsella bursa-pastoris* follows:

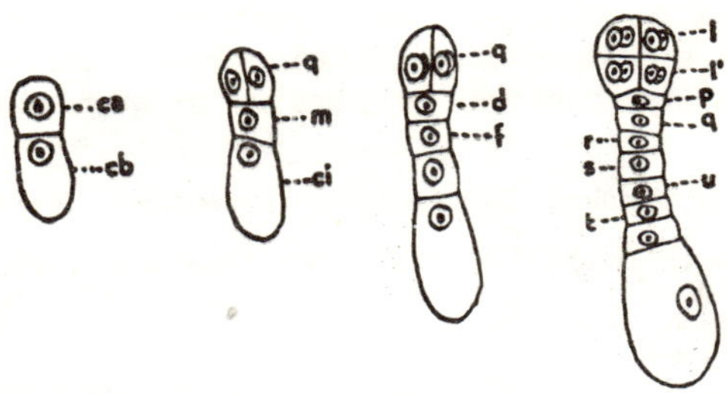

Recapitulatory Table for Capsella bursa-pastoris
(from left to right)

I. FIRST CELL GENERATION
Proembryo of two cells disposed in two tiers:
ca=*pco*+*pvt*+*phy*+*icc*
cb=*iec*+*co*+*s*

II. SECOND CELL GENERATION
Proembryo of four cells disposed in three tiers:
q=*pco*+*pvt*+*phy*+*icc*
m=*iec*+*co*+*s* (in part)
ci=*s* (in part)

III. THIRD CELL GENERATION
Proembryo of eight cells disposed in five tiers:
q=*pco*+*pvt*+*phy*+*icc*
d=*iec*+*co*+*s* (in part)
f+*g*+*v*=*s* (in part)

IV. FOURTH CELL GENERATION
Proembryo of sixteen (or fifteen) cells disposed in ten (or nine) tiers:
l =*pco*+*pvt*
l'=*phy*+*icc*
p=*iec*+*co*+*s* (in part)
q+*r*+*s*+*t*+*u*+*y*+*y'*=*s* (in part)

Alyssum Variation. — *Alyssum* is closely related to *Capsella* and in general has the same type of embryogeny, but appears to differ at two points. The first difference is that it is claimed that the basal cell of the three-celled proembryo does not divide further, all of the very long suspensor being derived from the middle cell (*m*). The suspicion immediately arises that this is a misinterpretation of exactly the same nature as happened in earlier accounts of the embryogeny of *Capsella*, but against this is the fact that the basal cell at the three-celled stage is extremely long and hence probably incapable of dividing. If the latter assumption is correct, then all the cells of the suspensor, totaling from eight to fifteen, are derived from cell *m*, as is the hypophysis cell. The second difference is that the first division in the hypophysis cell is vertical, as in the Ranunculus Variation; the next division is also vertical and is followed by transverse walls. All other histogens are delimited as in *Capsella*.

Assignment of Other Genera:- The following species conform to the Capsella Variation: *Capsella bursa-pastoris, Lepidium sativum, L. campestre, L. draba, Cochlearia officinalis, Ionopsidium savianum.* Those agreeing with the Alyssum Variation include *Alyssum macrocarpum*, and probably different varieties of *Brassica oleracea.*

However, the situation with respect to *Brassica*, and particularly concerning varieties of *B. oleracea*, is confused. In general, the embryogeny is claimed to resemble that of *Capsella bursa-pastoris* closely, but it is obvious from a consideration of the figures presented in several papers that this is not true with regard to the derivatives of the basal cell. The end cell of the suspensor is not swollen or vesicular as in *Capsella*; the suspensor is more like that of *Alyssum macrocarpum*. The intermediate cell, *m*, of the three-celled proembryo apparently does not function as a hypophysis initial but divides transversely at least twice. The cell adjacent to the embryo proper remains larger than the other cells of the suspensor and becomes the hypophysis cell. This cell divides by a typical curved wall. Since there is question as to the behavior of cell *m*, the disposition of *Brassica oleracea* must be held in abeyance. It seems to display characters intermediate between those of *Capsella* and *Alyssum*.

Corti, R. 1930. Embriologia del genere *Ionopsidium* Racb. Nuovo Giorn. Bot. Ital. 37: 510–526.
Pearson, O. H. 1933. Study of the life history of *Brassica oleracea*. Bot. Gaz. 94: 534–550.
Riddle, L. C. 1898. The embryology of *Alyssum*. Bot. Gaz. 26: 314–324.
Schaffner, M. 1906. The embryology of the Shepherd's Purse. Contrib. Bot. Lab. Ohio State Univ. 25: 1–8.
Souèges, R. 1914. Nouvelles recherches sur le développement de l'embryon chez les *Crucifères*. Ann. Sci. Nat., Bot. IX, 19: 311–339.
Souèges, R. 1919. Les premières divisions de l'oeuf et les différenciations du suspenseur chez le *Capsella bursa-pastoris* Moench. Ann. Sci. Nat., Bot. X, I: 1–28.
Thompson, R. C. 1933. A morphological study of flower and seed development in cabbage. Jour. Agr. Res. 47: 215–232.

Violales

Violaceae:- The embryonomy of *Viola tricolor* agrees in the main with that of the Geum Variation, Asterad Type, but there are certain essential differences. In *V. tricolor* the three subepiphyseal elements give rise to both the cotyledonary region proper and to the upper portion of the hypocotyledonary axis instead of producing the cotyledonary portion alone; the element *m* furnishes the lower portion of the hypocotyl in lieu of engendering the entire hypocotyl; the initials of the root cortex arise from derivatives of the hypocotyl instead of originating from cell *ci*; finally, no suspensor is developed, the lowest cell, *ci*, contributing only to the construction of the central portion of the root cap.

Viola fimbriatula and *V. cucullata* by inference conform to the description for *V. tricolor*. A second division is figured as being unequally transverse in the basal cell of the two-celled proembryo and the third is longitudinal in the terminal cell. A suspensor also is not formed in either species, the well developed embryo being perfectly globular in outline.

BLISS, M. C. 1912. A contribution to the life-history of *Viola*. Ann. Bot. 24: 155–163.
SOUÈGES, R. 1937. Embryogénie des *Violacées*. Développement de l'embryon chez le *Viola tricolor* L. C. R. Acad. Sci. Paris 205: 169–171.

Resedaceae:- *Reseda luteola* follows exactly the same type of embryonomy that prevails in the Onagrad Type. It is possible, however, that other species in the family do not maintain such a rigorous adherence to this developmental scheme.

SOUÈGES, R. 1935. Embryogénie des *Résédacées*. Développement de l'embryon chez le *Reseda luteola* L. C. R. Acad. Sci. Paris 201: 910–911.

Polygalales

There is a somewhat inconclusive report for one genus in the *Polygalaceae*, but none for either of the other two families, *Trigoniaceae* and *Vochysiaceae*.

Polygalaceae:- In *Salomonia* (*Epirrhizanthes*) two species, *S. elongata* and *S. cylindrica*, have been examined. The second division is transverse in the basal cell and the third is longitudinal in the terminal cell, following which the two daughters of the terminal cell are segmented longitudinally to form the quadrant stage. Next the quadrants are divided transversely to constitute the octant stage. Events in the root tip region are not at all clear.

Polygala vulgaris has an embryogeny decidedly similar to that occurring in *Ceanothus aureus* (*Rhamnaceae*) and may, like the latter, be assigned to the Geum Variation, Asterad Type. The species of *Salomonia* mentioned above apparently should be classified similarly.

SOUÈGES, R. 1941. Embryogénie des Polygalacées. Développement de l'embryon chez le *Polygala vulgaris* L. C. R. Acad. Sci. Paris 213: 446–448.
WIRZ, H. 1910. Beiträge zur Entwicklungsgeschichte von *Sciaphila* spec. und von *Epirrhizanthes elongata* Bl. Flora 101: 395–446.

Saxifragales

There is no report for the Australian family *Cephalotaceae*.

Crassulaceae:- There has been some divergence of opinion regarding the embryogeny of the *Crassulaceae*. A considerable number of species have been investigated, but only *Sedum acre* has been thoroughly studied. Despite statements to the contrary, the embryonomy of the family is essentially identical throughout and conforms to the Sedum Variation, Caryophyllad Type, with the superposition of certain characters of the Onagrad Type.

The differences of opinion result from a misinterpretation of the function of the basal cell of the two-celled proembryo. In *Sedum acre* it has been demonstrated beyond dispute that this cell never divides again. All of the embryo is derived solely from the terminal cell. After the first transverse division in the zygote, the basal cell becomes much swollen, occupying the entire apical portion of the embryo sac and becomes extensively haustorial. It is crowned by the very small, lenticular terminal cell. Cells later produced by the terminal cell were erroneously interpreted as derivatives of the basal cell.

The terminal cell divides transversely into two superposed cells, *cc* and *cd* (FIG. 41B). Shortly thereafter the lower cell, *cd*, divides transversely into the two cells *m* and *ci*, then the latter is partitioned by a longitudinal wall into two juxtaposed cells (FIG. 41C, D). The upper daughter cell, *cc*, of the original terminal cell of the two-celled stage now divides to form an octant stage precisely as in the Onagrad Type (FIG. 41E, F). The lower layer of octant cells presently divide by longitudinal walls to separate first the epidermal initials, then the periblem and plerome initials of the hypocotyl (FIG. 41G, H). In the developing plerome region the walls become more and more obliquely oriented, but this anomaly disappears during later stages. In the upper cells of the

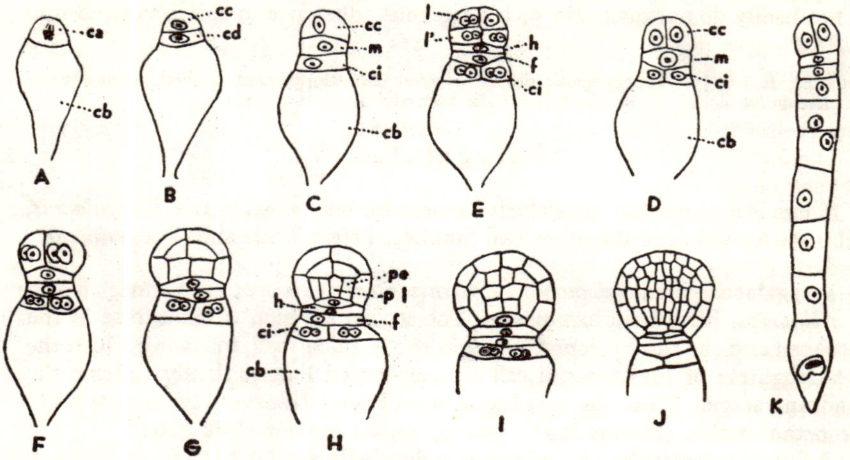

FIG. 41. — *Sedum acre.* A–J. Description in text. All × 400. *Sedum rupestre.* K. Proembryo with filamentous suspensor. ×230. (A–J *redrawn from* SOUÈGES 1927; K *redrawn from* MAURITZON 1933).

octant embryo curved walls delimit the epidermal initials (FIG. 41G–I). Each of the subepidermal cells is then partitioned by vertical walls. This development corresponds exactly with that in *Lepidium* (Capsella Variation, Onagrad Type).

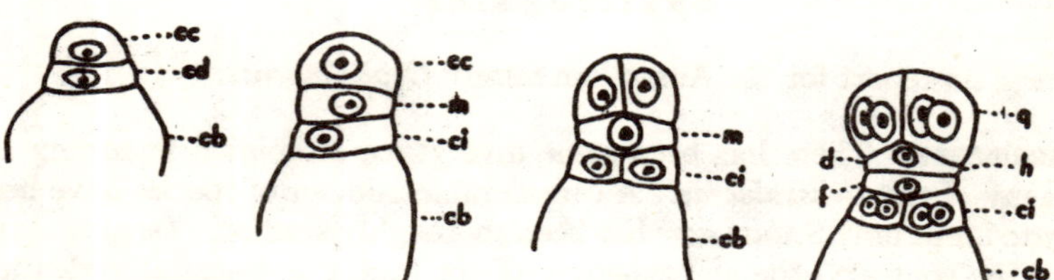

Recapitulatory Table for Sedum acre

(from left to right)

I. FIRST CELL GENERATION
Proembryo of two cells disposed in two tiers:

cc = *phy* + *pco* + *pvt*
cd = *iec* + *co* + *s*

II. SECOND CELL GENERATION
Proembryo of three cells disposed in three tiers:

cc = *phy* + *pco* + *pvt*
m = *iec* + *co* + *s* (in part)
ci = *s* (in part)

III. THIRD CELL GENERATION
Proembryo of five cells disposed in three tiers:

Stages and destinations are the same as in the second cell generation.

IV. FOURTH CELL GENERATION
Proembryo of ten cells disposed in four tiers:

cc (*q*) = *phy* + *pco* + *pvt*
d = *iec* + *co*
f = *s* (in part)
ci = *s* (in part)

The intermediate cell, *m*, of the four-celled proembryo behaves in a very regular manner. The first division is transverse (FIG. 41F): the daughter cell are designated *d* and *f* respectively. These cells are very much flattened in the horizontal plane. The upper cell *d* constitutes the hypophysis cell and behaves exactly as in the Onagrad Type (FIG. 41H). During the later stages of embryonal growth the basal region enlarges but little, so that the cells of the hypophyseal region are rather small. The epidermal cells of the embryo proper do not become extended to cover the root.

The lowest cell, *ci*, of the four-celled proembryo early becomes partitioned by vertical walls into four juxtaposed cells, which do not again divide, but become greatly flattened and stretched in the horizontal plane (FIG. 41D–I). These cells form a sort of suspensor.

The four cell generations and the regions of the embryo to which they give rise may be summarized as follows, leaving the basal cell out of consideration.

In certain species of *Sedum*, as in *S. rupestre* (FIG. 41K) and *S. anopetalum*, a filamentous suspensor of seven or more cells may be formed. The origin of the extra cells has not been explained, but it is reasonable to consider them as derivatives of cell *m* of the four-celled proembryo.

Penthorum sedoides has been placed by taxonomists in both the *Crassulaceae* and the *Saxifragaceae*, but embryonomically it is best included in the former. It develops precisely like *Sedum acre*. There is a large haustorial basal cell, with a row of four or five suspensor cells. *Rosularia sempervivum* and *Aeonium undulatum* also appear to agree in development with *S. acre*, as does *Cotyledon umbilicus*.

CRÉTÉ, P. 1946. Embryogénie des Crassulacées. Développement de l'embryon chez le *Cotyledon umbilicus* L. C. R. Acad. Sci. Paris 222: 1311–1313.

MAURITZON, J. 1933. Studien über die Embryologie der Familien *Crassulaceae* und *Saxifragaceae*. Lund, Ohlsson.

ROCEN, T. 1928. Beitrag zur Embryologie der *Crassulaceen*. Svensk Bot. Tidskr. 22: 368–376.

SOUÈGES, R. 1927. Développement de l'embryon chez le *Sedum acre* L. Bull. Soc. Bot. France 74: 234–251.

SOUÈGES, R. 1936. Modifications au tableau récapitulatif des lois de développement chez le *Sedum acre* L. Le type embryonomique de cette espèce chez les autres *Crassulacées*. Bull. Soc. Bot. France 83: 13–18.

SOUÈGES, R. 1936. Les relations embryogéniques des *Crassulacées*, *Saxifragacées* et *Hypericacées*. Bull. Soc. Bot. France 83: 317–329.

Saxifragaceae:- *Saxifraga granulata* follows the Sedum Variation, Caryophyllad Type, closely, with certain minor variations: (*1*) in the construction of the two subepidermal layers in the circumaxial zone of the cotyledonary region; (*2*) in the mode of construction of the hypophyseal region, and (*3*) in the formation of large cells by the descendants of the basal cell. The basal cell of the two-celled proembryo divides rather promptly and does not become swollen. It produces from four to eight large cells which become the lower portion of the suspensor. At the time of development of the cotyledons the somewhat swollen basal suspensor region is compressed and completely resorbed by the endosperm. The terminal cell is first segmented transversely to produce two superposed cells, *cc* and *cd*. The cell *cc* engenders the embryo proper in the same manner as in the Onagrad Type. The cell *cd* divides transversely into two superposed cells, *m* and *ci*. The latter, as a general rule, is divided vertically into two juxtaposed cells and produces the middle region of the suspensor. The element *m* is invariably segmented transversely; the lower daughter cell, *f*, which may be further divided by one or two transverse walls, develops the upper portion of the suspensor. The upper daughter cell, *d*, is transformed directly into a hypophysis cell. Occasionally, however, cell *d* may divide transversely and the upper daughter cell then becomes the hypophysis cell. The hypophysis cell next behaves exactly as in the Onagrad Type.

Many genera have been described as forming a linear four-celled proembryo; the further development is decidedly variable according to the genus. In no species has a sufficiently detailed study been made to allow a definite assignment to a given Type. From the crude diagrams and meager descriptions available it would appear that most of them follow the Sedum Variation. The behavior of the basal cell is extremely variable and irregular. Sometimes it does not divide and at the other extreme a massive structure surmounted by a comparatively small embryo (*Ribes aureum*) is formed. Filamentous suspensors are occasionally developed, as in *Bergenia hessiana* and *Mitella nuda*. In *Deutzia longiflora* and *Vahlia oldenlandioides* the basal cell becomes excessively elongated.

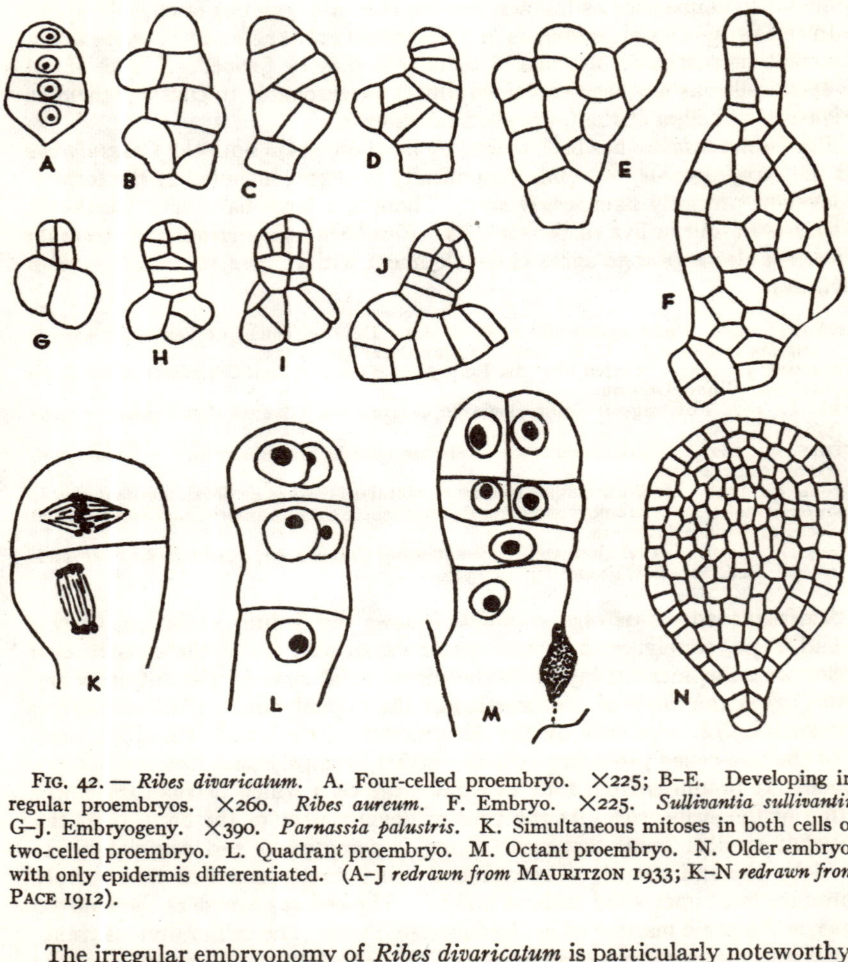

FIG. 42. — *Ribes divaricatum*. A. Four-celled proembryo. ×225; B–E. Developing irregular proembryos. ×260. *Ribes aureum*. F. Embryo. ×225. *Sullivantia sullivantii*. G–J. Embryogeny. ×390. *Parnassia palustris*. K. Simultaneous mitoses in both cells of two-celled proembryo. L. Quadrant proembryo. M. Octant proembryo. N. Older embryo, with only epidermis differentiated. (A–J *redrawn from* MAURITZON 1933; K–N *redrawn from* PACE 1912).

The irregular embryonomy of *Ribes divaricatum* is particularly noteworthy. As in the other genera, a row of four cells is first formed (FIG. 42A–E), after which each cell divides in the most peculiar fashion. Daughter cells may project laterally, but most often derivatives become extended in the direction of the chalaza. As the later development was not followed out, the ultimate fate of the lateral outgrowths is not known.

The embryonomy of *Sullivantia sullivantii* (FIG. 42G–J) and *Chrysosplenium alternifolium* is similar and also irregular. The basal cell appears to divide vertically, followed by additional more or less vertical divisions so that the suspensor region becomes broad and somewhat flattened. Development of the embryo proper appears to conform with that of the Sedum Variation.

In *Peltiphyllum peltatum* both basal and terminal cells of the two-celled proembryo divide transversely.

A proembryo of six cells in a linear row, topped by two juxtaposed cells, is figured for *Mitella pentandra*. Behavior in *M. diphylla* is different, especially for the derivatives of the basal cell. At the four-celled proembryonic stage, the upper two cells are derived by transverse segmentation of the terminal cell and the lower two by a similar division of the basal cell. The latter two cells are unevenly swollen in such a manner that the lower one is turned to one side. Further divisions in both cells presently result in a broad suspensor. There is a row of three to five cells above and later the uppermost one divides vertically.

Boykinia tellimoides develops a linear five-celled proembryo. *Vahlia viscosa* clearly conforms to the Lythrum Variation, Onagrad Type; *V. oldenlandioides* is said to follow the same scheme of development as in the former species.

Parnassia palustris and *Saxifraga crassifolia* inferentially conform to the Sedum Variation, but there is some indication that they might follow the Lythrum Variation since in the first species the first division in the terminal cell is longitudinal (FIG. 42K–M).

DAHLGREN, K. V. O. 1930. Zur Embryologie der *Saxifragoideen.* Svensk Bot. Tidskr. 24: 429–448.

GÄUMANN, E. 1919. Studien über die Entwicklungsgeschichte einiger Saxifragales. Rec. Trav. Bot. Néerl. 16: 285–322.

MAURITZON, J. 1933. Studien über die Embryologie der Familien *Crassulaceae* und *Saxifragaceae.* Lund, Ohlsson.

PACE, L. 1912. *Parnassia* and some allied genera. Bot. Gaz. 54: 306–329.

RAGHAVAN, T. S., and V. K. SRINIVASAN. 1942. A contribution to the life-history of *Vahlia viscosa* Roxb. and *Vahlia oldenlandioides* Roxb. Proc. Indian Acad. Sci., B, 15: 83–105.

SOUÈGES, R. 1936. Embryogénie des *Saxifragacées.* Développement de l'embryon chez le *Saxifraga granulata* L. C. R. Acad. Sci. Paris 202: 240–242.

SOUÈGES, R. 1936. Les relations embryogéniques des *Crassulacées, Saxifragacées* et *Hypericacées.* Bull. Soc. Bot. France 83: 317–329.

Sarraceniales

Droseraceae:- The embryonomy of *Drosera rotundifolia* is simple and constant, and offers certain peculiar characteristics. It is a little difficult to assign this species to a definite Type, since the embryogeny partakes of characters from several different Types. For instance, the filamentous embryo conforms to the Solanad Type; the disposition of the quadrants resembles that in both the Solanad and Chenopodiad Types; the terminal cell of the proembryonic tetrad, which alone develops into the body of the embryo, is the daughter cell of the terminal cell, as in the Sedum Variation, Caryophyllad Type; the basal cell does not participate in the formation of the embryo proper but gives rise directly to the suspensor, as again is the case in the Caryophyllad Type; finally, the two tiers *h* and *h'*, which engender respectively the initials of the root cortex and the primordium of the root cap, behave according to the Asterad Type. Altogether, it appears most logical to assign this species to the Caryophyllad Type as a Variation.

To proceed with the detailed discussion of the embryonomy: the terminal cell of the two-celled proembryo always divides in advance of the basal cell and is partitioned transversely into two superposed cells, *cc* and *cd*. The element *cc* next divides transversely into two new cells, *l* and *l'*. These two cells thereupon segment vertically; the four cells thus resulting represent quadrants disposed in a vertical plane. They then undergo meridional partitionings into eight octants; the upper four octants correspond to the cotyledonary region while the four lower octants represent the hypocotyledonary region.

The element *cd* divides transversely into two superposed elements *h* and *h'*, which next divide by vertical walls which are almost but not quite meridional. Through new vertical partitionings, the two juxtaposed elements of *h* give rise to four circumaxial cells, then to a plateau of eight to twelve cells of which only four may be observed in longitudinal sections of an embryo at this stage. The four elements adjacent to the axis constitute the initials of the root cortex, while the four or eight others contribute to the formation of the lateral portion of the root cap. The two juxtaposed elements of *h'* divide very tardily by means of two meridional walls to engender four circumaxial cells that next segment horizontally and finally give rise to the central portion of the root cap.

The basal cell, *cb*, divides transversely into *m* and *ci*. These two cells usually divide transversely, but the walls may be vertical or oblique, and their descendants constitute a very short suspensor *sensu stricto*.

The zygote of *Dionaea muscipula* divides transversely, whereupon the basal cell segments similarly. This species therefore differs from *Drosera rotundifolia* in this respect. The terminal cell segments longitudinally, then each daughter cell is partitioned transversely. A later stage clearly reveals the presence of an epiphysis initial, but it is not clear how it originated. It is very difficult to assign this species also; it doubtless belongs under the Onagrad Type: if an epiphysis is present (a hypophysis is lacking), then it may be assigned to the Lotus Variation, but if an epiphysis is lacking it conforms to the Trifolium Variation.

SMITH, C. M. 1929. Development of *Dionaea muscipula*. I. Flower and seed. Bot. Gaz. 87: 507–530.

SOUÈGES, R. 1936. Embryogénie des *Droséracées*. Développement de l'embryon chez le *Drosera rotundifolia* L. Cr. R. Acad. Sci. Paris 202: 1457–1459.

Sarraceniaceae:- To judge from the very brief account of the development of the proembryo of *Sarracenia purpurea*, it would seem to agree substantially with the Drosera Variation, Caryophyllad Type.

SHREVE, F. 1906. The development and anatomy of *Sarracenia purpurea*. Bot. Gaz. 42: 107–126.

Podostemonales

No report is available for the *Hydrostachyaceae*.

Podostemonaceae:- Conditions within the family are rather interesting, but unfortunately no detailed description of a given species is available.

In all investigated species the zygote divides transversely. The basal cell thereupon segments transversely and the lower daughter cell (the equivalent of *ci*) undergoes mitosis but no wall is subsequently formed. The terminal cell appears to divide transversely several times, then all cells save the apical derivative segment longitudinally. The suspensor, at least in *Dicraea elongata*, consists of not over four cells arranged in a row. The binucleate cell at the distal end sends out haustorial-like protuberances which grow through the nucellus in a chalazal direction.

It is not possible to determine the assignment of species in the Podostemonaceae. Those studied include *Podostemon subulatus*, *Dicraea elongata*, *Lawia zealanica* and *Farmeria metzgerioides*.

MAGNUS, W., und E. WERNER. 1913. Die atypische Embryonalentwicklung der Podostema ceen. Flora 105: 275–336.

Caryophyllales

Elatinaceae:- *Bergia ammanioides*, *B. capensis* and *Elatine* sp. appear to conform to the Solanad Type, and possibly to the Nicotiana Variation.

FRISENDAHL, A. 1927. Über die Entwicklung chasmo- und kleistogamer Blüten bei der Gattung *Elatine*. Meddel. Göteborgs Bot. Trägd. 3: 99–142.
KAJALE, L. B. 1940. A contribution to the life-history of *Bergia ammanioides* Roxb. Jour. Indian Bot. Soc. 18: 157–167.
RAGHAVAN, T. S., and V. K. SRINIVASAN. 1940. A contribution to the life-history of *Bergia capensis* Linn. Jour. Indian Bot. Soc. 19: 283–291.

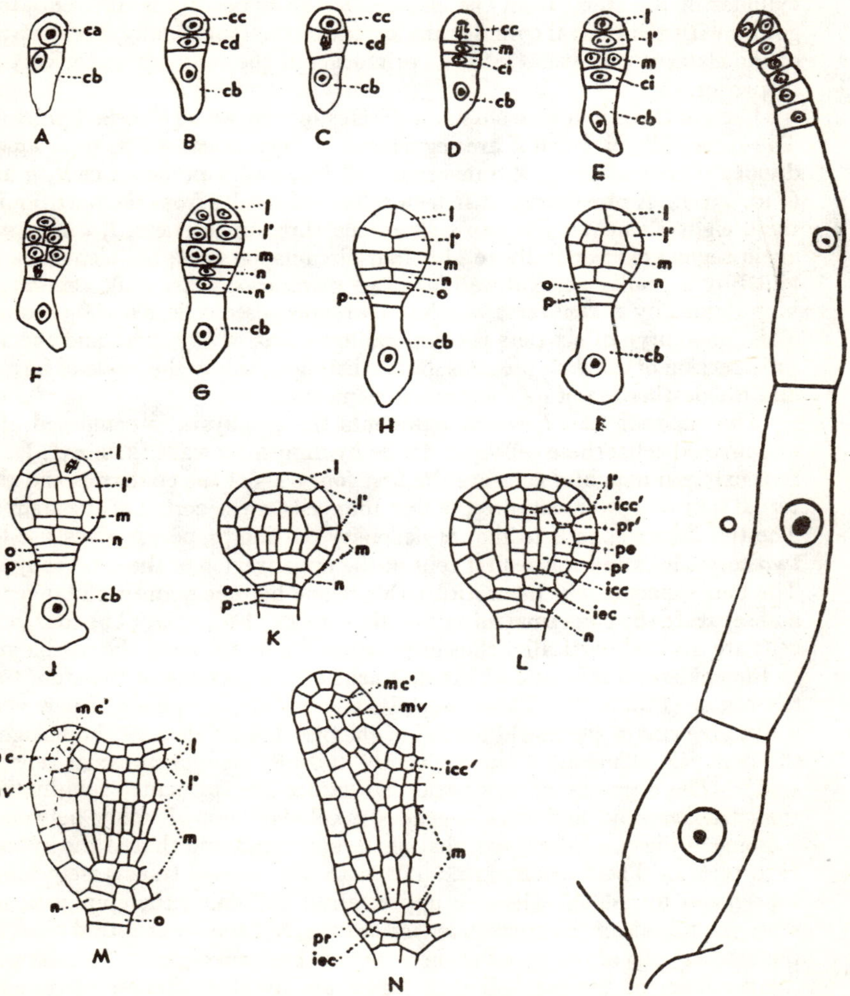

FIG. 43. — *Sagina procumbens*. A–N. Description in text. ×420. *Vaccaria vaccaria*. O. Filamentous proembryo. (A–N *redrawn from* SOUÈGES 1922; O *redrawn from* COOK 1909).

Caryophyllaceae:- The first transverse division in the zygote of *Sagina procumbens* partitions it into a terminal and a basal cell. The latter (FIG. 43A *et seq.*) does not again divide, but eventually becomes the lowermost cell of the suspensor, assuming an irregular form. The terminal cell becomes segmented transversely to engender two superposed elements, *cc* and *cd* (FIG. 43B). Shortly the element *cd* in turn is divided by a transverse wall into two new elements, *m* and *ci*; the other element *cc* undergoes a similar division to give rise to two new cells, *l* and *l'* (FIG. 43D–F). There thus is formed a tetrad of cells comparable to that found in *Chenopodium bonus-henricus* (Chenopodiad Type). It differs (1) in that all cells are engendered by the terminal cell alone, and (2) by the eventual destination of the four elements of which it is composed. In effect, the element *l* gives rise in a unique manner to the stem apex, repre-enting an epiphysis initial corresponding to that in *Geum* and *Myosotis* but already individualized at the tetrad stage; the element *l'* engenders the cotyledonary regions — that is to say, the cotyledonary tissues plus the central cylinder of the stem; from the element *m* the hypocotyl is differentiated together with the central cylinder and cortex of the root; finally, the derivatives of the element *ci* develop into the primordia of the root cap and a very short suspensor.

Each of these four elements now divides to form an eight-celled proembryo. The three cells *m*, *l* and *l'* are segmented by meridional walls; the element *ci* divides shortly after by a transverse wall into two superposed cells, *n* and *n'* (FIG. 43G). A proembryo of sixteen cells next results from the partitioning of these eight elements: the two cells of the three upper tiers, *l*, *l'* and *m*, are again segmented vertically to give four circumaxial cells; the element *n* is divided by a similar vertical wall into two juxtaposed cells, while the sister cell *n'* is divided by a transverse wall into two superposed cells, *o* and *p* (FIG. 43H). This proembryo of six tiers is comparable to that in *Sagittaria* and the *Persicaria* section of the *Polygonaceae*, but is distinguished by the mode of formation and the destination of its constituent elements.

The superior tier, *l*, which represents the epiphysis, is composed of four circumaxial cells: these cells now divide by tangential walls (FIG. 43I, J). The four external daughter cells are the first four cells of the epidermis and the internal four daughter cells are the first four cells of the cortex of the stem apex. The tier *l'* corresponds to the cotyledonary portion proper; it gives rise to the two cotyledons and to the first cells of the upper region of the central cylinder. The four elements which constitute this region become segmented tangentially and separate the epidermal initials on the outside (FIG. 43I). The four internal cells are divided vertically, thus originating the first cells of the periblem next to the epidermal initials and the first cells of the plerome of the stem toward the center (FIG. 43J). These four plerome initials, by means of new vertical walls, give rise in the neighborhood of the periblem to elements homologous to the pericycle elements of the hypocotyl and four new circumaxial cells (FIG. 43K). The elements of this pericyclic nature are segmented longitudinally, then transversely, and thus engender a cellular core of union between the plerome of the cotyledon (*mv*), that of the stem apex and that of the hypocotyl (FIG. 43M). The four circumaxial cells are partitioned transversely into two superposed tetrads of cells: the upper tetrad, adjoining the epiphysis, represents the initials of the stem stele (*icc'*, FIG. 43N); the lower tetrad constitutes the mother cells of the vascular meristem of the same organ. The first cells of the periblem of the cotyledonary region are divided oftenest by tangential walls, which, in a singular manner, appear to be inserted at the lower end to the wall of the epidermal cell of the hypocotyl. Anticlinal walls succeed these tangential divisions and occasionally precede them. The vascular meristems of the cotyledons are differentiated from the innermost elements of this cell group (*mv*); the other elements of the same group contribute to the formation of the internal and external periblem of the cotyledons (*mc* and *mc'*, FIG. 43M, N).

The tier *m* gives rise to the hypocotyl. The four elements which compose this layer in the twelve-celled proembryo are the first to become segmented. Tangential walls first delimit the epidermal initials; then, in the four internal cells, as also occurs in *Nicotiana*, transverse walls separate the initials of the root cortex in the lower region and four new circumaxial cells in upper region (FIG. 43H–J). The latter divide longitudinally to give rise to the periblem and plerome. The initials of the plerome, by new vertical walls, contribute the first cells of the pericycle toward the outside and four new cells adjoining the axis; the latter divide transversely into two new superposed cells (FIG. 43K, L). The lower tetrad represents the initials of the central cylinder of the root tip, while the upper tetrad engenders the vascular meristems of the root. The initials of the periblem are segmented transversely, then longitudinally in two directions.

The tier *n* is composed of two cells: these divide vertically (FIG. 43H); then, in the four elements thus formed, there is established, either immediately or following the formation of new vertical walls, tangential walls which separate the calyptrogen initials (FIG. 43L–N). The cells *o* and *p* go into the formation of the suspensor.

The embryonomy of *Sagina procumbens*, briefly, is distinguished by the method of origin of the epiphysis, the order of segmentation in the cotyledonary region and the method of differentiation of the root cortex. The recapitulatory table follows:

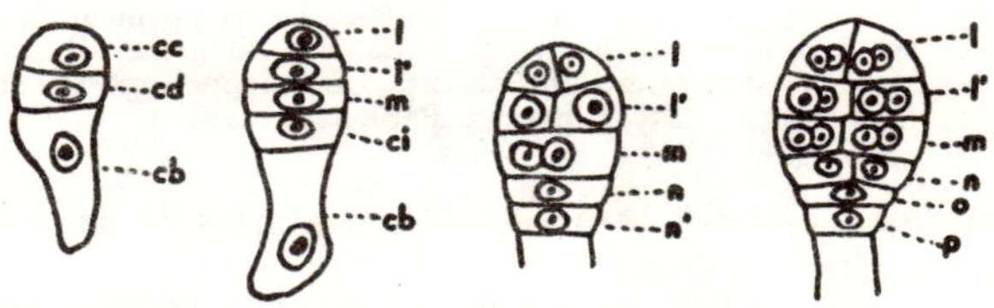

Recapitulatory Table for Sagina procumbens
(from left to right)

I. FIRST CELL GENERATION
(The basal cell is omitted)
Proembryo of two cells disposed
in two tiers:

cc=*pco*+*pvt*
cd=*phy*+*icc*+*iec*+*co*+*s*

II. SECOND CELL GENERATION
Proembryo of four cells disposed
in four tiers:

l =*pvt*
l'=*pco*
m=*phy*+*icc*+*iec*
ci=*co*+*s*

III. THIRD CELL GENERATION
Proembryo of eight cells disposed
in five tiers:

l =*pvt*
l'=*pco*
m=*phy*+*icc*+*iec*
n=*co*
n'=*s*

IV. FOURTH CELL GENERATION
Proembryo of sixteen cells disposed
in six tiers:

l =*pvt*
l'=*pco*
m=*phy*+*icc*+*iec*
n=*co*
o+*p*=*s*

Agrostemma (*Lychnis*) *githago* appears to conform to the same developmental scheme as does *Sagina procumbens*. *Silene conoides* apparently develops similarly, while the related *Vaccaria vaccaria* differs in that an elongated suspensor is developed (FIG. 43O). The early development of the proembryo in the latter species was not followed out, but the impression is obtained that the basal cell of the two-celled proembryo does not divide again, so that all of the suspensor cells, whatever their diverse dimensions, are derived from the terminal cell.

Arenaria serpyllifolia likewise conforms to the Sagina Variation, differing in minor details: in that the initials of the root cortex are tardily differentiated, and in the even more tardy individualization of the initial of the root cap.

Tunica saxifraga and *Saponaria cerastioides* apparently conform to the Vaccaria Variation, Caryophyllad Type, whereas *Stellaria media*, *Polycarpon tetraphyllum*, *Drymaria cordata* and *Moehringia trinervia*, although imperfectly described and illustrated, would seem to agree most closely with the Sagina Variation, Caryophyllad Type.

Claytonia perfoliata, on the other hand, has an embryogeny radically different from that of any of the species mentioned above. The embryonomy represents a superposition of those of the Polygonum and Senecio Variations of the Asterad Type. It differs from both in that tier *m* alone forms all of the hypocotyledonary region, but in other details it might conform to either Variation.

The disposition of *Montia rivularis* is difficult; it appears to agree more closely with the Senecio Variation, Asterad Type, than with any other.

COOK, M. T. 1903. The development of the embryo-sac and embryo of *Agrostemma githago*. Ohio Nat. 3: 365–369.
COOK, M. T. 1909. Notes on the embryology of the *Caryophyllaceae*. Ohio Nat. 9: 477–479.
ROCEN, T. 1927. Zur Embryologie der Centrospermen. Diss., Uppsala.
SOUÈGES, R. 1922. Embryogénie des *Caryophyllacées*. Les premiers stades du développement de l'embryon chez le *Sagina procumbens* L. Les derniers stades du développement chez le *Sagina procumbens* L. C. R. Acad. Sci. Paris 175: 709–711; 894–896.
SOUÈGES, R. 1924. Développement de l'embryon chez le *Sagina procumbens* L. Bull. Soc. Bot. France 71: 590–614.
SOUÈGES, R. 1945. Embryogénie des Caryophyllacées. Développement de l'embryon chez *L'Arenaria serpyllifolia* L. C. R. Acad. Sci. Paris 221: 320–322.
SOUÈGES, R. 1945. Embryogénie des Portulacacées. Développement de l'embryon chez le *Claytonia perfoliata* Donn. C. R. Acad. Sci. Paris 221: 111–113.

Molluginaceae:- *Giesekia pharnaceoides* conforms rather closely to the Sagina Variation, Caryophyllad Type.

A linear proembryo of five or six cells is formed in *Mollugo nudicaulis*, but the exact derivation of these cells was not described. The cell at the apex of the proembryo is divided vertically when the six- or seven-celled stage is reached. Assignment is probably also to the Sagina Variation. However, there is considerable resemblance to the Linum Variation, Solanad Type.

BHARGAVA, H. R. 1934. Contribution to the morphology of *Mollugo nudicaulis* Lamk. Proc. Indian Acad. Sci., B, 1: 271–278.
JOSHI, A. C., and V. R. RAO. 1936. The embryology of *Gisekia pharnaceoides* Linn. Proc. Indian Acad. Sci., B, 3: 71–92.

Aizoaceae (Ficoidaceae):- The proembryo of *Sesuvium portulacastrum* divides repeatedly in the transverse plane to form a linear row of about eight cells. It is claimed that the two terminal cells alone participate in the formation of the embryo proper, while the third cell functions as a hypophysis. It is highly probable that this species belongs under the Linum Variation, Solanad Type, though the suspensor is much longer than in *Linum*.

Trianthema decandra is very similar in development to *S. portulacastrum*. On the other hand, the first division in the zygote of *T. monogyna* is said to be transverse but might be slightly oblique, and the terminal cell is claimed to divide longitudinally, but it is not at all clear whether the terminal cell of the two-celled proembryo or the terminal cell of a much older proembryo consisting of a linear row of about eleven cells is meant. It is so highly improbable that all of the cells of such a well developed proembryo were derived from the basal cell alone that the only conclusion which can be reached is that the first division in the terminal cell of the two-celled proembryo is transverse. Assignment to the Linum Variation is probable. The suspensor becomes quite massive and is larger than the embryo proper until after cotyledon development commences.

For *Mesembryanthemum crystallinum* there is depicted a proembryo of two juxtaposed cells occupying the terminal end, then three superposed basal cells. Assignment of this species is problematical, but might be like that of the preceding species.

BHARGAVA, H. R. 1935. The life history of *Trianthema monogyna* L. Proc. Indian Acad. Sci., B, 2: 49–58.
RAGHAVAN, T. S., and V. K. SRINIVASAN. 1940. Studies in the Indian *Aizoaceae*. Ann. Bot., n.s. 4: 651–661.
WOODCOCK, E. F. 1930. Morphological studies on the seed of *Mesembryanthemum crystallinum* L. Papers Mich. Acad. Sci., Arts and Letters 13: 221–226.

Portulacaceae:- *Portulaca oleracea* conforms to the Myriophyllum Variation, Caryophyllad Type. *P. tuberosa*, on the contrary, appears to conform to the Drosera Variation of the same Type. *Claytonia virginiana* probably belongs to the Myriophyllum Variation.

COOK, M. T. 1903. The development of the embryo-sac and embryo of *Claytonia virginiana*. Ohio Nat. 3: 349–353.
COOPER, D. C. 1940. Macrosporogenesis and embryology of *Portulaca oleracea*. Amer. Jour. Bot. 27: 326–330.
RAGHAVAN, T. S., and V. R. SRINIVASAN. 1941. Cyto-morphological features of *Portulaca tuberosa* Roxb. Proc. Indian Acad. Sci., B, 14: 472–488.
SOUÈGES, R. 1938. Embryogénie des *Portulacacées*. Développement de l'embryon chez le *Portulaca oleracea* L. C. R. Acad. Sci. Paris 20: 768–770.

Polygonales

Polygonaceae:- The embryogeny of the *Polygonaceae* (Polygonum Variation) closely resembles that of *Urtica pilulifera* (Urtica Variation). The differences are in the development of the derivatives of cell *m*; otherwise, the eventual destination of each tier of proembryonic cells is the same in the two Variations.

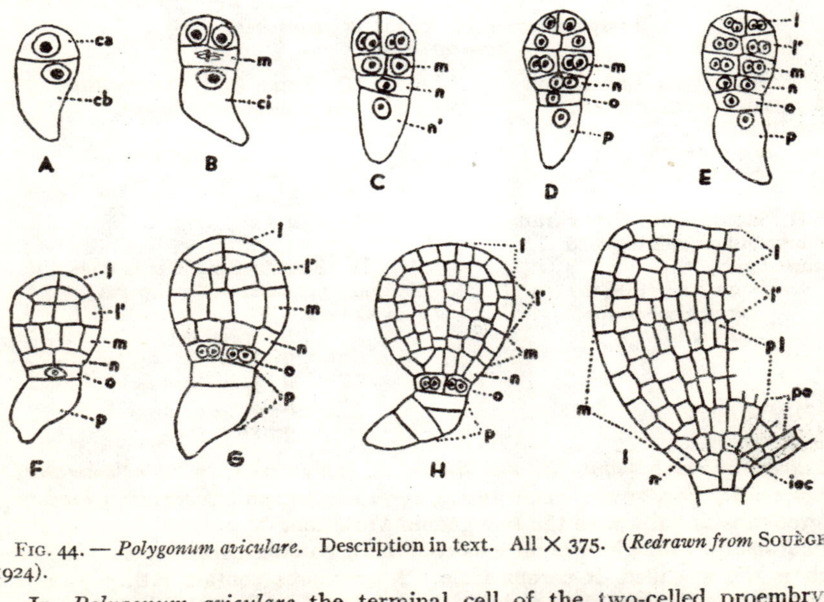

FIG. 44. — *Polygonum aviculare*. Description in text. All × 375. (*Redrawn from* SOUÈGES 1924).

In *Polygonum aviculare* the terminal cell of the two-celled proembryo divides longitudinally into two juxtaposed cells, while the basal cell is partitioned transversely into two superposed cells, *m* and *ci* (FIG. 44A, B). The two upper cells are next divided longitudinally to form the quadrant stage (FIG. 44C). The cell *m*, either slightly before or at about the same time, is divided vertically; then the lower cell *ci* divides transversely (FIG. 44C). The proembryo at this stage consists of eight cells in all; each of these eight cells is bipartitioned to form a sixteen-celled proembryo: the quadrant group of cells, by transverse segmentation of each, engenders four upper octants, *l*, and four lower octants, *l′*, while the two juxtaposed cells of tier *m* are separated by meridional walls into four circumaxial cells and finally the two superposed cells *n* and *n′* respectively divide longitudinally and transversely (FIG. 44D, E). During the later stages of embryo development, the tier *l*, which is reduced to two cell layers, gives rise to the cotyledonary portion; more precisely, to the epidermal initials and the cortex of the stele, to the internal cortical meristems and the vascular meristems of the cotyledons. The tier *l′* gives rise to the short upper portion of the hypocotyl, whereas the tissues derived from *m* make up by far the larger portion of the hypocotyl. The region *n* gives rise to the initials of the root cortex (in contrast to giving rise to the central portion of the root cap in other genera of the *Polygonaceae*), while the derivatives of an upper daughter cell of *n′* constitute the primordia of the root cap, and the derivatives of *p* (the other daughter cell) form the suspensor (FIG. 44F–I).

Polygonum persicaria agrees substantially with *P. aviculare* in most details, but cell *o* may be divided by a vertical wall and cell *p* generally remains very small and is conical in shape, but these two cells may eventually give rise to a small suspensor.

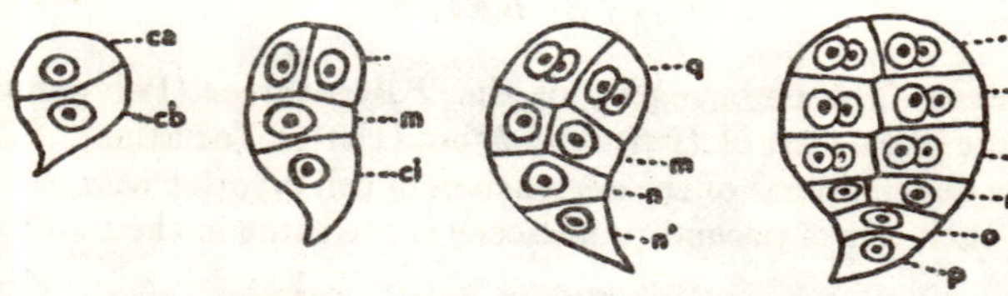

Recapitulatory Table for the Polygonaceae
(from left to right)

I. FIRST CELL GENERATION
Proembryo of two cells disposed in two tiers:

$ca = pco + pvt + \frac{1}{2}phy$
$cb = \frac{1}{2}phy + icc + iec + co + s$

II. SECOND CELL GENERATION
Proembryo of four cells disposed in three tiers:

$q = pco + pvt + \frac{1}{2}phy$
$m = \frac{1}{2}phy + icc + iec$
$ci = co + s$

III. THIRD CELL GENERATION
Proembryo of eight cells disposed in four tiers:

$q = pco + pvt + \frac{1}{2}phy$
$m = \frac{1}{2}phy + icc + iec$
$n = co$
$n' = s$

IV. FOURTH CELL GENERATION
Proembryo of sixteen cells disposed in six tiers:

$l = pco + pvt$
$l' = \frac{1}{2}phy$
$m = \frac{1}{2}phy + icc + iec$
$n = co$
$o + p = s$

Rumex hydrolapathum, *R. maritimus*, *R. sanguineus*, *R. conglomeratus*, *Rheum emodi*, *Fagopyrum esculentum*, *Oxyria digyna*, and *Koenigia islandica* all agree substantially with the Polygonum Variation.

In *Rumex acetosa* the egg nucleus contains 7 haploid chromosomes, one of which is the so-called X-chromosome. Microspores contain either 7 or 8 chromosomes. In developing embryos, some have been found with 14 somatic chromosomes and others with 15. Those with 15 chromosomes appear to follow the Polygonum Variation closely, but those with 14 chromosomes are far less regular in construction.

The laws of embryonomy for *Polygonum persicaria* may be recapitulated as shown in the table on the opposite page.

EDMAN, G. 1929. Zur Entwicklungsgeschichte der Gattung *Oxyria* Hill, nebst zytologischen, embryologischen und systematischen Bemerkungen über einige andere *Polygonaceen*. Acta Hort. Bergiani 9: 165–291.
HAGERUP, O. 1926. Könsdelenes bygning og udvikling hos *Koenigia islandica* L. Meddelelser om Grönland 58: 199–204.
LONAY, H. 1923. L'ovule du *Polygonum aviculare*. Bull. Soc. Roy. Bot. Belgique 55:175–177.
MAHONY, K. L. 1936. Morphological and cytological studies on *Fagopyrum esculentum*. II. Embryogeny. Amer. Jour. Bot. 23: 129–133.
ONO, T. 1935. Chromosomen und Sexulität von *Rumex acetosa*. Sci. Rep. Tohoku Imp. Univ., IV, Biol. 10: 41–210.
SOUÈGES, R. 1919. Embryogénie des *Polygonacées*. Développement de l'embryon chez le *Polygonum persicaria*. C. R. Acad. Sci. Paris 168: 791–793.
SOUÈGES, R. Recherches sur l'embryogénie des *Polygonacées*. Bull. Soc. Bot. France 66: 168–199, 1919; 67: 1–11, 75–85. 1920.
SOUÈGES, R. 1924. Embryogénie des *Polygonacées*. Le développement de l'embryon chez le *Polygonum aviculare* L. C. R. Acad. Sci. Paris 178: 409–412.

Illecebraceae:- *Scleranthus perennis* and *Herniaria glabra* show profound analogies to the Sagina Variation, Caryophyllad Type, but differ somewhat in minor details. The former has a very late differentiation of the initials of the root cortex and a still later individualization of the cell which gives rise to the primordium of the root cap. *H. glabra* differs from both *S. perennis* and *Sagina procumbens* in the destination of the proembryonic tetrad elements; in this respect this species approaches *Papaver rhoeas*.

The basal cell of the two-celled proembryo of *H. glabra* segments transversely a few times to produce the lower portion of the suspensor. These derivatives take no part in the construction of the embryo proper. From the derivatives of the terminal cell there is developed a proembryonic tetrad composed of four superposed elements, *l*, *l'*, *m* and *ci*. Element *l* is converted into an epiphysis; element *l'* produces the cotyledonary region proper, the hypocotyl and the initials of the root cortex; element *m* gives rise, in a very general way, to the central portion of the root cap and *ci* contributes to the construction of the upper portion of the filamentous suspensor.

The embryogeny of *Polycarpon tetraphyllum* is far more regular than that of the species mentioned above. Altogether, the evidence indicates that all three species should be assigned to the Sagina Variation.

SOUÈGES, R. 1938. Embryogénie des *Illécébracées*. Développement de l'embryon chez le *Herniaria glabra* L. Bull. Soc. Bot. France 85: 353–363.
SOUÈGES, R. 1938. Embryogénie des *Illécébracées*. Développement de l'embryon chez le *Scleranthus perennis* L. C. R. Acad. Sci. Paris 206: 1404–1406.
SOUÈGES, R. 1945. Embryogénie des Illécébracées. Développement de l'embryon chez le *Polycarpon tetraphyllum* L. C. R. Acad. Sci. Paris 221: 41–44.

Chenopodiales

There is no report for the *Batidaceae*.

Basellaceae:- The earliest stages for *Basella alba* are neither described nor figured; early embryos are depicted with very irregularly arranged cells. An assignment is impossible.

ROCEN, T. 1927. Zur Embryologie der Centrospermen. Diss., Uppsala.

Phytolaccaceae:- The description of the embryonomy of *Phytolacca decandra* agrees closely with the Myosurus Variation, Onagrad Type, and this seems to be true also for *Rivinia humilis*. In *R. brasiliensis* a linear proembryo of about nine cells is produced; development may possibly conform to that of some Type other than the Onagrad Type.

LEWIS, I. F. 1905. Notes on the development of *Phytolacca decandra* L. Johns Hopkins Univ. Circ. No. 178, 35–43.
MAURITZON, J. 1934. Ein Beitrag zur Embryologie der *Phytolaccaceen* und *Cactaceen*. Bot. Notiser 1934: 111–135.

Cynocrambaceae:- The embryogeny of *Thelygonum cynocrambe* appears to belong under the Chenopodiad Type.

SCHNEIDER, H. 1914. Morphologische und Entwicklungsgeschichtliche Untersuchungen an *Thelygonum cynocrambe* L. Flora 106: 1–41.

Chenopodiaceae:- The zygote of *Chenopodium bonus-henricus* is segmented transversely, whereupon each daughter cell in turn divides transversely, with the basal cell segmenting a little in advance of the terminal cell (FIG. 45A, B). The destination of the four cells of the proembryonic tetrad is as follows: the upper cell *l* gives rise to the cotyledonary region; the next cell, *l′*, engenders

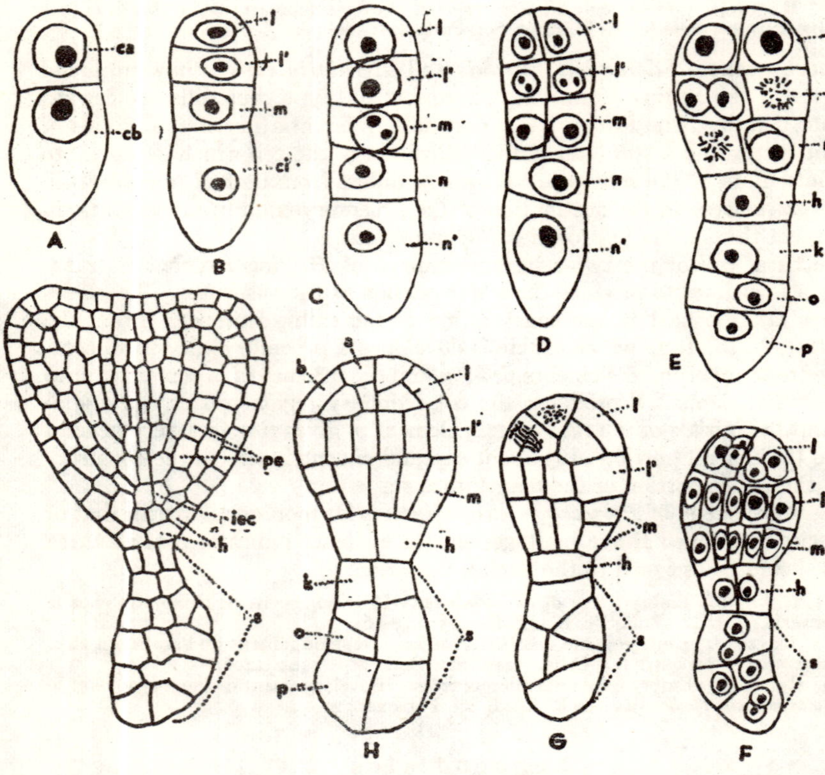

FIG. 45. — *Chenopodium bonus-henricus*. Description in text. A–E × 650; F–H × 460; I × 200. (*Redrawn from* SOUÈGES 1920).

the upper portion of the hypocotyledonary region, with the lower portion originating from the next cell, *m*; finally, the lowest cell, *ci*, develops the hypophysis cell and the suspensor proper. It is thus apparent that a considerable proportion of the embryo proper is derived from the basal cell.

Each of the four elements is segmented to engender the eight-celled proembryo: the lowest cell, *ci*, is divided transversely into *n* and *n′*, while the other three cells are partitioned vertically (FIG. 45C, D). The wall in *m*, however, may occasionally be oblique; so much so, indeed, that it sometimes appears to be almost transverse. The sixteen-celled proembryo is created by cytokinesis in each cell of the eight-celled proembryo (FIG. 45E). The cell *m* is generally divided transversely into two superposed cells, *h* and *k*. The cell *h* constitutes the hypophysis cell; while the derivatives of *k* go into the construction of the suspensor. The cell *n'* also divides transversely (into *o* and *p*) and the derivatives join with those from *k* to form the suspensor. Next the two elements of tier *l'* divide vertically to engender four circumaxial cells which correspond to four "lower octants." At the same time these four cells become greatly enlarged and give the upper portion of the embryo a globular shape. Similar divisions occur next in region *m*; these four cells together with those in tier *l'* constitute the hypocotyledonary region. Finally the two elements of tier *l* are in turn segmented vertically into four cells which might be designated the "upper octants" and which represent the cotyledonary region.

The order of the segmentations during the further development in tier *l* is somewhat irregular (FIG. 45F, G). In a very general way, the first wall in any one of the four cells is inserted exteriorly on the peripheral membrane and then follows a horizontal direction more or less parallel to the wall separating

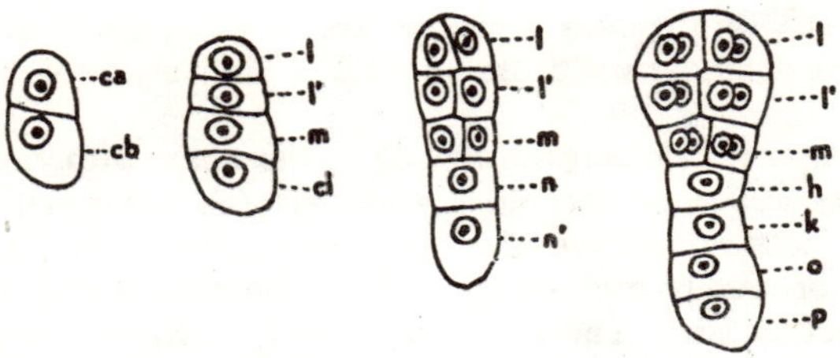

Recapitulatory Table for Chenopodium bonus-henricus
(from left to right)

I. FIRST CELL GENERATION
Proembryo of two cells disposed in two tiers:

$ca = pco + pvt + \frac{1}{2}phy$
$cb = \frac{1}{2}phy + icc + iec + co + s$

II. SECOND CELL GENERATION
Proembryo of four cells disposed in four tiers:

$l = pco + pvt$
$l' = \frac{1}{2}phy$
$m = \frac{1}{2}phy + icc + iec$
$ci = co + s$

III. THIRD CELL GENERATION
Proembryo of eight cells disposed in five tiers:

$l = pco + pvt$
$l' = \frac{1}{2}phy$
$m = \frac{1}{2}phy + icc + iec$
$n = co + s$ (in part)
$n' = s$ (in part)

IV. FOURTH CELL GENERATION
Proembryo of sixteen cells disposed in seven tiers:

$l = pco + pvt$
$l' = \frac{1}{2}phy$
$m = \frac{1}{2}phy + icc + iec$
$h = co$
$k + o + p = s$

the two tiers *l* and *l'*. Two dissimilar cells are produced: an upper one, *b*, with a triangular aspect, and a lower one, *a*, with a quadrangular aspect (FIG. 45H). Exceptionally, the wall may be almost vertical, parallel to one of the two meridional walls. Again, the wall may be diagonally curved to follow the rounding of the external wall. The next divisions, whatever the position of the preceding wall, individualize the epidermal initials. In cell *a*, the wall is generally tangential. Eventually, when the terminal depression appears and the protuberances indicating the positions of the two cotyledons become formed, region *l* consists of two cell layers in addition to the epidermal layer. In tier *l'* each of the four elements is segmented vertically (FIG. 45G, H).

Further development is very similar to the corresponding growth in the Onagrad Type. The first transverse walls in this tier generally appear at a rather late stage of development.

Tier *m* develops much like tier *l'*, but transverse walls may appear at an earlier stage (FIG. 45G, H). At the moment of formation of the cotyledons tiers *m* and *l'* both consist of two layers of cells which correspond respectively to the upper and lower portions of the hypocotyledonary region. The periblem, plerome and initials of the cortex all originate in tier *m*.

Tier *h*, as has already been indicated, becomes transformed directly into the hypophysis cell. This cell segments longitudinally into two juxtaposed cells, and the latter in turn are divided similarly to form four circumaxial cells which represent hypophyseal quadrants (FIG. 45H, I). Next each quadrant cell is divided more or less horizontally, then, in a general way, the upper daughter cell is again segmented horizontally and the derivatives constitute the two innermost layers of the root cap.

The suspensor proper takes its origin from the three elements *k*, *o*, and *p*. Each, as a general rule, is partitioned vertically to give rise to two juxtaposed cells (FIG. 45F–H). Subsequent divisions are very irregular and are most numerous in the regions derived from *o* and *p*. The cells soon acquire large dimensions.

The embryogeny of *Beta vulgaris* agrees in the main with that of *Chenopodium bonus-henricus*. One account states that the basal cell of the two-celled proembryo does not divide and becomes haustorial, but this is either rather inaccurate or merely represents an exceptional phenomenon associated with environmental factors. There are certain irregularities in the planes and sequence of the divisions in the four cells of the tetrad proembryo, so that no rigid laws appear to be applicable. Cell *l'* always divides vertically and cell *m* usually does so. During the third cell generation the proembryo may thus consist of either five, six or seven tiers. The disposition of the cells during the fourth cell generation is therefore dependent upon the disposition during the preceding generation. Thus, if the proembryo is five-tiered during the third cell generation, the proembryo may have eight tiers during the succeeding generation.

Chenopodium album is said to produce a proembryonic row of eight cells with subsequent longitudinal divisions in all but the second and third cells from the terminal end.

Events are not very clear in *Haloxylon ammodendron*, but it apparently conforms to the Chenopodiad Type.

ARTSCHWAGER, E. 1927. Development of flowers and seed in the sugar beet. Jour. Agr. Res. 34: 1–25.

ARTSCHWAGER, E., and R. C. STARRETT. 1933. The time factor in fertilization and embryo development in the sugar beet. Jour. Agr. Res. 47: 823–843.

BHARGAVA, H. R. 1937. The life-history of *Chenopodium album* Linn. Proc. Indian Acad. Sci., B, 4: 179–200.

BOCANZEVA, Z. P. 1944. (An investigation into the biology of flowering and embryological development in *Haloxylon ammodendron* Bge.) Jour. Bot. URSS 29: 36–48. (Russian with English summary)

OKSIJUK, P. 1927. Entwicklungsgeschichte der Zuckerrübe (*Beta vulgaris*). Jard. Bot. Kieff 5-6: 1–19.

SOUÈGES, R. 1920. Embryogénie des *Chenopodiacées*. Développement de l'embryon chez le *Chenopodium bonus-henricus* L. C. R. Acad. Sci. Paris 170: 467–469.

SOUÈGES, R. 1920. Développement de l'embryon chez le *Chenopodium bonus-henricus* L. Bull. Soc. Bot. France 67: 233–257.

Amarantaceae:- *Amarantus retroflexus* embryonomically represents a transition form between the Solanad and Chenopodiad Types, but since the basal cell of the two-celled proembryo contributes to the formation of the embryo, it most logically belongs under the latter Type. Otherwise, the orientation of the first wall in region *l'* of the proembryonic tetrad determines which Type the embryo concerned follows: the wall may be either longitudinal (Solanad) or transverse (Chenopodiad). In most respects, however, the embryonomy very closely parallels that of *Chenopodium bonus-henricus*.

Amarantus caudatus is identical in development with *A. retroflexus*, and *Digera arvensis*, *Alteranthera sessilis* and *Achryanthus asper* are also remarkably similar.

JOSHI, A. C., and L. B. KAJALE. 1937. Fertilization and seed development in *Amarantaceae*. Proc. Indian Acad. Sci., B, 5: 91–100.

JOSHI, A. C., and C. V. RAO. 1934. A contribution to the anatomy, morphology and cytology of the flower of *Digera arvensis* Forsk. Jour. Indian Bot. Soc. 13: 201–236.

KAJALE, L. B. 1937. Embryology of *Achyranthes asper* Linn. Proc. Indian Acad. Sci., B, 5: 195–205.

SOUÈGES, R. 1937. Embryogénie des *Amarantacées*. Développement de l'embryon chez l'*Amarantus retroflexus* L. C. R. Acad. Sci. Paris 204: 892–894.

SOUÈGES, R. 1937. L'embryon chez les *Amarantus*. Relations embryologiques entre les *Solanacées* et les *Centrospermales*. Bull. Soc. Bot. France 84: 242–255.

Geraniales

There is no record for the *Limnanthaceae*.

Linaceae:- By transverse partitioning of both the basal and the terminal cells of the two-celled proembryo of *Linum cathcarticum*, a tetrad of four superposed cells results (FIG. 46A–C). These cells are designated, in order, as *l*, *l'*, *m* and *ci*. The upper cell, *l*, always divides vertically (FIG. 46D–P), and the lowest cell, *ci*, always transversely (FIG. 46I–U). Sometimes the element *l'* is divided longitudinally (FIG. 46E, F, J, M, N, P, Q), sometimes transversely (FIG. 46G, H, I, L, O, R, T, V), or occasionally obliquely (FIG. 46K). The element *m* is oftenest divided transversely (FIG. 46K), but is sometimes segmented longitudinally (FIG. 46M). These variations in the behavior of the divisions in these two central elements of the tetrad proembryo thus present three types of eight-celled proembryos: the first consists of five tiers of cells (FIG. 46M), the second of six cells (FIG. 46N), and the third of seven tiers (FIG. 46O). The last type appears to be the one most commonly encountered.

Region *l* engenders the cotyledons. The four proembryonic elements of which it is composed become segmented obliquely; the cells later very probably multiply in a manner similar to that occurring in the Urtica Variation. The destination of the two elements *l'* and *m* of the proembryonic tetrad is variable. In the general case of the eight-celled proembryo of seven tiers, the element *l'* is divided into two superposed cells, from the upper one of which the hypocotyl arises, while the lower derivative is divided transversely into two superposed cells. The upper daughter cell becomes a hypophysis cell which behaves as in the Onagrad Type, while the lower daughter cell becomes part of the suspensor proper.

The element *m* of the eight-celled proembryo of seven tiers, which is likewise divided into two superposed cells, and its descendants go entirely into the construction of the suspensor (FIG. 46L–U). In the case of proembryos of eight cells arranged in six tiers, the stage *l'*, here composed of two juxtaposed cells, gives rise to the hypocotyl; the upper daughter cell of *m* (*d*), after one or two additional transverse segmentations, becomes the hypophysis cell and the lower daughter cell (*f*) goes into the construction of the suspensor. Finally, in the rare instances of eight-celled proembryos arranged in five tiers, the tier *l'* gives rise to the hypocotyl and the tier *m* becomes the hypophysis cell but is exceptional in that the first wall is vertical (FIG. 46M).

The first division in the hypophysis initial or cell, as a general rule, occurs exactly as in the Onagrad Type (FIG. 46AA–AC). The upper daughter cell, by two cruciate walls, gives rise to the four initials of the root cortex, while the lower cell engenders the central portion of the root cap (FIG. 46AD).

Radiola linoides agrees substantially with *Linum cathcarticum* but displays a greater regularity in the sequence of divisions.

MAURITZON, J. 1934. Zur Embryologie einiger *Gruniales*. Svensk Bot. Tidskr. 28: 84–102.
SOUÈGES, R. 1924. Embryogénie des *Linacées*. Développement de l'embryon chez le *Linum cathcarticum* L. C. R. Acad. Sci. Paris 178: 1307–1310.
SOUÈGES, R. 1937. Développement de l'embryon chez le *Radiola linoides* Roth. Bull. Soc. Bot. France 84: 297–306.

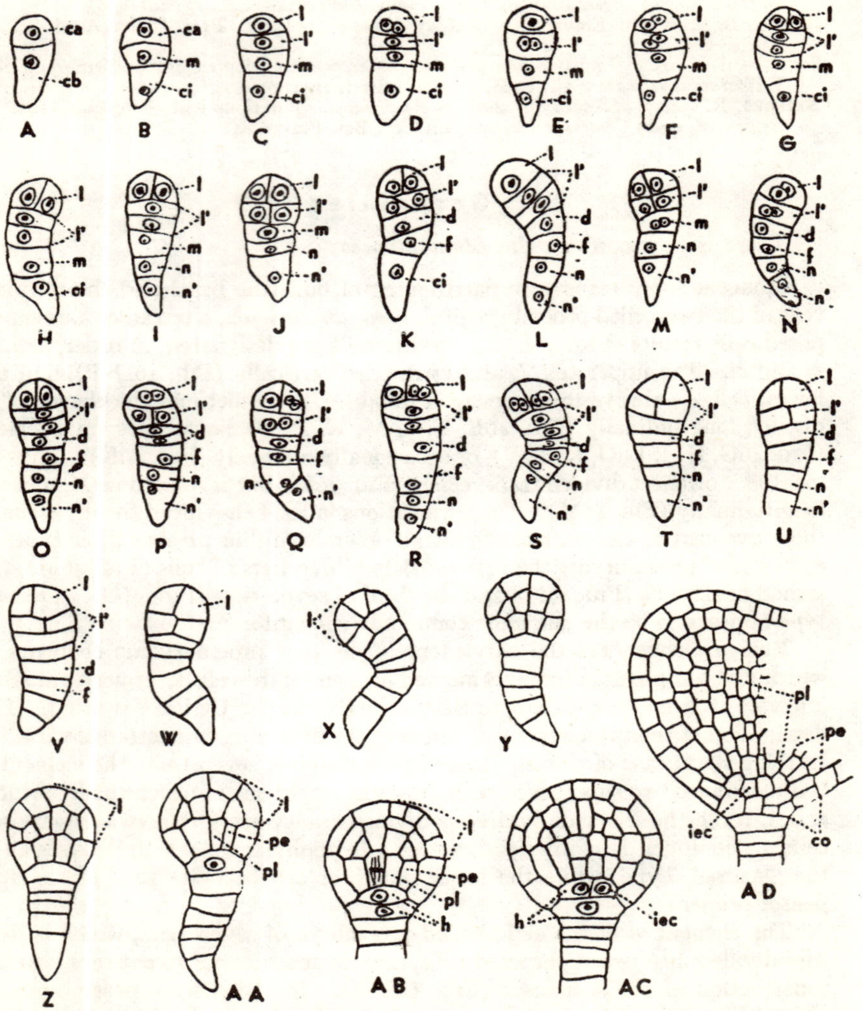

FIG. 46. — *Linum cathcarticum*. Description in text. All × 420. (*Redrawn from* SOUÈGES 1924).

Zygophyllaceae:- For *Tribulus terrester* there is depicted a linear proembryo of five cells of which all but the terminal one are much longer than wide. A satisfactory assignment is impossible.

MAURITZON, J. 1934. Etwas über die Embryologie der *Zygophyllaceen* sowie einige Fragmente über die der *Humiriaceen*. Bot. Notiser 1934: 407–422.

Geraniaceae:- The embryonomy of *Geranium molle* is a curious mixture of those of *Myosurus minimus* and *Erodium cicutarium*. The four- and eight-celled proembryonic stages are constructed as in *M. minimus*; the cotyledonary and hypocotyledonary regions originate from the terminal cell and the hypophysis region develops from the intermediate cell, but there is one singular difference in that only two of the four quadrants are segmented transversely to engender two cells which appear to occupy the summit of the proembryo. These indicate a manifest tendency to differentiate an epiphysis resembling that in *Erodium cicutarium*. This epiphysis region never becomes as fully developed as in the latter species but disappears by the time the cotyledonary protuberances appear.

It would appear that *G. molle* should be placed under the Onagrad Type, despite the fact that the intermediate cell, *m*, contributes a substantial portion of the entire embryo, but adds nothing to the body proper.

The embryonomy of *Erodium cicutarium* closely parallels that of *Geum urbanum* (Geum Variation, Asterad Type), the most significant difference residing in the fact that region *n* in the former species does not contribute to the construction of the suspensor. In the two species, an epiphysis absolutely identical in origin and functions is differentiated.

Souèges, R. 1923. Embryogénie des *Geraniacées*. Développement de l'embryon chez l'*Erodium cicutarium*. C. R. Acad. Sci. Paris 176: 1565–1567.
Souèges, R. 1923. Embryogénie des *Geraniacées*. Développement de l'embryon chez le *Geranium molle* L. C. R. Acad. Sci. Paris 177: 556–559.

Tropaeolaceae:- There are several old accounts of the embryogeny of various species of *Tropaeolum*, but no conclusive evidence for the disposition of these species can be adduced. The terminal cell in the same species is said to divide in either plane. The most noteworthy feature is that the basal cell is claimed to segment longitudinally, especially in *T. majus*, whereupon each daughter divides separately to form two unequal divergent branches, one of which attains a length of three or four cells, the other of as many as 25 cells.

Brunotte, C. 1900. Recherches embryogéniques sur quelques espèces des genres *Impatiens* et *Tropaeolum*. Diss., Paris et Nancy.
Dickson, A. 1876. On the embryogeny of *Tropaeolum peregrinum* (L.) and *T. speciosum* (Endl. et Poepp.). Trans. Soc. Edinburgh 27: 223–235.
Hegelmaier, F. 1878. Vergleichende Untersuchungen über die Entwicklung dicotyledoner Keime mit Berücksichtigung der Pseudo-monokotyledonen. Stuttgart.
Leidicke, J. W. 1903. Beiträge zur Embryologie von *Tropaeolum majus*. Diss., Breslau.

Oxalidaceae:- In *Oxalis corniculata* the rules which prevail in the construction of the proembryonic tetrad, of the eight-celled proembryo, and of the twelve-celled pro-embryo, and those which later govern the processes of divisions in the eight elements which compose the upper region of the sixteen-celled proembryo are similar to those in the Asterad Type. Slight divergencies permit the erection of the Oxalis Variation under that Type.

The four circumaxial cells composing region *m* in the sixteen-celled proembryo segment vertically (Fig. 47A–D); the external daughter cells are epidermal initials. In the four internal cells transverse walls are erected, in a very constant manner; they separate, internally, the initials of the root cortex (*iec*), and, toward the upper part, those elements which by means of vertical divisions give the first elements of the periblem and plerome (Fig. 47E). The cells of the plerome segment longitudinally and transversely, or, inversely, first transversely and then longitudinally, and thus produce the first elements of the pericycle. The processes of differentiation of the initials of the root cortex and the mother cells of the two internal histogens are, in sum, comparable to those in *Nicotiana*.

The two-celled region *n* of the sixteen-celled proembryo gives rise to the primordium of the root cap (Fig. 47C–G). By longitudinal divisions they form four circumaxial cells which then segment transversely once or twice, to give the elements of the calyptrogen layer above and below these those elements which enter into the construction of the culminating portions of the cap. The elements originating from *o* enter wholly into the construction of the same portions. The cell *p* is converted, through oblique divisions, into several large cells which together constitute a short suspensor. These become resorbed by the time the cotyledons have originated.

Biophytum dendroides and *B. sensitivum* both conform to the Oxalis Variation.

Mauritzon, J. 1934. Zur Embryologie einiger *Gruniales*. Svensk Bot. Tidskr. 28: 84–102.
Noll, W. 1935. Embryonalentwicklung von *Biophytum dendroides* DC. Planta 24: 609–648.
Souèges, R. 1939. Embryogénie des *Oxalidacées*. Développement de l'embryon chez l'*Oxalis corniculata* L. C. R. Acad. Sci. Paris 209: 698–700.

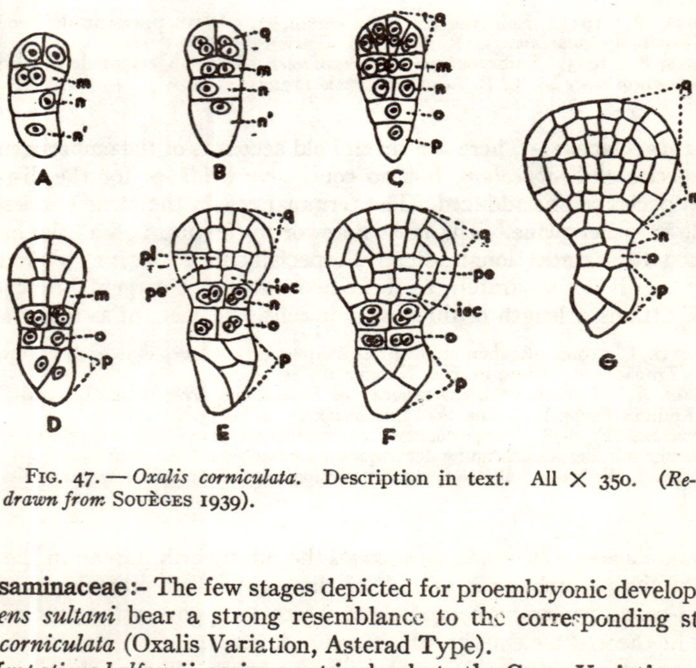

Fig. 47. — *Oxalis corniculata*. Description in text. All × 350. (*Redrawn from* Souèges 1939).

Balsaminaceae:- The few stages depicted for proembryonic development in *Impatiens sultani* bear a strong resemblance to the corresponding stages in *Oxalis corniculata* (Oxalis Variation, Asterad Type).

In *Impatiens balfourii*, assignment is clearly to the Geum Variation, Asterad Type. Since the development of this species was followed more carefully than in the preceding species, the classification of the latter may possibly be in the Geum rather than in the Oxalis Variation.

Events in *Impatiens roylei* are not so clear.

Dahlgren, K. V. O. 1934. Die Embryologie von *Impatiens roylei*. Svensk Bot. Tidskr. 28: 103–125.
Ottley, A. M. 1918. A contribution to the life history of *Impatiens sultani*. Bot. Gaz. 66: 289–317.
Souèges, R. 1945. Embryogénie des Balsaminacées. Développement de l'embryon chez l'*Impatiens balfourii* Hook. C. R. Acad. Sci. Paris 220: 837–840.

Lythrales

Four of the eight families included in the *Lythrales* have never been investigated: *Crypteroniaceae*, *Punicaceae*, *Oliniaceae* and *Callitrichaceae*.

Lythraceae:- The embryonomy of *Lythrum salicaria* bears a remarkable resemblance to that of the Onagrad Type. As a matter of fact, the sole difference resides in the mode of origin of the hypophysis cell. In the Onagrad Type this cell is derived directly from the middle cell (*m*) of the proembryonic tetrad, while in *Lythrum* this cell divides transversely and the upper daughter cell becomes the hypophysis cell. It is on this basis that the Lythrum Variation differs from the main Onagrad Type.

The following species all conform to the Lythrum Variation: *Peplis patula*, *Nesaea syphilitica*, *N. myrtiflora*, *Ammania senegalensis*, *A. latofilia*, *A. baccifera*, *A. multiflora*, *A. pentandra*, *A. peploides* and *Woodfordia floribunda*. *Cuphea lanceolata* digresses in that there are two superposed basal cells, then two tiers of two cells each. The basal cell derivatives apparently again divide once each, transversely, then the lowermost cell divides vertically several times and the cells become spread out and flattened. The development of the embryo proper, however, is much like that of *Lythrum*.

The suspensor is variable in structure. It remains uniseriate in *Peplis*, *Nesaea* and species of *Ammania* belonging to the subgenus *Euammania*, while it becomes multiseriate in *Cuphea* and species of *Ammania* in the subgenus *Rotala*.

JOSHI, A. C., and J. VENKATESWARLU. 1936. Embryological studies in the *Lythraceae* — III. Proc. Indian Acad. Sci., B, 3: 377–400.

MAURITZON, J. 1934. Zur Embryologie einiger *Lythraceen*. Meddel. Göteborgs Bot. Trädg. 9: 1–21.

SOUÈGES, R. 1925. Embryogénie des *Lythracées*. Développement de l'embryon chez le *Lythrum salicaria* L. C. R. Acad. Sci. Paris 180: 1417–1418.

Sonneratiaceae:- *Dusbasya sonneratioides* conforms closely to the Lythrum Variation, Onagrad Type.

VENKATESWARLU, J. 1937. A contribution to the embryology of *Sonneratiaceae*. Proc. Indian Acad. Sci., B, 5: 206–223.

Onagraceae:- Perhaps the most fundamental type of angiospermic embryonomy is that exhibited by the great majority of species belonging to the *Onagraceae*. The family constitutes a remarkably homogeneous embryological unit. The genus *Trapa*, placed in the *Onagraceae* by many taxonomists, must be excluded since it does not agree with any species in the family in either megagametogeny or embryogeny.

The terminal cell of the two-celled proembryo is bisected longitudinally and the basal cell transversely; the second cell generation thus consists of two juxtaposed upper cells, *q*, a middle cell, *m*, which is also the hypophysis initial, and a lower cell, *ci*. The next division is simultaneous in both cells of tier *q* and bisects each longitudinally to form the quadrant stage. Each quadrant is next segmented transversely to give the octant stage; the upper tier is designated as *l* and the lower tier as *l'*. Region *l* corresponds to the cotyledonary portion and the lower region *l'* represents the hypocotyledonary portion.

The middle cell *m* is the hypophysis initial and gives rise to the root tissues and to the root cap. The division in this cell, with exceptions in a few species, is characteristic. The wall which originally separated the terminal cell from the basal cell in the two-celled proembryo becomes somewhat diagonal and higher at the center than at the periphery; in other words, cell *m* has what might be called a "gabled roof." The ends of the wall in cell *m* are usually attached to this transverse roof about one-third of the way in from the peripheral wall. The wall is curved, so that the lower daughter cell has the shape of a watch glass and the upper daughter cell resembles a short cone with a curved basal wall. While one end of the wall is always attached to the primary zygotic wall, the other end may not join the latter wall but becomes extended in a straight line to meet the wall separating cell *m* from cell *ci*. This variation occurs rather commonly in *Eulobus sceptrostigma*, much less so in other species. When the wall is curved, the next division is in the lower daughter cell and bisects it vertically; when the wall is diagonal, the next division is in the cell with the broadest width of its upper portion adjacent to the embryo proper and segments it diagonally. In either case, it is apparent that the immediate effect is the same. Both upper and lower daughter cells become transformed into two groups of four cells each through the erection of cruciately meridional walls. The derivatives of the upper daughter cell of *m* represent the initials of the root cortex (*iec*); the derivatives of the lower daughter cell are the initials of the

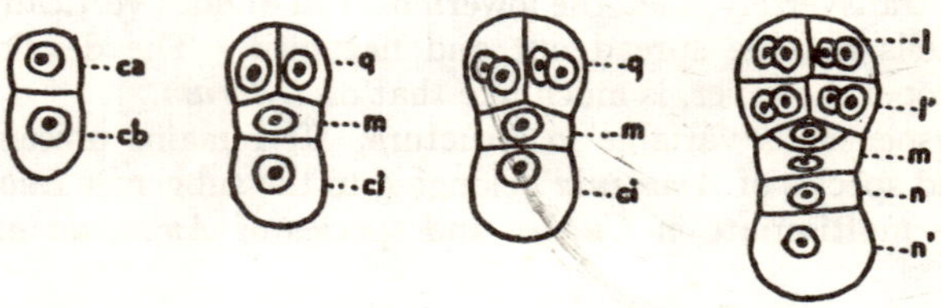

Recapitulatory Table for the Onagrad Type
(from left to right)

I. First Cell Generation
Proembryo of two cells disposed in two tiers:
$ca = pco + pvt + phy + icc$
$cb = iec + co + s$

II. Second Cell Generation
Proembryo of four cells disposed in three tiers:
$q = pco + pvt + phy + icc$
$m = iec + co$
$ci = s$

III. Third Cell Generation
Proembryo of six cells disposed in three tiers:
The destinations of the three tiers are the same as in the preceding cell generation.

IV. Fourth Cell Generation
Proembryo of twelve cells disposed in five tiers:
$l = pco + pvt$
$l' = phy + icc$
$m = iec + co =$ (hypophysis)
$n + n' = s$

root cap. In the latter, vertical walls are first formed, followed by tangential walls, and the cells become extended by tangentially centripetal divisions of the innermost layer to develop the calyptrogen. Cell *m* may rarely be divided by a purely vertical meridional wall, as in *Oenothera biennis*, followed by a second wall in each daughter cell and oriented perpendicularly to the first so that a tier of hypophyseal quadrants is produced.

It appears that a diagonal wall in cell *m* is characteristic of primitive species in the family and of those which are more or less irregular in their embryonomy, while the peculiar curved wall alone occurs in the more highly evolved species. The oblique wall represents a partial transformation or transition from the transverse or longitudinal wall first formed in the hypophysis initial of many species in other families, and the curved wall is the completion of the transition.

The lowest cell, *ci*, divides transversely into two superposed cells, *n* and *n'*. Cell *n* does not ordinarily divide again, but cell *n'* may do so, either transversely or longitudinally. Save in a few species in which development is more or less anomalous, the suspensor rarely includes more than three evanescent cells.

Among the species which conform exactly to the Onagrad Type are: *Circaea pacifica*, *Gayophytum ramosissimum*, *Godetia amoena*, *Stenosiphon linifolium*, *Hartmannia tetraptera*, *Galpinsia hartwegi*, *Epilobium angustifolium*, *E. franciscanum*, *E. paniculatum* (and probably all species included in the genus), *Oenothera biennis*, *Ludwigia palustris* and *Jussieua octonervia*.

Circaea lutetiana differs from the other species in the family in three important respects: in the destinations of the derivatives of the basal cell (*cb*), in the origin and the mode of differentiation of the initials of the root cortex. In the Type, $cb = iec + co + s$, but in *C. lutetiana*, $cb = co + s$. At the fourth cell generation there are but twelve cells in place of the usual sixteen. Cell *m* is the upper daughter derivative of cell *cb* (as is the case in the Lythrum Variation); it divides twice vertically to form a circumaxial tier of four cells which eventually provide the root cap. This behavior represents such a radical departure, though it is in no way questionable, from the normal proembryonal development in the family that it is desirable to await its discovery in other species before erecting a Variation to accommodate *C. lutetiana*.

Stenosiphon linifolium is peculiar in that development of the embryo proceeds after the green ovaries become abscissed from the stems and drop to the ground.

There is considerable irregularity in the embryonomy of *Clarkia elegans*. Evidence of fasciation may be detected during the development of the embryo; the cells are irregularly arranged and are commonly binucleate and even trinucleate. Growth of the embryo is often unilateral with resultant production of only one cotyledon. Tricotyledonous embryos are also frequent.

Embryos are never formed in *Anogra pallida*, and are rare in both species of *Burragea* (*B. fruticulosa* and *B. frutescens*); occasionally they develop in normal fashion in the two latter species, but are frequently very irregular and in this respect resemble similar embryos in *Clarkia elegans*.

The genus *Fuchsia* is predominantly parthenocarpic; the eggs are not fertilized and the megagametophytes slowly degenerate.

Zauschneria latifolia exhibits numerous embryological peculiarities. Seven distinct types of embryos have been described:

A. Perfectly normal diploid embryos. The endosperm is coenocytic, sometimes fairly abundant.

B. Apparently normal embryos which, however, lack the typical regularity of cell organization. Cotyledons, a radicle or a plumule have never been found in any such embryos. The endosperm, if present, consists of from three to six very tiny nuclei in a mere vestige of cytoplasm.

C. Atypical embryos which at first are columnar, then ovoidal and finally ellipsoidal. Endosperm is generally lacking.

D. Embryos resulting from fertilized synergids. Endosperm is absent.

E. Monocotyledonous embryos. Endosperm is absent.

F. Embryos of apomictic origin. Endosperm is never present.

G. Irregular embryos, including instances of budding. Endosperm is sometimes present.

The normal embryos included in group *A* are most numerous and are easy to identify as such from the bicellular proembryonic stage onward. All the stages of normal embryonomy follow the typical Onagrad Type closely. The quasinormal embryos included in group *B* represent the second most numerous group. The earlier stages are much bulkier than the corresponding stages in normal embryos and the planes of division are scarcely regular. No hypophysis is formed; consequently, mature embryos of this type lack a radicle.

Unless carefully scrutinized, intermediate stages superficially resemble normal embryos; from the beginning the shape of the embryo is either ellipsoidal, ovoidal or urceolate, never globular as in normal embryos. The sizes and shapes of the individual cells, as well as their positions, vary greatly and the cell walls are quite thin in comparison with those of normal embryos. The suspensor is irregular and often comprises only two cells. Cotyledons have never been found on these embryos.

Embryos belonging to group *G* may be merely variations of the preceding group. The earlier proembryonic stages consist of a linear row of three or four cells, of which at least the lower two belong to the suspensor. These cells, which would ordinarily develop into a typical massive suspensor, early degenerate and disappear, leaving a somewhat globular structure which henceforth develops in much the same manner as in the preceding group. No embryonal organs of any nature have been recognized, save possibly for the epidermis.

Group *D* embryos, derived from fertilized synergids, are discussed in the chapter on apomictic embryogeny. In Group *E*, several nearly mature pseudomonocotyledonous embryos were observed; they are probably aberrant forms belonging to some other group. Budding of the suspensor to give rise to two or more embryos was often observed; such embryos are included in group *G*. These instances may represent cases of true polyembryony among angiosperms.

When *Oenothera muricata* is pollinated with pollen from *O. biennis*, irregular and sickly embryos are the result, or embryo-like outgrowths from the nucellus may be produced at the chalazal end of the embryo sac. Either normal-appearing or sickly embryos arise when *O. lamarckiana* is fertilized by pollen from *O. biennis*.

Observations on species not mentioned in the following citations are original.

JOHANSEN, D. A. 1930. Studies on the morphology of the *Onagraceae*. II. Embryonal manifestations of fasciation in *Clarkia elegans*. Bot. Gaz. 90: 75–91.

JOHANSEN, D. A. 1930. Studies on the morphology of the *Onagraceae*. IV. *Stenosiphon linifolium*. Bull. Torrey Bot. Club 57: 315–326.

JOHANSEN, D. A. 1931. Studies on the morphology of the *Onagraceae*. V. *Zauschneria latifolia*, typical of a genus characterized by irregular embryology. Annals New York Acad. Sci. 33: 1–26.

JOHANSEN, D. A. 1931. Studies on the morphology of the *Onagraceae*. VI. *Anogra pallida*. Amer. Jour. Bot. 18: 854–864.

JOHANSEN, D. A. 1932. Studies on the morphology of the *Onagraceae*. VII. *Gayophytum ramosissimum*. Bull. Torrey Bot. Club 60: 1–8.

JOHANSEN, D. A. 1934. Studies on the morphology of the *Onagraceae*. VIII. *Circaea pacifica*. Amer. Jour. Bot. 21: 508–510.

RENNER, O. 1915. Befruchtung und Embryobildung bei *Oenothera lamarckiana* und einigen verwandten Arten. Flora 107: 115–150.

SOUÈGES, R. 1920. Embryogénie des *Oenothéracées*. Développement de l'embryon chez l'*Oenothera biennis*. C. R. Acad. Sci. Paris 170: 946–949.

SOUÈGES, R. 1935. Embryogénie des *Oenothéracées*. Les principaux termes du développement de l'embryon chez le *Ludwigia palustris* Elliott. C. R. Acad. Sci. Paris 200: 1626–1628.

SOUÈGES, R. 1946. Embryogénie des Oenothéracées. Développement de l'embryon chez le *Circaea lutetiana* L. C. R. Acad. Sci. Paris 223: 700–702.

Trapaceae:- The zygote of *Trapa natans* segments transversely, whereupon both derivative cells are divided in the same plane. A series of both transverse and longitudinal walls are next erected, resulting in a proembryo of about six tiers of cells with at least two cells in each tier and a unicellular basal tier. One account states that the suspensor always has a single cell at its distal end, but this is not borne out by the figures in the other account. This cell may occasionally divide longitudinally. As the proembryo grows, the suspensor cells closest to the embryo proper increase considerably in number but not in size and presently develop a collar-like protuberance which encircles the embryo proper; the side opposite the funiculus develops more extensively than does the other and attains a length almost exceeding the embryo. The suspensor cells below this region become irregularly inflated and convoluted, bearing an amazing resemblance to the suspensor system of certain Coniferophyta. The whole suspensor system eventually collapses and disappears. Growth of the embryo proper is comparatively slow at first. The basal cell apparently contributes nothing to the construction of the embryo proper. The latter becomes cylindrical in shape, with the cells arranged in rather definite tiers. The epidermis is early delimited, but the other histogens do not become recognizable until late in the development of the embryo. The two cotyledons are very unequal in size.

Assignment is undoubtedly to the Solanad Type and the development closely parallels that of the Sherardia Variation, except in the construction of the suspensor system. It should be observed that the genus *Trapa* emphatically does not belong under the *Onagraceae,* in which it has been placed by many taxonomists, since neither its megagametogeny nor its embryonomy conforms to the types of each characteristic of that family.

GIBELLI, G., and F. FERRERO. 1891. Intorno allo sviluppo dell'ovolo e del seme della *Trapa natans* L. Malpighia 5: 186–218.
TISON, A. 1919. Sur le suspenseur du *Trapa natans* L. Rev. Gén. de Bot. 31: 219–228.

Halorrhagaceae:- The two genera belonging to this family that have been investigated each belong to a different Type; the family therefore is not embryonomically homogeneous.

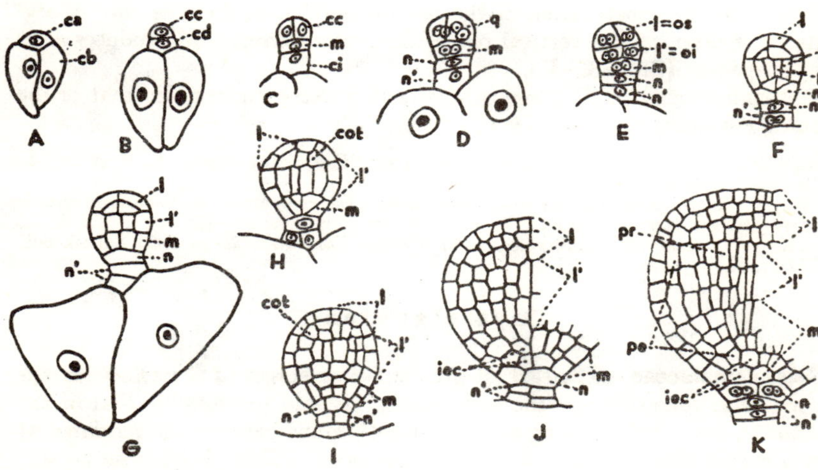

FIG. 48. — *Myriophyllum alternifolium.* Description in text. (*Redrawn from* SOUÈGES 1940).

The zygote of *Myriophyllum alternifolium* divides into two very unequal cells: a flattened, small terminal cell and a vesicular basal cell (FIG. 48A). The latter divides first, in the vertical meridional plane, then the terminal cell is segmented transversely into two superposed cells, *cc* and *cd* (FIG. 48B). Cell *cc* is next partitioned vertically and cell *cd* transversely into cells *m* and *ci* (FIG. 48C). The two upper juxtaposed elements then divide longitudinally to produce four quadrants, and next transversely to engender four upper octants (*os*) and four lower octants (*oi*) (FIG. 48E). The upper octants produce the cotyledonary region (*l*); they divide by tangential walls, the outer daughter cells becoming epidermal initials and the inner daughters then dividing by generally vertical walls to separate the mother cells of the cotyledons exteriorly and, interiorly, the elements which enter into the construction of the stem apex. The lower octants give rise to the upper portion of the hypocotyledonary axis (*l′*); they undergo vertical segmentations which separate the epidermal initials, then the periblem and plerome (FIG. 48H–K). Transverse walls then follow these first vertical walls.

The processes of division and the functions of the element *m* of the tetrad proembryo constitute a delicate point and perhaps the most debatable one in the course of development. This element presently produces four circumaxial cells (FIG. 48D, E). These cells then become partitioned by walls which are inserted at the lower end either on the vertical meridional walls or on the horizontal wall which separates region *m* from region *n* (FIG. 48F). There thus results certain irregularities in the processes of cell construction in that portion of the embryo originating from region *m*. The second mode of division of the four circumaxial cells occurs more frequently and therefore might be considered as being the general rule. The epidermal initials are separated by the first walls and succeeding walls delimit the periblem. In the central cells transverse walls separate, toward the base, the initials of the root cortex, and, toward the apex, those elements which enter into the construction of the central cylinder. There appear to be two superposed tetrads of root initials (FIG. 48J, K).

The element *ci* of the proembryonic tetrad is presently divided into two superposed cells, *n* and *n′* (FIG. 48D): *n* produces four circumaxial cells which contribute to the construction of the central portion of the root cap; *n′* segments most often in the vertical plane, then transversely and produces a reduced suspensor (FIG. 48F–K).

The embryogeny of *Hippurus vulgaris* very closely resembles that of the Veronica Variation, Onagrad Type.

SOUÈGES, R. 1922. Recherches embryogéniques sur l'*Hippurus vulgaris*. C. R. Acad. Sci. Paris 175: 529–532.
SOUÈGES, R. 1940. Embryogénie des *Haloragacées*. Développement de l'embryon chez le *Myriophyllum alternifolium* DC. C. R. Acad. Sci. Paris 211: 185–187.
STOLT, K. A. H. 1928. Die Embryologie von *Myriophyllum alterniflorum* DC. Svensk Bot. Tidskr. 22: 305–319.

Thymelaeales

Geissolomataceae:- The earliest proembryonic stages of *Geissoloma marginata* were not followed out; the embryonomy is said to resemble that of the *Penaeaceae*, save that the long and linear cotyledons become differentiated at a much earlier stage and are far better developed. There is likewise no suspensor; all of the proembryo enters into the construction of the embryo.

This species is doubtless assignable to the Penaea Variation, Asterad Type.

STEPHENS, E. L. 1909. The embryo-sac and embryo of *Geissoloma marginata*. New Phytol. 8: 345–348.

Thymelaeaceae:- In *Daphne kiusiana* and *D. pseudomezereum* the terminal cell of the two-celled proembryo divides longitudinally and the basal cell transversely. Older proembryos and apparently mature embryos are depicted as becoming globular and there is no evidence of a suspensor, which leads to the conclusion that the basal cell goes completely into the formation of the embryo proper. It is highly probable that these species conform to the Penaea Variation, Asterad Type.

In *Daphne mezereum*, on the other hand, the terminal and basal cells both segment longitudinally, though the wall in the basal cell is frequently very oblique. As with the two species mentioned above, there is no suspensor; consequently, this species also conforms to the Penaea Variation. *D. laureola* appears to agree with the other species.

Events are decidedly different in *Thymelaea arvensis*, which conforms very closely to the Erodium Variation of the Asterad Type, including the formation of a suspensor and an epiphyseal region.

Osawa, I. 1913. On the development of the pollen-grain and embryo-sac of *Daphne*, with special reference to the sterility of *Daphne odora*. Jour. Coll. Agr. Imp. Univ. Tokyo 4: 237–264.
Fuchs, A. 1938. Beiträge zur Embryologie der *Thymelaeaceae*. Oesterr. Bot. Zeit. 87: 1–41.
Souèges, R. 1942. Embryogénie des Thyméléacées. Développement de l'embryon chez le *Daphne mezereum* L. C. R. Acad. Sci. Paris 214: 569–571.
Venkateswarlu, J. Development of the embryo of *Thymelaea arvensis* Lamk. Bot. Gaz. (in press).

Penaeaceae:- In no species of the *Penaeaceae* has a suspensor been found at any stage of embryonal development. The first zygotic division is transverse, after which longitudinal walls are erected in both basal and terminal cells, and the whole of this spherical proembryo enters into the formation of the mature embryo. This early proembryonic behavior indicates adherence to the Asterad Type, but the lack of a suspensor and other differences necessitate the creation of a special Variation, the Penaea Variation.

The embryo has a broad root tip, completely lacks a root cap, the hypocotyl is massive and the cotyledons are rudimentary.

Species investigated include *Penaea mucronata, Sarcocolla squamosa, S. formosa* and *Brachysiphon imbricatus*.

Stephens, E. L. 1909. The embryo-sac and embryo of certain *Penaeaceae*. Ann. Bot. 23: 363–378.

Nyctaginaceae:- *Oxybaphus viscosus* agrees with the Polygonum Variation, Asterad Type, in all essential details.

Boerhaavia diffusa and *B. repanda* are very close in their embryonomy; from the description it appears that both are also referable to the Polygonum Variation. The disposition of *Mirabilis jalapa* is far less clear.

Kajale, L. B. 1938. Embryo and seed development in the *Nyctaginaceae*. I. Studies in the genus *Boerhaavia*. Jour. Indian Bot. Soc. 17: 243–254.
Rocen, T. 1927. Zur Embryologie der Centrospermen. Diss., Uppsala.
Souèges, R. 1938. Embryogénie des *Nyctaginacées*. Développement de l'embryon chez l'*Oxybaphus viscosus* L'Hérit. C. R. Acad. Sci. Paris 206: 1830–1832.

Proteales

Proteaceae:- The four-celled proembryo of *Grevillea banksii* consists of two tiers of two juxtaposed cells each; no suspensor is formed, all of the derivatives of both basal and terminal cells going into the construction of the embryo. Conformation to the Penaea Variation, Asterad Type, is therefore indicated. *Grevillea robusta* similarly seems to conform to the Penaea Variation.

Brough, P. 1933. The life-history of *Grevillea robusta* Cunn. Proc. Linn. Soc. N. S. Wales 58: 33–73.
Kausik, S. B. 1939. Studies in the *Proteaceae*. III. Embryology of *Grevillea banksii* R. Br. Ann. Bot., n.s. 3: 815–824.
Kausik, S. B. 1942. Studies in the *Proteaceae*. VII. The endosperm of *Grevillea robusta* Cunn., with special reference to the structure and development of the vermiform appendage. Proc. Indian Acad. Sci., B, 16: 121–140.

Dilleniales

Neither of the two families, *Dilleniaceae* and *Crossosomataceae*, has ever been investigated embryologically.

Coriariales

The sole family, *Coriariaceae*, has been so superficially presented that no conclusions are possible.

MAURITZON, J. 1936. Zur Embryologie und systematischen Abgrenzung der Reihen *Terebinthales* und *Celastrales*. Bot. Notiser 1936: 161–212.

Pittosporales

All three families (*Pittosporaceae*, *Byblidaceae* and *Tremandraceae*) remain unstudied.

Bixales

Only one of the seven families, the *Cistaceae*, has been investigated. The others include the *Bixaceae*, *Cochlospermaceae*, *Flacourtiaceae*, *Samydaceae*, *Canellaceae* and *Frankeniaceae*.

Cistaceae:- Embryologically, the *Cistaceae* display a remarkably close resemblance to the *Linaceae*. *Helianthemum guttatum*, for instance, is almost identical with *Radiola linoides* embryonomically; this species, together with *Cistus monspelliensis*, is referable to the Linum Variation, Solanad Type.

CHIARUGI, A. 1925. Embriologia della *Cistaceae*. Nuovo Giorn. Bot. Ital., n.s. 32: 223–316.
SOUÈGES, R. 1937. Développement de l'embryon chez l'*Helianthemum guttatum* Mill. Bull. Soc. Bot. France 84: 400–407.

Tamaricales

The *Malesherbiaceae* are still uninvestigated.

Tamaricaceae:- In *Tamarix chinensis* a linear proembryo of four to seven cells is produced, but nothing is said as to their derivation.

A linear proembryo of six to seven cells is formed in *Tamarix ericoides*, but the derivation again was not described. Beginning with the uppermost cell, all cells appear to be segmented longitudinally. This species perhaps should be referred to the Sherardia Variation, Solanad Type.

PURI, V. 1939. Studies in the order *Parietales*. I. A contribution to the morphology of *Tamarix chinensis* Lour. Beih. Bot. Centralbl. 59: 335–349.
SHARMA, Y. M. L. 1939. Gametogenesis and embryogeny of *Tamarix ericoides* Rottl. Ann. Bot., n.s. 3: 861–870.

Fouquieriaceae:- *Fouquieria peninsularis* conforms to the Polygonum Variation, Asterad Type.

JOHANSEN, D. A. 1936. Morphology and embryology of *Fouquieria*. Amer. Jour. Bot. 23: 95–99.

Passiflorales

Neither the *Passifloraceae* nor the *Achariaceae* have ever been investigated.

Cucurbitales

All the families save the *Datiscaceae* have been reported upon.

Cucurbitaceae:- Although considerable work has been done on various species of the *Cucurbitaceae*, the embryological affinities of the family are still in a very dubious condition. The embryogeny of *Bryonia dioica*, although followed out in precise detail, does not by itself, as the original investigator remarks, furnish the solution of the problem. It is conjectured that the secret is to be found among the primitive groups following the embryological type of the *Ranunculaceae.*

The tetrad proembryo of *Bryonia dioica* presents two aspects: one whose lower cells, *m* and *ci*, are separated by an oblique wall more or less approaching the vertical; and a second in which these two cells are separated by a clearly transverse wall. From these two types of tetrads are derived two series of proembryonic forms.

The first series is derived from the tetrad with an oblique basal wall. Cell *ci* divides, usually transversely, into *n* and *n'*; while cell *m* usually divides obliquely, the wall being more or less parallel to the peripheral wall. Cell *n'* engenders a short suspensor, while *n* together with *m* enters into the construction of a true hypophyseal region which later produces the initials of the root cortex and the primordium of the root cap.

The second series is derived from the tetrad with a basal horizontal wall. Cell *ci* divides transversely into two superposed cells, *n* and *n'*, but here both cells enter into the construction of a short suspensor. Cell *m* divides by vertically cruciate walls to form four circumaxial hypophyseal quadrants in which the later divisions occur as in *Myosurus.*

During the later growth stages it is impossible to determine whether an embryo was derived from the first or the second of proembryonic stages.

The two upper cells of the tetrad, which are always juxtaposed, give rise by means of vertical walls to four quadrants which, however, do not always appear to be disposed in a horizontal plane. Eight octants arise as derivatives of the quadrant cells; these are more or less disposed in two tiers, *l* and *l'*, of which the former engenders the cotyledonary region and the latter the hypocotyl. In the octants the walls next formed are sometimes tangential, and at other times inclined to diverse planes. The irregularity of these divisions renders internal differentiation of the embryo very tardy and the internal histogens are scarcely distinguishable.

B. dioica may provisionally be assigned to the Myosurus Variation since it clearly belongs under the Onagrad Type. It is possible that the two series of embryogenic types as found during the four- and eight-celled stages may represent separate male and female forms, because in species with hermaphroditic flowers the same series of forms are encountered.

Fevillea cordifolia appears to differ considerably from the other genera in its early embryogeny. The first few proembryonic divisions are apparently transverse; a later stage consisting of about ten cells has the lower cells flattened against the nucellus while the upper cells are separated by longitudinal walls. There is a strong suspicion that an epiphysis initial may be present. Two somewhat later stages, as depicted, do not show the appressed basal cells, but again reveal evidence of an epiphysis region.

The earlier proembryonic stages in *Cucumis sativus* clearly reveal that an epiphysis initial is formed. The embryonomic relationship is probably with the Trifolium Variation, Onagrad Type.

Melothria pendula, *Apodanthera undulata*, *Bryonopsis laciniosa*, *Trichosanthes anguina*, *Momordica charantia*, *Luffa acutangula*, *Cucumis myriocarpus*, *Lagenaria lagenaria* (*L. leucantha*), *Benincasa hispida*, *Citrullus vulgaris*, *Cucurbita pepo*, *Sicyos angulata* and *Micrampelis lobata* appear to be all more or less alike embryologically; it is said that there are two or three transverse divisions of the zygote. The suspensor is short and ephemeral.

KIRKWOOD, J. E. 1904. The comparative embryology of the *Cucurbitaceae*. Bull. New York Bot. Gard. 3: 313–402.

SOUÈGES, R. 1939. Embryogénie des *Cucurbitacées*. Développement de l'embryon chez le *Bryonia dioica* Jacq. C. R. Acad. Sci. Paris 208: 227–229.

TILLMANN, O. J. 1906. The embryo sac and embryo of *Cucumis sativus*. Ohio Nat. 6: 423–430.

Begoniaceae:- *Begonia semperflorens* follows the Lythrum Variation, Onagrad Type.

SOUÈGES, R. 1939. Embryogénie des *Bégoniacées*. Développement de l'embryon chez le *Begonia semperflorens* Link èt Otto. C. R. Acad. Sci. Paris 208: 534–536.

Caricaceae:- The embryonomy of *Carica papaya* is difficult to determine because of proembryonic irregularities. The zygote presumably divides transversely, after which both the basal and the terminal cells are segmented longitudinally. Divisions thereafter are rather irregular and it is not at all clear whether the basal cell contributes to the construction of the embryo proper. On the supposition that it contributes but little, then there is considerable resemblance to the Trifolium Variation, Onagrad Type. The basal cell, on the other hand, produces a peculiar lateral outgrowth reminiscent of that in *Erythronium* or *Tulipa*.

FOSTER, L. T. 1943. Morphological and cytological studies on *Carica papaya*. Bot. Gaz. 105: 116–126.

Cactales

Cactaceae:- A linear tetrad is produced in *Rhipsalis cassytha*, but an irregular five-celled proembryo and a linear one of six cells are also depicted. Further illustrations show proembryos of three tiers of two cells each, of two tiers of two juxtaposed cells at the basal end plus two superposed upper cells. The latest stage figured shows an upper tier of four cells, then one of five cells (probably only four cells, one of which has segmented prematurely by an oblique wall), then a linear suspensor of eight cells.

The description is too inadequate to allow determination of the Type to which *R. cassytha* conforms.

MAURITZON, J. 1934. Ein Beitrag zur Embryologie der *Phytolaccaceen* und *Cactaceen*. Bot. Notiser 1934: 111–135.

Theales

There is no report for any of the following families: *Theaceae* (*Ternstroemiaceae*), *Medusagynaceae*, *Marcgraviaceae*, *Caryocaraceae*, *Saurauiaceae*, *Ochnaceac* (except for *Ochna serrulata*), *Ancistrocladaceae*, *Dipterocarpaceae* and *Chlaenaceae*.

Actinidiaceae:- Two different embryogenic schemes occur in *Actinidia chinensis*: one is referable to the Drosera Variation, Caryophyllad Type and the other closely resembles that occurring in *Hypericum perforatum* (Onagrad Type, in general).

Souèges, R. 1943. Embryogénie des Actinidiacées. Développement de l'embryon chez l'*Actinidia chinensis* Planch. C. R. Acad. Sci. Paris 217: 430–432.

Myrtales

Combretaceae:- *Poivrea coccinea* conforms to the Erodium Variation, Asterad Type.

Venkateswarlu, J. Development of the embryo of *Thymelaea arvensis* Lamk. Bot. Gaz. (in press).

Lecythidaceae:- In *Barringtonia vriesei* there is formed a flattish proembryo whose derivation was not followed out. Later a suspensor, flatter and wider than the embryo proper, is distinguishable.

Treub, M. 1884. Notes sur l'embryon, le sac embryonnaire et l'ovule. 5. L'embryon du *Barringtonia vriesei* T. et B. Ann. Jard. Bot. Buitenzorg 4: 101–106.

Myrtaceae:- The embryogeny of *Myrtus communis* agrees most closely with that of *Clematis* among the *Ranunculaceae*, with the earlier proembryonic stages resembling those of *Myosurus* and *Ranunculus*. Assignment of this species is perhaps best made to the Myosurus Variation, Onagrad Type, since the later divisions in the developing embryo are equally as irregular as in *Myosurus*.

Souèges, R. 1940. Embryogénie des *Myrtacées*. Développement de l'embryon chez le *Myrtus communis* L. C. R. Acad. Sci. Paris 210: 548–550.

Melastomaceae:- A linear proembryo of about six cells is produced in *Monochaetum ensiferum*. The lower four or five cells comprise the suspensor and apparently divide no further. All of the embryo appears to develop from the uppermost cell of the filamentous row; presumably the best assignment of this species is to the Linum Variation, Solanad Type.

Development is essentially similar in *Centradenia floribunda*.

In *Sonerila wallichii* the basal cell segments transversely and the terminal cell longitudinally. The lower derivative of the basal cell (cell *ci*) is divided longitudinally, then the two juxtaposed daughter cells quickly produce a tier of eight somewhat inflated and more or less disjointed cells. The hypophysis initial (cell *m*) may divide either transversely or longitudinally; the derivatives develop in much the same manner as in the Capsella Variation, Onagrad Type. However, since the peculiar suspensor development is not known in any other variation of that Type, *Sonerila* constitutes a variation by itself.

Subramangan, K. 1944. A contribution to the life-history of *Sonerila wallichii* Benn. Proc. Indian Acad. Sci., B, 19: 115–120.
Ziegler, A. 1925. Beiträge zur Kenntnis des Androeciums und der Samenentwicklung einiger *Melastomaceen*. Bot. Archiv 9: 398–467.

Rhizophoraceae:- The early embryogeny of *Rhizophora mangle* was not followed out in detail, but is said to be "probably of the Capsella Type," although the evidence is unconvincing. The growth of the embryo is divided into three periods: (1) first growth of the cotyledons, (2) growth of the hypocotyl, (3) second growth of the cotyledons. During the first period of growth about one-third of the embryo and the greater part of the endosperm are forced out of the embryo sac and lie in the ovarian chamber. The second period of growth is confined largely to the hypocotyl, but during this time the structure of the two cotyledons undergoes considerable modification. During the third period of growth the cotyledons elongate and the point of junction between them and the hypocotyl is forced beyond the apex of the ovary. At this time the embryo is approaching maturity. The cotyledons and the hypocotyl then separate and the hypocotyl drops, carrying the stem apex with it; while the ovary, containing the outer integument and cotyledons, remains hanging on the tree.

In *Rhizophora mucronata* the youngest embryo found consisted of an undifferentiated mass of tissue with a massive suspensor. The suspensor of this species thus differs from that in *R. mangle*, in which it is described as being a linear series of cells. The two cotyledons are fused throughout the greater part of their length, with the free tips closely appressed to one another. The growth of the cotyledons pushes the hypocotyl through the micropyle into the ovarian cavity; the hypocotyl then elongates and pierces the ovarian wall. The cotyledonary bend, which conceals the plumule, enlarges and when fully developed resembles a "Phrygian cap" and occupies almost all of the space within the ovule. In the final stage of development, the cotyledonary sheath elongates and the base of this sheath passes through the aperture of the fruit so that the concealed plumule and hypocotyl are free to drop into the mud in which the trees grow, there to continue development.

Ceriops candolleana develops essentially as does *Rhizophora mucronata*, save that the cotyledonary head is cylindrical at maturity.

Carey, G. 1934. Further investigations on the embryology of viviparous seeds. Proc. Linn Soc. N. S. Wales 59: 392–410.

Cook, M. T. 1907. The embryology of *Rhizophora mangle*. Bull. Torrey Bot. Club 34: 271–277.

Hypericales (Guttiferales)

The as yet uninvestigated families include the *Eucryphiaceae* and *Quiinaceae.*

Hypericaceae:- *Hypericum perforatum* and *Androsaemum officinale* have the identical embryonomy; the latter species has been more carefully investigated and the following account is based on it.

The terminal and basal cells of the two-celled proembryo are both segmented transversely to give a tetrad of four superposed elements (Fig. 49A, B). The uppermost cell, *cc*, is divided longitudinally into two juxtaposed cells, whereupon each daughter cell is partitioned by a longitudinal wall oriented perpendicular to the first to form the quadrant stage (Fig. 49C–F). By transverse segmentations of each cell of the quadrant, the octant stage is produced (Fig. 49G). The sequence of divisions in the terminal cell and the histogenic role of the resultant octant cells are comparable to those in the Onagrad Type. The element *cd*, sister cell to *cc*, is divided horizontally into two superposed cells, *m* and *ci* (Fig. 49D); these are segmented, the first much sooner than the latter, transversely into new elements situated one above the other and more or less flattened (Fig. 49G–I). Occasionally the cell *ci* is divided longitudinally, as in *Saxifraga granulata.* It is ordinarily the upper daughter cell of *m* (*d*) which becomes differentiated as the hypophysis cell (Fig. 49F), which in its mode of individualization, its processes of division and its functions is comparable to that characterizing the Onagrad Type.

The cell *f*, sister of *d*, contributes, together with the derivatives of *ci* and all the elements produced by the basal cell, *cb*, to the construction of a considerably elongated suspensor (FIG. 49B–I). Despite the presence of longitudinal walls in a few of the cells derived from *cb*, the suspensor should be classified in the filamentous category.

There are several exceptions in the embryogeny of *Hypericum perforatum.*

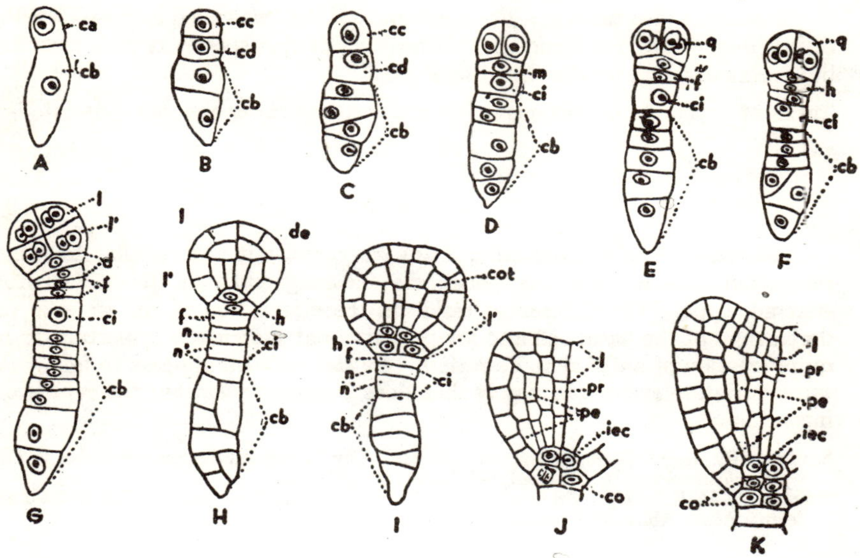

FIG. 49. — *Androsaemum officinale.* Description in text. All × 350. (*Redrawn from* SOUÈGES 1936).

For example, the terminal cell, *ca*, of the two-celled proembryo may divide longitudinally; in this case, the embryo proper necessarily takes its origin not from *cc* but from its mother cell, as in the Onagrad Type. Again, the cells *m* and *ci*, derivatives of *cd*, often divide vertically — sometimes both, sometimes the one or the other alone —; when *m* divides vertically, it becomes the hypophysis initial directly. Yet again, the hypophysis is frequently individualized not from *d*, but from its upper daughter cell; this most often occurs in those embryos with extensively elongated suspensors.

The following table recapitulates the cell generations as found in *H. perforatum*, leaving the basal cell out of consideration after the first cell generation:

Recapitulatory Table for Hypericum perforatum

I. FIRST CELL GENERATION
Proembryo of two cells disposed in two tiers:
cc = *pvt* + *pco* + *phy*
cd = *icc* + *co* + upper portion of *s*

II. SECOND CELL GENERATION
Proembryo of four cells disposed in three tiers:
cc = *pvt* + *pco* + *phy*
m = *icc* + *co* + *s* (in part)
ci = *s* (in part)

III. THIRD CELL GENERATION
Proembryo of six cells disposed in three tiers:
The dispositions and destinations are the same as in the preceding generation.

IV. FOURTH CELL GENERATION
Proembryo of twelve cells disposed generally in six tiers:
l = *pvt* + *pco*
l' = *phy*
d = *iec* + *co*
f + *n* + *n'* = *s* (upper portion)

SOUÈGES, R. 1925. Embryogénie des *Hypericacées*. Développement chez l'*Hypericum perforatum* L. C. R. Acad. Sci. Paris 180: 949–951.
SOUÈGES, R. 1936. Embryogénie des *Hypericacées*. Développement de l'embryon chez l'*Androsaemum officinale* All. C. R. Acad. Sci. Paris 202: 679–681.
SOUÈGES, R. 1936. Les relations embryogéniques des *Crassulacées*, *Saxifragacées* et *Hypericacées*. Bull. Soc. Bot. France 83: 317–329.

Clusiaceae (Guttiferae):- For *Garcinia kydia* a two-celled and two several-celled proembryos are depicted. Disposition of this species is problematical, but seems to be under the Onagrad Type.

TREUB, M. 1911. Le sac embryonnaire et l'embryon dans les Angiospermes. Ann. Sci. Nat., Bot. II, 9: 1–17.

Tiliales

Tiliaceae:- Embryonomically, *Tilia platyphyllos* is very similar to *Malva rotundifolia* and, as with that species, may be assigned to the Urtica Variation, Asterad Type. The differences reside in the orientation of the walls in the derivatives of the basal cell and are probably not sufficiently important to require erection of a different Variation. *Entelea palmata* appears to follow the same Variation even more closely than does *T. platyphyllos*, but the evidence is inconclusive.

SOUÈGES, R. 1941. Embryogénie des Tiliacées. Développement de l'embryon chez le *Tilia platyphyllos* Scop. C. R. Acad. Sci. Paris 212: 998–1000.
STENAR, A. H. S. 1925. Embryologische Studien. I. Zur Embryologie einiger Columniferen. Akad. Abhandl. Uppsala 1: 1–197.

Sterculiaceae:- The zygote of *Theobroma cacao* does not divide until some 45 days after syngamy. Only a single stage is figured and described for this species; the orientation of all walls is markedly diagonal. The evidence is insufficient to determine the assignment of *T. cacao*. The occurrence of parthenogenesis, frequently described, is very doubtful.

CHEESMAN, E. E. 1927. Fertilization and embryogeny in *Theobroma cacao* L. Ann. Bot. 41: 107–126.

Malvales

Malvaceae:- The embryonomy of *Malva rotundifolia* agrees very closely with that of the Urtica Variation, Asterad Type.

Lavatera thuringiana apparently also follows the Urtica Variation.

SOUÈGES, R. 1922. Embryogénie des *Malvacées*. Développement de l'embryon chez le *Malva rotundifolia* L. C. R. Acad. Sci. Paris 175: 1435–1436.
STENAR, A. H. S. 1925. Embryologische Studien. I. Zur Embryologie einiger Columniferen. Akad. Abhandl. Uppsala 1: 1–197.

Malpighiales

Nothing has been described for the *Humiriaceae* and *Erythroxylaceae*. In the *Malpighiaceae* five genera (*Hiptage*, *Aspicarpa*, *Banisteria*, *Heteropteris* and *Stigmatophyllum*) have all been described as reproducing solely by means of nucellar embryos.

RAO, A. M. S. 1940. Studies in the *Malpighiaceae*. I. Embryo-sac development and embryogeny in the genera *Hiptage*, *Banisteria* and *Stigmatophyllum*. Jour. Indian Bot. Soc. 18: 145–156.

Euphorbiales

Euphorbiaceae:- The terminal cell of the bipartitioned zygote of *Euphorbia exigua* is divided longitudinally and the basal cell is segmented transversely. The two upper cells, designated together as *q*, are next partitioned vertically by walls oriented at right angles to the first longitudinal wall and shortly thereafter each quadrant, usually at a different time, is divided transversely into two layers, *l* and *l'*. The upper four cells of the octant eventually give rise to the cotyledonary portion and the lower four cells to the hypocotyledonary region. The subsequent order of divisions in the octant cells displays little regularity. The intermediate cell *m* gives rise to the hypophysis: two cruciate walls are first erected to give four circumaxial cells and each is shortly divided transversely. The upper four cells thus resulting give rise to the initials of the root cortex and the lower four represent the primordium of the root cap. The lowest cell, *ci*, divides in a rather indeterminate manner to form a few-celled, conical suspensor.

The embryogeny of *Euphorbia* greatly resembles that of the *Ranunculaceae*; the tetrad is erected in the same manner; the quadrants and octants are disposed similarly, take their origin from the same elements and give rise to the same embryonal regions; furthermore, the intermediate cell *m* acts as a hypophysis initial and segments by essentially comparable processes, and finally the lowest cell, *ci*, produces a suspensor whose origin, form and dimensions are manifestly similar. There are, however, minor differences which suffice to set the Euphorbia Variation apart from the Myosurus Variation: the lower cells segment much less actively than the upper cells, and again the cells at the root tip end of the developing embryo divide more rapidly in *Euphorbia*.

Euphorbia procera, *E. esula*, *E. splendens* and *Acalypha australis* conform to the Euphorbia Variation. *Euphorbia mauritanica* and *E. preslii* appear to develop differently. The former is said to have no suspensor, which leads to the inference that it develops much like *E. preslii*. The zygote of *E. preslii* is claimed to divide longitudinally, then each daughter cell is segmented transversely. Further divisions result in the formation of a globular embryo lacking a suspensor. Later the embryo becomes elongated and cotyledons and a well developed root cap are produced. These two species, therefore, should be assigned to the Scabiosa Variation, Piperad Type.

The description of the embryogeny of *Aleurites fordi* is incomplete and puzzling. A figure of a late proembryo stage shows no apparent suspensor, though one is found on the older embryo; this figure also clearly reveals an apparent epiphysis initial. Such an initial is not known for any other species in the Euphorbiaceae; *A. fordi*, therefore, requires reinvestigation.

McCann, L. P. 1945. Embryology of the Tung tree. Jour. Agr. Res. *71*: 215–229.

Modilewski, J. 1909. Zur Embryobildung von *Euphorbia procera*. Ber. D. Bot. Ges. 27: 21–26.

Sateishi, S. 1927. On the development of the embryo sac and fertilization of *Acalypha australis* L. Bot. Mag. Tokyo 41: 477–485. (In Japanese)

Souèges, R. 1924. Embryogénie des *Euphorbiacées*. Développement de l'embryon chez l'*Euphorbia esula* L. C. R. Acad. Sci. Paris 179: 989–991.

Souèges, R. 1925. Développement de l'embryon chez l'*Euphorbia exigua* L. Bull. Soc. Bot. France 72: 1018–1031.

Ventura, M. 1934. Sviluppo del gametofito femminile di *Euphorbia mauretanica* L. Ann. di Bot. 20: 267–273.

Weniger, W. 1917. Development of embryo sac and embryo in *Euphorbia preslii* and *E. splendens*. Bot. Gaz. 63: 266–281.

Cunoniales

All the families — *Cunoniaceae, Brunelliaceae, Escalloniaceae, Greyiaceae,* and *Hydrangeaceae* — have never been studied embryologically.

Grossulariaceae:- Certain species in *Ribes*, which apparently properly belong in the *Grossulariaceae*, are mentioned under the *Saxifragaceae*, in which family they were placed by the original investigator.

Rosales

The families which remain uninvestigated include the *Chailletiaceae* (*Dichapetalaceae*), *Calycanthaceae* and *Malaceae.*

Rosaceae:- The zygote of *Geum urbanum* is divided transversely into two very unequal cells (FIG. 50A). The basal cell is the larger one and is segmented first, in the transverse plane, into two superposed cells, *m* and *ci* (FIG. 50B). Shortly thereafter, the terminal cell in turn is divided by a wall which is always oblique. The two daughter cells, *a* and *b*, are very dissimilar. (FIG. 50C). The destination of the four cells of the proembryonic tetrad is as follows: the two upper cells give rise to the cotyledonary portion, the intermediate cell *m* engenders the hypocotyledonary region and the lower cell, *ci*, develops the initials of the root cortex, the root cap and the suspensor.

The plane of division in the terminal cell may appear to be transverse if the embryo is microtomed either too thinly or in an incorrect plane.

Bipartitioning of each element of the tetrad leads to the formation of the eight-celled proembryo. The intermediate cell *m* generally divides first, by a vertical meridional wall, to give two juxtaposed cells (FIG. 50C, D). Later the lowest cell, *ci*, divides transversely into *n* and *n'*. Of the two upper cells, cell *a* is divided by an oblique wall oriented perpendicularly to the preceding one and thus gives rise to two new elements with one slightly below the other; while cell *b* is divided vertically by a meridional wall to give two juxtaposed elements. Segmentation of cell *b* generally precedes that of cell *a*. The four elements issuing from these two cells appear to occupy the four summits of a tetrahedron bounded on one of its faces by the horizontal wall limiting the upper side of tier *m*. The summit of the proembryo is occupied by a triangular cell (derived from cell *a*) which by its subsequent sequence of division and by its histogenic role is an epiphysis initial (*e*, FIG. 50D *et seq.*). It resembles that of *Myosotis hispida* and other plants, but in most of these the epiphysis initial is a unit of the fourth cell generation, whereas in *G. urbanum* it is a unit of the third cell generation. Nothing corresponding to octants is formed in *G. urbanum*; the two upper tiers *l* and *l'*, of other types are not differentiated.

The entire embryonic region derived from the original terminal cell is designated as *q* (FIG. 50D *et seq.*). The daughter cells of cell *b* usually divide first; in Fig. 50E are shown four nuclei separated by walls which are not visible in the plane of the figure. These walls may be tangential, thus isolating the first initials of the epidermis toward the exterior (FIG. 50J, O–Q); but most often they are oriented parallel or perpendicular to the preceding wall, thus separating from the derivatives of the first two elements issuing from cell *b* a quadrilateral cell and a triangular cell (this is best shown in transverse sections, and is diagrammed in FIG. 50L, M, in which the two daughter cells issuing from derivatives of *b'* and *b''* are shown).

The lower daughter cell of cell *a*, sister to the epiphysis initial, is divided by a meridional wall into two juxtaposed elements, *a′* and *a″* (FIG. 50G–J, L, M). These sometimes segment by the same processes that occur in cells

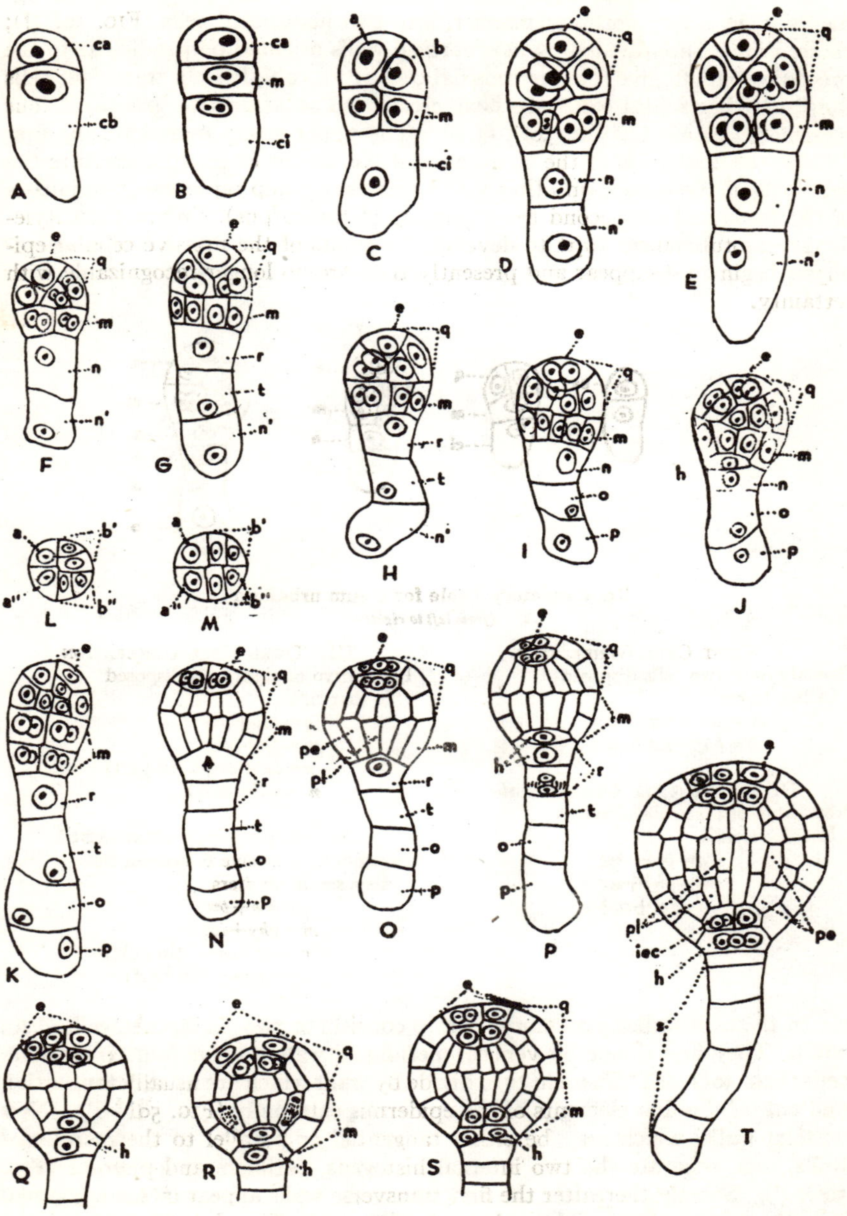

FIG. 50. — *Geum urbanum.* Description in text. A–E × 530; F–K × 380; L, M × 380; N–T × 350. (*Redrawn from* SOUÈGES 1923).

b′ and *b″*, but more frequently are divided by tangential walls, separating outwardly new initials of the epidermis. The central subepidermal cells presently divide vertically by tangential or rectangular walls to engender a cellular plateau of which the most external elements give rise to the cortical and vascular meristems of the cotyledon, while the cells closest to the embryonic axis become the mother cells of the central cylinder of the stem.

The epiphysis initial divides very tardily. A vertical meridional wall segments it, in a very constant manner, into two juxtaposed cells (FIG. 50I, J); these, in turn, are divided by new vertical walls oriented perpendicular to the preceding one, to give four circumaxial cells (FIG. 50N, P). In these four cells tangential horizontal walls are next erected to separate two groups of four superposed elements (FIG. 50O, Q–S). The upper group represents the first cells of the epidermis at the stem apex, while the lower group constitute the first cells of the cortex (in other words, the first group represents the initials of the torus and the second the beginning of the corpus). When the cotyledonary protuberances start to develop, the limits of the massive cellular epiphysis begin to disappear and presently they are no longer recognizable with certainty.

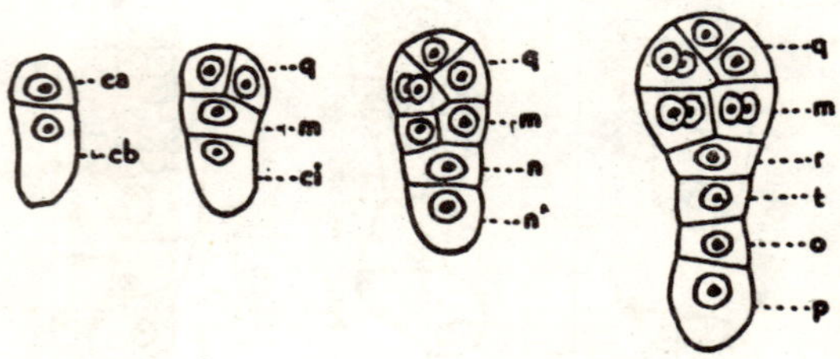

Recapitulatory Table for Geum urbanum
(from left to right)

I. FIRST CELL GENERATION
Proembryo of two cells disposed in two tiers:
ca=*pco*+*pvt*
cb=*phy*+*icc*+*iec*+*co*+*s*

II. SECOND CELL GENERATION
Proembryo of four cells disposed in three tiers:
q=*pco*+*pvt*
m=*phy*+*icc*
ci=*iec*+*co*+*s*

III. THIRD CELL GENERATION
Proembryo of eight cells disposed in four tiers:
q=*pco*+*pvt*
m=*phy*+*icc*
n=*iec*+*co*+*s* (in part)
n'=*s* (in part)

IV. FOURTH CELL GENERATION
Proembryo of twelve to sixteen cells disposed in six tiers:
q=*pco*+*pvt*
m=*phy*+*icc*
r=*iec*+*co*+*s* (in part)
t+*o*+*p*=*s* (in part)

In the eight-celled proembryo tier *m* consists of two juxtaposed cells (FIG. 50C). They first divide by vertical meridional walls to give four circumaxial cells (FIG. 50D–K). These in turn divide by walls which are usually tangential, and cut off the first elements of the epidermis externally (FIG. 50N–P). New vertical walls, which may be either tangential or parallel to the meridional walls, soon separate the two internal histogens, periblem and plerome (FIG. 50O, P). Shortly thereafter the first transverse walls appear in the epidermal initials, then in the periblem elements (FIG. 50R–T), while the cells of the plerome segment longitudinally to give exteriorly the first elements of the pericycle. The primordia of the three histogens thus delimited continue to multiply according to the ordinary rules to erect the hypocotyledonary axis. The initials of the central cylinder at the root end take their origin from the lowest elements of the plerome. The epidermal cells of region *m* become divided tangentially and much later contribute to the construction of the lateral portion of the root cap.

The cell which represents region *n* in the eight-celled proembryo becomes divided transversely into two superposed cells, *r* and *t* (FIG. 50H, K–P). Occasionally, the wall which divides these two cells has one end attached to the upper wall which separates tiers *m* and *n* (FIG. 50J); in this case, cell *n* is converted directly into a hypophysis cell and the suspensor is derived solely from cell *n'*. Most often, however, the wall is disposed in a purely horizontal plane; in such instances, the lower daughter cell, *t*, enters into the construction of the suspensor and the upper cell, *r*, behaves in its turn like the element *n* from which it was derived. It may be transformed directly into a hypophysis cell or segmented horizontally into two new cylindrical elements (FIG. 50N), of which the upper one necessarily differentiates as a hypophysis cell, while the other adds further units to the suspensor. Processes of segmentation in the hypophysis cell are identical with those occurring in *Capsella* (FIG. 50P–S).

Cell *n'* always divides transversely into two superposed cells, *o* and *p* (FIG. 50I *et seq.*). Later derivatives, together with those from element *t* or the lower daughter cell of *r*, enter into the construction of a filamentous suspensor of very variable dimensions.

The laws governing the embryonomy of *Geum urbanum* may be recapitulated as shown in the table on the opposite page.

The tribe *Potentilleae* of the *Rosaceae* has a very uniform embryogeny. The embryonomy of *Potentilla reptans*, for instance, agrees exactly with that of *Geum urbanum.*

Cultivated varieties of strawberries (*Fragaria* spp.) have frequently been described in the older literature as producing adventitious embryos, but an examination under precise modern conditions of several hundreds of seeds of three separate varieties failed to reveal other than embryos derived from regularly fertilized eggs. The embryonomy of *Fragaria* is precisely like that of *Geum urbanum.*

SOUÈGES, R. 1922. Embryogénie des *Rosacées*. Les premiers stades du développement de l'embryon chez le *Geum urbanum* L. C. R. Acad. Sci. Paris 174: 1070–1072.
SOUÈGES, R. 1922. Embryogénie des *Rosacées*. Les derniers stades du développement de l'embryon chez le *Geum urbanum* L. C. R. Acad. Sci. Paris 174: 1197–1199.
SOUÈGES, R. 1923. Développement de l'embryon chez le *Geum urbanum* L. Bull. Soc. Bot. France 70: 645–660.
SOUÈGES, R. 1935. Observations embryologiques sur quelques *Fragaria* de culture. Bull. Soc. Bot. France 82: 458–461.
SOUÈGES, R. 1935. Embryogénie des *Rosacées*. Développement de l'embryon chez le *Potentilla reptans* L. C. R. Acad. Sci. Paris 200: 1972–1974.

Amygdalaceae:- The fruits of various edible species of *Prunus* develop in three clearly defined stages. In Stage I there is a period of rapid development of the pericarp beginning at about the time of full bloom. State II is a period of retarded pericarp development during midseason. Stage III is a period of rapid pericarp development just prior to fruit ripening.

The length of each of these periods varies with the species, the variety of a given species, and the season. In the latitude of Geneva, New York, Stage I in *Prunus cerasus* has a duration of seventeen days; in *P. avium* it continues for twenty-one days, and in *P. persica* for forty-nine days. Stage II is directly correlated with the season of fruit ripening. Early-ripening varieties have a short period of retarded development, in contrast to a longer period for late-ripening varieties. Stage III is variable and continues until final ripening.

Development of the embryo also progresses in three distinct stages, but not necessarily parallel with those of the pericarp. While the pericarp, nucellus and integuments are in Stage I following full bloom, the embryo is in a state of arrested development which terminates abruptly in a rapid increase in growth at just the time that Stage II of development of the pericarp commences. The initiation of rapid increase of the embryo is further correlated with the initiation of hardening of the stony pericarp and with attainment of maximum size by the nucellus and integuments. In late-ripening varieties the rapid development of the embryo continues to maximum size during Stage II. Seed formation proceeds to maturity, and late-ripening varieties are characterized by viable seed. In the case of early-ripening varieties, on the other hand, Stage III of pericarp development begins before the embryo attains full size, and such varieties are characterized by numerous abortive or non-viable embryos.

In *Prunus persica* the embryo develops slowly during Stage I and is still "microscopic" fifty days after full bloom; at the beginning of Stage III it enters on a period of rapid development which continues until it attains maximum size. In the case of very early-ripening varieties, Stage III is initiated while the embryo is in this period of rapid development, but the embryo fails to reach maturity and aborts. In the case of late-ripening varieties, the beginning of Stage III does not occur until the embryo is nearly mature, and the seed is viable.

No detailed studies on the embryonomy of any species in the *Amygdalaceae* are available; it is said that in *Prunus persica* the zygote does not divide until about twelve days after fertilization and embryo development is extremely slow. A linear four-celled proembryo is produced, but nothing is said of its origin or subsequent development. Young and developing embryos of *P. avium* and *P. cerasus* are ellipsoid to pyriform in shape and show no suspensors and no resemblance to the Geum Variation so prevalent in the related *Rosaceae*.

BRADBURY, D. 1929. A comparative study of the developing and aborting fruits of *Prunus cerasus*. Amer. Jour. Bot. 16: 525–542.
HARROLD, T. J. 1935. Comparative study of the developing and aborting fruits of *Prunus persica*. Bot. Gaz. 96: 505–520.
TUKEY, H. B. 1933. Embryo abortion in early-ripening varieties of *Prunus avium*. Bot. Gaz. 94: 433–468.
TUKEY, H. B. 1936. Development of cherry and peach fruits as affected by destruction of the embryo. Bot. Gaz. 98: 1–24.
TUKEY, H. B., and F. A. LEE. 1937. Embryo abortion in the peach in relation to chemical composition and season of fruit ripening. Bot. Gaz. 98: 586–597.

Malaceae:- Development of the pericarp, nucellus and integuments, and embryo is similar in *Pyrus malus* to some details in the stages of development in similar parts in species of *Prunus*, described above, but here there are no such stages in the development of the fruit since it is accessory (stem) tissue.

TUKEY, H. B., and J. O. YOUNG. 1942. Gross morphology and histology of developing fruit of the apple. Bot. Gaz. 104: 3–25.

Leguminosales

Caesalpiniaceae:- *Caesalpinia mimosoides* resembles *Cercis siliquastrum* (*Mimosaceae*) in its embryogeny, but here the sequence of divisions is somewhat more regular.

GUIGNARD, L. 1881. Recherches sur l'embryogénie des *Légumineuses*. Ann. Sci. Nat., Bot. VI, 12: 5–166.

Mimosaceae:- In *Acacia baileyana* the zygote divides transversely, whereupon both basal and terminal cells divide longitudinally. The division in the basal cell is decidedly oblique and the larger of the two daughter cells retains its individuality, whether it divides again or not, until the cotyledonary protuberances appear. After the four-celled stage, the proembryo grows rather irregularly. There is no root cap. The best diagnosis that can be made for this species is that it follows the Trifolium Variation, Onagrad Type, save that a suspensor is not formed.

Proembryo development in *Acacia retinoides* agrees with that of *A. baileyana*, save that the larger of the two basal cells soon loses its identity. The embryo of *A. farnesiana* at the stage just before the cotyledons appear is very similar to the corresponding stages of *A. retinoides*, as are those of *Mimosa pudica* and *M. denhartii* at somewhat earlier stages.

Cercis siliquastrum differs somewhat from *Acacia* and *Mimosa* at the six-celled stage. Here there are three juxtaposed cells separated by vertical walls in the basal region and three similar cells in the terminal half. One of the walls in the latter region is oblique and is suggestive of the formation of an epiphysis initial. This cell preserves its identity for several subsequent divisions but presently becomes indistinguishable. The basal cell forms a massive suspensor.

It is evident that there are two types of embryogeny in the *Mimosaceae*, one with and the other lacking a suspensor.

Guignard, L. 1881. Recherches sur l'embryogénie des *Légumineuses*. Ann. Sci. Nat. Bot. VI, 12: 5–166.

Newman, I. V. 1934. Studies in the Australian Acacias IV. The life history of *Acacia baileyana* F. v. M., Part. 2. Gametophytes, fertilization, seed production and germination, and general conclusions. Proc. Linn. Soc. N. S. Wales 59: 277–313.

Papilionaceae (Fabaceae):- The embryonomy of species belonging to the *Papilionaceae* is perplexing; a great deal of careful study of all the genera is required before the situation can be fully elucidated. Both the Onagrad (Trifolium and Lotus Variations) and Caryophyllad (Medicago Variation) Types are represented, and it appears that species in the same genus may follow different Types (*Trifolium*). The family is further characterized by the greatest diversity in the structure and formation of the suspensor. Very often it is exceedingly difficult to determine whether the basal cell contributes to the construction of the embryo proper, so that the assignment of certain species can only be considered as provisional pending more critical examination.

A rough separation of the genera into seven groups, based on the presence or absence of a suspensor, and on its structure when present, was proposed long ago by Guignard (who also included species properly belonging to the *Mimosaceae*) and may be of some value in classifying embryological types:

1. Suspensor lacking: *Hedysarum coronarium* (and probably most species in the tribe *Hedysareae*).
2. Suspensor rudimentary, three- or four-celled: *Soja*, *Amphicarpaea*, *Trifolium* (some species, at least).
3. Suspensor composed of two pairs of cells lying irregularly crosswise one above the other. The micropylar pair are much lengthened, the other globose; both contain numerous coenocytic nuclei: most genera in the tribe *Viceae* (with the marked exception of *Cicer arietinum*), such as *Lathyrus*, *Pisum*, *Orobus* and *Vicia*.
4. Suspensor composed of a filament of a variable number of cells: *Ononis*.
5. Suspensor formed of a greater or lesser number of cell pairs, which may be arranged one over the other in the same plane: most species of *Lupinus*; or in regular alternation: *Cicer arietinum*.
6. Suspensor in the form of a lengthened cellular body whose elements are very similar to those of the embryo proper: *Medicago*, *Trigonella*; or very little differentiated: *Galega*, *Lotus*; or usually blended indistinguishably: *Phaseolus*.
7. Suspensor in the form of a pyriform or globular cellular body whose cells differ from those of the body of the embryo in form, number, plane of division and contents: *Trifolium minus*.

However, the most logical procedure is to classify the embryos according to the Type and Variation followed, describing the most completely known examples first, then citing those species which appear to conform to the same embryonomic laws.

Lotus Variation, Onagrad Type. — *Lotus corniculatus* in the earlier proembryonic stages follows the typical Onagrad Type of development but after the four-celled stage it is almost impossible to determine the sequences of division with certainty. There are certain resemblances to the Myosurus Variation; and again, there are other points of resemblance to the Datura Variation of the Solanad Type.

A fairly large suspensor composed of very irregularly shaped cells is produced; it is almost indistinguishably connected to the derivatives of cell *m* of the tetrad embryo, which have the appearance of being a stalace.

Trifolium Variation, Onagrad Type. — Whereas the Trifolium Variation appears to be well grounded in fact, the assignment to the Type is somewhat problematical. The Variation is based upon the embryonomy of *T. minus*, but the assignment to the Onagrad Type is based upon other species and upon the eventual destination of each cell of the four-celled proembryo.

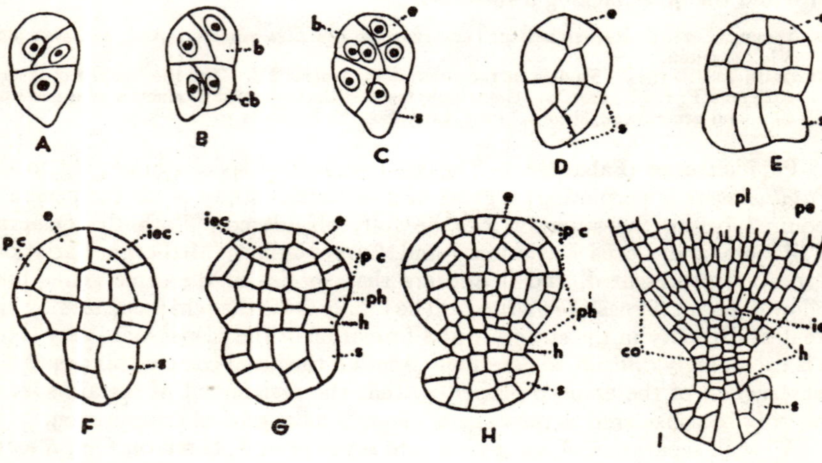

FIG. 51. — *Trifolium minus.* Description in text. A–G × 500; H × 350; I × 180. (*Redrawn from* SOUÈGES 1927).

In *Trifolium minus* the terminal cell of the two-celled proembryo divides by a vertically oblique wall to form two dissimilar juxtaposed cells, *a* and *b* (FIG. 51A). Shortly afterwards the basal cell is divided by an equally oblique wall into two juxtaposed cells (FIG. 51B). The succeeding walls in the daughter cells of the basal cell are erected in very variable planes; they result in the formation of a globular or oblate suspensor and do not contribute to the construction of the embryo proper.

Cell *a* is bisected by an oblique wall oriented perpendicular to the first wall, and cell *b* by a median, more nearly vertical wall (FIG. 51C). The upper daughter cell resulting from the division of cell *a* is the epiphysis initial (*e*), identical with that of *Geum* and other species. The three cells below the epiphysis initial divide as a general rule by vertical walls and presently six cells are recognizable in this horizontal layer (FIG. 51D, E). Each of these cells becomes partitioned transversely, with the result that two distinct tiers of cells are formed. Soon the elements of the upper tier (*pc*), situated directly beneath the epiphysis, commence to divide either vertically or tangentially; while the elements of the lower tier (*pc'*), situated directly above the suspensor, become segmented transversely, and give rise to two new tiers, *ph* and *h* (FIG. 51F, G).

The epiphysis initial divides by a vertical wall into two juxtaposed cells, which in turn are partitioned by new vertical walls, then by transverse walls, thus giving rise to two superposed tetrads of cells (FIG. 51F, G). The upper tetrad is considered as representing epidermal initials, while the lower tetrad represents the initials of the cortex of the stem apex (or, in other words, the tunica and corpus initials, respectively). The cells of the two tetrad groups multiply but little during the earlier growth of the embryo; at the time of origin of the cotyledons they still form a distinct group but shortly afterwards the distinction is completely effaced.

The cell tier, *pc*, immediately below the epiphyseal region, becomes extended to the right and left by both vertical and periclinal walls (FIG. 51D–G). The central cells thus resulting become part of the central cylinder of the stem, while the lateral derivatives contribute to the development of the cotyledons. All the elements of the subepiphyseal layer, *pc*, therefore, represent the cotyledonary portion of the embryo.

The tier *ph*, on the contrary, corresponds to the hypocotyledonary portion of the embryo. The transverse walls produced in both the internal and peripheral regions of this tier form two new layers. A third is soon developed toward the upper portion adjoining the cotyledonary region (FIG. 51H). At the same time vertical or tangential partitions are established throughout the parts corresponding to the hypocotyledonary region, which thereby becomes massive. It is only at the time of origin of the cotyledons that the epidermal initials are definitely developed; shortly after, the periblem and plerome are in turn differentiated. At this moment the equatorial plane which separates the cotyledonary and hypocotyledonary regions disappears (FIG. 51I).

The tier *h* adjoining the suspensor proper gives rise to the hypophyseal region. There is a certain degree of regularity in the segmentation of this tier; when first recognizable, it consists of four cells, two peripheral and two central (FIG. 51G). Each becomes divided transversely to give rise to two new tiers (FIG. 51H). By successive transversal segmentations, the lower tier gives rise to four longitudinal series of elements which are always distinct. In the upper layer the peripheral elements are segmented tangentially and form the lateral portion of the root cap; the circumaxial elements, after several transverse divisions precede the longitudinal ones, engender the initials of the root cortex toward the upper end and the initials of the calyptrogen toward the base (FIG. 51I).

Other genera which probably conform to the Trifolium Variation include the following: *Spartium*, *Cytisus*, *Genista*, *Crotalaria*, *Ononis* (?), *Tetragonolobus*, *Phaseolus* and *Onobrychis*. Some of these may be discussed further.

Trifolium pratense, *T. hybridum*, *T. resupinatum*, *T. repens*, *Spartium junceum*, *Cytisus laburnum*, *C. alschingeri*, *Tetragonolobus purpureus*, *Onobrychis petrea* and *Phaseolus multiflorus* are among the species that have been studied. In *Anthyllis tetraphylla* the early proembryonic development is like that in *Trifolium minus*. After the first longitudinal division of the basal cell of the two-celled proembryo, a second vertical wall is erected in each daughter cell at right angles to the first wall, whereupon all four cells are segmented transversely. None of these eight cells divides further, as a rule, but each grows in an outward direction to become a sort of inflated vesicle. Together they occupy the entire space at the micropylar end of the embryo sac. The embryo proper begins as a single cell which first divides in the longitudinal plane, then again in the same plane but with the walls oriented perpendicular to the first one and next all four quadrant cells are segmented transversely. Certain species of *Ononis* appear to follow the Trifolium Variation, others the Medicago Variation.

Medicago Variation, Caryophyllad Type. — In *Medicago lupulina* the terminal cell of the two-celled proembryo divides transversely into two superposed cells, *cc* and *cd*, while the basal cell, *cb*, generally remains undivided (FIG. 52A). If it should divide it does not produce more than a few cells which are attached to those of the suspensor. This behavior therefore indicates that this species belongs under the Caryophyllad Type; there are many points of resemblance to the Sagina Variation, but the differences are such that another Variation is necessary.

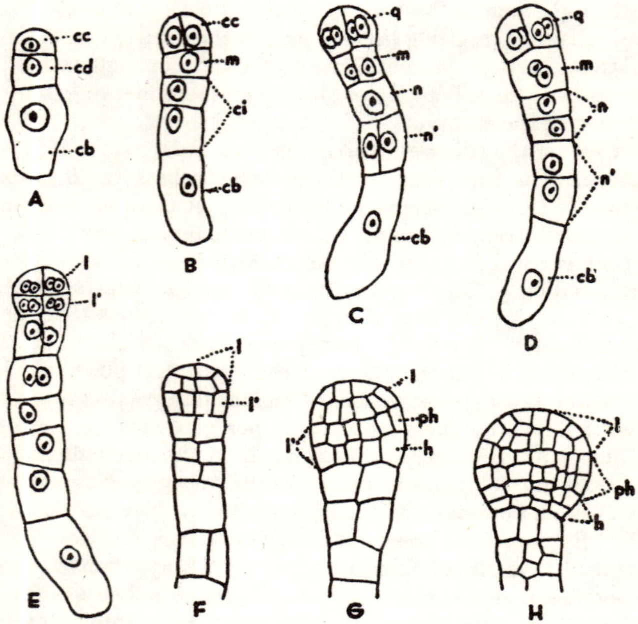

FIG. 52. — *Medicago lupulina.* Description in text. A–C × 390; F–H × 330. (*Redrawn from* SOUÈGES 1927).

Cell *cc* of the three-celled proembryo divides vertically and its sister cell, *cd*, horizontally (FIG. 52B). The two upper cells next are segmented vertically, to form quadrants, then each quadrant is divided transversely to engender eight octant cells (FIG. 52C–E). There is little regularity in the divisions that occur in cell *m*; it most often divides vertically (FIG. 52C–E), but it may divide transversely. Cell *ci* generally divides transversely into two superposed elements, *n* and *n'* (FIG. 52C). All the elements issuing from *m*, *ci* and *cb* give rise to a large, elongated suspensor composed of cells whose form and dimensions clearly distinguish them from those of the embryo proper.

The four upper octants sometimes segment by means of vertical walls perpendicular to the meridional walls, but most often by horizontal walls parallel to the equatorial wall. It is extremely difficult to determine the order and orientation of the succeeding walls. All these divisions eventually result in the formation of three or four layers of cells of which the outermost are differentiated as the epidermis. Later the cotyledons and stem apex originate from this hemispherical mass.

In the four lower octants (l') the first walls are perpendicular to the meridional walls, then transverse walls appear so that the region consists of two tiers of cells (*ph* and *h*, FIG. 52E–G). The upper tier *ph* represents the hypocotyl; by vertical, then transverse segmentations, the limits of the three histogens are tardily distinguishable. The lower tier *h* corresponds to the hypophyseal region; it engenders the initials of the root cortex and the central portion of the root cap (FIG. 52G, H). Development of regions *ph* and *h* is much like that in *Trifolium minus*.

Orobus angustifolius and *O. aureus* (possibly both are species of *Lathyrus*?), *Pisum sativum*, *Lathyrus heterophyllus*, *L. odoratus*, *L. aphaca*, *Ervum ervilia*, *Vicia narbonnensis* and *V. americana* — in fact, probably all species assigned to the tribe *Vicieae* — form a most peculiar "middle piece" and suspensor, both with numerous coenocytic nuclei. The detailed account for *Pisum sativum* as given below will serve for all these species.

The zygote of *P. sativum* divides transversely, whereupon the basal cell is segmented longitudinally and the terminal cell transversely into *m* and *cc*. Cell *cc* is the equivalent of the original terminal cell and all of the embryo proper is derived from it; cell *m* divides longitudinally and becomes the "middle piece," while the two juxtaposed elements of the basal cell do not divide again. The first segmentation in cell *cc* is longitudinal; the orientation of this wall is at right angles to that in cell *m* below. The nuclei in the two juxtaposed basal cells multiply at a rapid rate without cytokinesis, at the same time that these cells elongate and broaden. The nuclei in the "middle piece" likewise proliferate but not to such an extent as do those in the two suspensor cells. The total number of nuclei in each of the two suspensor cells of *P. sativum* eventually reaches 64, while each cell of the "middle piece" has half that number.

Growth of the embryo proper is rather slow. After the first division in cell *cc*, each derivative is segmented transversely, but it is said that in *P. sativum* the derivatives divide by longitudinal walls perpendicular to the first wall and then all four quadrants are segmented transversely to produce octants. It is entirely conceivable that there may be variations in the sequence of divisions in the two cells derived from *cc*, even in the same species.

The following species also conform to the Medicago Variation: *Melilotus alba*, *M. arvensis*, *M. officinalis*, *Vicia sepium*, *Orobus tuberosus*, *Lathyrus* spp.; *Lupinus* spp.; *Galega* spp.; *Hedysarum nutans*, *Erythrina cristagalli*, *Medicago sativa*, *M. falcata*.

Ervum hirsutum represents a transition form between the Medicago and Myriophyllum Variations, in that development in the derivatives of cell *ca* develop as in the former, while cell *cb* is segmented longitudinally as in the latter, but the cells do not become appreciably inflated. *Orobus vernus*, unlike *O. tuberosus*, conforms very closely to the Fumaria Variation, Caryophyllad Type.

In *Lupinus polyphyllus* the earlier proembryonic development has been described and figured, but matters are a little puzzling. At the eight-celled stage, the embryo proper consists of two juxtaposed and very small cells, with six globular and more or less detached cells below. Apparently these cells are part of the suspensor, but it is difficult to conceive of their assuming a globular form and then becoming more or less separated. A similar situation is found in *L. hartwegi* and *L. varius*, while in *L. mutabilis*, *L. cruikshankii* and *L. truncatus* there is an even greater number of globular cells, and the extreme development occurs in *L. succulentus* in which fourteen pairs of superposed cells were found. In almost all species the embryo is pushed around the side of the embryo sac cavity until it reaches a position near the chalaza.

COOK, M. T. 1924. Development of seed of *Crotalaria sagittalis*. Bot. Gaz. 77: 440–445.
COOPER, D. C. 1933. Macrosporogenesis and embryology of *Melilotus*. Bot. Gaz. 95: 143–155.
COOPER, D. C. 1935. Macrosporogenesis and embryology of *Medicago*. Jour. Agr. Res. 51: 471–477.
COOPER, D. C. 1938. Embryology of *Pisum sativum*. Bot. Gaz. 100: 123–132.
GUIGNARD, L. 1881. Recherches sur l'embryogénie des *Légumineuses*. Ann. Sci. Nat., Bot., VI, 12: 5–166.
MARTIN, J. N. 1914. Comparative morphology of some *Leguminosae*. Bot. Gaz. 58: 154–167.
SOUÈGES, R. 1927. Embryogénie des *Légumineuses*. Développement du proembryon chez le *Trifolium minus* Rehl. C. R. Acad. Sci. Paris 184: 1018.
SOUÈGES, R. 1927. Embryogénie des *Légumineuses*. Les derniers stades du développement de l'embryon chez le *Trifolium minus* Rehl. C. R. Acad. Sci. Paris 184: 1196.
SOUÈGES, R. 1927. Embryogénie des *Légumineuses*. Les premiers stades du développement de l'embryon chez le *Medicago lupulina* L. C. R. Acad. Sci. Paris 185: 1062–1064.
SOUÈGES, R. 1927. Embryogénie des *Légumineuses*. Les derniers stades du développement de l'embryon chez le *Medicago lupulina* L. C. R. Acad. Sci. Paris 185: 1206–1208.
SOUÈGES, R. 1929. Recherches sur l'embryogénie des *Légumineuses*. Bull. Soc. Bot. France 76: 93–112; 338–346; 527–540.
SOUÈGES, R. 1946. Embryogénie des *Papilionacées*. Développement de l'embryon chez le *Melilotus arvensis* Vallr. C. R. Acad. Sci. Paris 222: 1361–1363.
SOUÈGES, R. 1946. Embryogénie des *Papilionacées*. Développement de l'embryon chez l'*Orobus vernus* L. (*Lathyrus vernus* Bernh.). C. R. Acad. Sci. Paris 223: 60–62.
SOUÈGES, R. 1946. Embryogénie des *Papilionacées*. Développement de l'embryon chez le *Vicia sepium* L. C. R. Acad. Sci. Paris 223: 389–391.
SOUÈGES, R. 1946. Embryogénie des *Papilionacées*. Développement de l'embryon chez l'*Orobus tuberosus* L. (*Lathyrus macrorrhizus* Wimm.). C. R. Acad. Sci. Paris 223: 493–495.
SOUÈGES, R. 1946. Embryogénie des *Papilionacées*. Développement de l'embryon chez l'*Ervum hirsutum* L. (*Vicia hirsuta* S. F. Gray). C. R. Acad. Sci. Paris 223: 838–840.

Hamamelidales

The *Bruniaceae*, *Eucommiaceae* and *Myrothamnaceae* have never been investigated embryologically, while in the *Hamamelidaceae*, *Buxaceae* and *Platanaceae* the available information is very scanty and of practically no service in determining assignments of the species concerned.

Hamamelidaceae:- *Hamamelis virginiana* is said to produce a linear proembryo of five or six cells; there are two enlarged terminal cells and three or four small flattened cells in between. Nothing is said of their origin. The embryonic terminal cell is next segmented vertically.

SHOEMAKER, D. N. 1905. On the development of *Hamamelis virginiana*. Bot. Gaz. 39: 248–266.

Buxaceae:- *Buxus sempervirens* conforms to the Onagrad Type, but the illustrations do not bear out the investigator's opinion that it follows the Capsella Variation. *Simmondsia californica* forms a slightly irregular embryo; while it may also conform to the Onagrad Type, this is doubtful since sufficiently early stages were not available.

WIGER, J. 1935. Embryological studies on the families *Buxaceae*, *Meliaceae*, *Simarubaceae* and *Burseraceae*. Thesis, Lund.

Platanaceae:- All that has been described of the embryonomy of *Platanus acerifolia* (?) is that a linear tetrad is produced.

BRETZLER, E. 1924. Beiträge zur Kenntnis der Gattung *Platanus*. Bot. Archiv 7: 388–417.

Salicales

Salicaceae:- The embryonomy of *Salix triandra* agrees with the Polygonum Variation, Asterad Type. The following other species also appear to agree with the same Variation: *S. petiolaris*, *S. glaucophylla*, *S. cordata* and *S. tristis*.

CHAMBERLAIN, C. J. 1897. Contribution to the life history of *Salix*. Bot. Gaz. 23: 147–179.
SOUÈGES, R. 1923. Embryogénie des *Salicacées*. Développement de l'embryon chez le *Salix triandra* L. C. R. Acad. Sci. Paris 177: 1234–1237.

Garryales

Garryaceae:- The zygote of *Garrya elliptica* becomes an elongated tubular structure, then a transverse wall cuts off a cell at the outer tip and two additional walls are erected in the same plane and close to the first wall. Unfortunately, however, it is not at all clear in which cells the latter two walls are developed; hence it can only be surmised that the terminal cell segments transversely. The lower half of the long suspensor region then divides repeatedly by approximately transverse walls. From ten to fourteen cells are produced. Later those cells lying nearest the embryonal end, in addition to those in the center of the filamentous row, divide longitudinally. Still later longitudinal segmentations may occur in any of the cells toward the embryonal end of the suspensor, so that the suspensor region ultimately consists of two definite parts: (*1*) the distal half, which is a long tube with the end next to the micropyle slightly swollen, and (*2*) the proximal half, which is a cellular filament. Each segment of this filament consists of one or two cells or a plate of four cells. The suspensor henceforth remains in approximately this condition. The three cells of the proembryo proper each segments lengthwise to form three horizontal tiers of two cells each. Almost immediately a second longitudinal wall is erected in each daughter cell, in a plane perpendicular to that of the preceding wall, so that each tier now consists of four cells. The cells of the terminal tier are then segmented by periclinal walls to delimit exteriorly the epidermal initials. The cells of the tier adjacent to the suspensor do not undergo a corresponding segmentation.

That *G. elliptica* is assignable to the Solanad Type is evident, but so many essential points are obscure that the classification cannot be carried further. However, there are certain resemblances to the Sherardia Variation. It is noteworthy that a hypophysis is apparently not developed.

HALLOCK, F. A. 1930. The relationships of *Garrya*. The development of the flowers and seeds of *Garrya* and its bearing on the phylogenetic position of the genus. Ann. Bot. 44: 771–812.

Leitneriales

Leitneriaceae:- The first wall formed in the zygote of *Leitneria floridana* may be either transverse or longitudinal. Divisions follow in rapid order, but there is said to be no fixed sequence. The information is so scanty that it is impossible to assign this species with certainty; if the longitudinal segmentation of the zygote is a regular occurrence, disposition would be under the Piperad Type.

PFEIFFER, W. M. 1912. The morphology of *Leitneria floridana*. Bot. Gaz. 53: 189–203.

Myricales

The single family, *Myricaceae*, is embryologically unknown.

Balanopsidales

The New Caledonian family *Balanopsidaceae* has never been investigated.

Fagales

There are no reports for two of the families — *Betulaceae* and *Corylaceae*.

Fagaceae:- The four-celled proembryo of *Quercus velutina* is depicted as having a large basal cell and three juxtaposed cells above it, the latter presumably being derived from the terminal cell. At about the time the epidermal initials are being delimited, the basal cell divides once vertically; the suspensor remains two-celled at all stages of further growth of the embryo. Subsequent divisions in the upper cells of the proembryo are extremely irregular.

Q. velutina clearly belongs under the Onagrad Type. Since there is no perceivable hypophysis, and as the behavior of the terminal cell is very similar to that in *Trifolium minus*, it is probably best placed in the Trifolium Variation.

CONRAD, A. H. 1900. A contribution to the life history of *Quercus*. Bot. Gaz. 29: 408–418.

Casuarinales

Casuarinaceae:- Interpretation of the embryonomy of *Casuarina equisetifolia* depends upon whether or not a hypophysis occurs; in the original account this important feature is not mentioned. One figure, however, indicates the presence of a hypophysis initial which segments vertically; on this basis, therefore, assignment is definitely to the Alyssum Variation, Onagrad Type.

SWAMY, B. G. L. 1944. A preliminary note on the embryology of *Casuarina equisetifolia* Forst. Proc. Indian Acad. Sci. 20, B: 187–191.

Urticales

The uninvestigated families include the *Barbeyaceae*, *Scyphostegiaceae* and *Cannabinaceae*.

Ulmaceae:- The terminal cell of the two-celled proembryo in both *Ulmus americana* and *U. fulva* divides transversely. The basal cell apparently does not divide again. The upper daughter of the terminal cell next divides longitudinally and is followed by a transverse segmentation in the cell next below. Both species certainly belong in the Solanad Type, and probably to the Nicotiana Variation.

In *Holoptelea integrifolia* the basal cell divides transversely and the terminal cell longitudinally. Quadrants and octants are next produced in the typical Onagrad Type fashion. There is no indication of the formation of a hypophysis.

CAPOOR, S. P. 1937. The life history of *Holoptelea integrifolia* Planch. Beih. Bot. Centralbl. 57: 233–249.
SHATTUCK, C. H. 1905. A morphological study of *Ulmus americana*. Bot. Gaz. 40: 209–223.
WALKER, R. I. 1938. Macrosporogenesis and embryo development in *Ulmus fulva*. Bot. Gaz. 99: 592–598.

Moraceae:- The proembryo of *Ficus carica* is a linear tetrad; subsequent development is stated to resemble that of *Urtica pilulifera*, but since the terminal cell divides transversely (in contrast to the longitudinal segmentation in the latter species), this is hardly correct.

The description of the embryogeny of *Artocarpus integrifolia* is confusing. To judge from the illustrations, some of which represent oblique sections, it is clear that the basal cell divides transversely and the terminal cell longitudinally. Subsequent events render it probable that this species conforms to the Urtica Variation, Asterad Type.

The proembryonic tetrad of *Streblus asper* is similar to that of the preceding species, but the structure of the indistinct suspensor precludes assigning *S. asper* to the Urtica Variation, although it certainly belongs under the Asterad Type.

RAO, A. M. S. 1940. Cytology and embryogeny in *Artocarpus integrifolia* Linn. Jour. Mysore Univ. 2, B: 63–73.
RAU, M. A. 1942. Development of the embryo-sac and embryo in *Streblus asper* Lour. Jour. Mysore Univ. 2, B: 109–114.
CONDIT, I. J. 1931. The structure and development of flowers in *Ficus carica* L. Diss., Stanford Univ.

Urticaceae:- The embryogeny of *Urtica pilulifera* agrees very closely with that of the Asterad Type. The differences are all secondary: (*1*) two supepidermal layers are differentiated in the cotyledonary region; (*2*) two superposed groups of four initials are formed at the radicular summit of the cortex; (*3*) the suspensor consists of a single vesicular cell or a small number of such cells, and (*4*) the form of the cotyledons is sharply obcordate at the base.

The four-celled proembryo is formed exactly as in the Onagrad Type (FIG. 53A), but the destination of each cell is different: the two upper cells engender the cotyledonary portion, the intermediate cell *m* gives rise to the hypocotyledonary axis, and from the derivatives of the basal cell *ci* are developed the initials of the cortex, the central portion of the root cap and the suspensor. The eight-celled proembryo arises as a result of the division of each of the four tetrad cells. The lowest cell, *ci*, divides transversely into *n* and *n'*; the element *m* divides meridionally into two juxtaposed cells; and each of the two upper cells is divided vertically to engender four circumaxial cells which are customarily called quadrants (FIG. 53B). This eight-celled proembryo closely resembles that of the *Ranunculaceae.*

The sixteen-celled proembryo arises by the bipartitioning of each cell of the eight-celled proembryo. At this period of development important variations in the planes of cell division occur. The three elements of the three lower tiers, *m*, *n* and *n'*, are segmented in the ordinary manner: cell *n'* is divided transversely into *o* and *p* (FIG. 53C, D); the element *n* is divided meridionally into two juxtaposed cells, and the two cells of tier *m* are divided by vertical walls to engender four circumaxial cells.

The four elements of the upper layer, now designated as *q*, are divided not by transverse but by oblique walls: each wall has its outer insertion in the center of the peripheral wall and the inner insertion at a point about three-fourths of the way in on the horizontal wall which separates the two tiers *q* and *m* (FIG. 53 C). This wall isolates two cells which present different aspects when viewed in median longitudinal sections of the proembryo. The one adjacent to the axis is generally the larger of the two and appears essentially quadrilateral, while the other, outer one is triangular. These cells may be designated as *a* and *b* respectively (FIG. 53C). Tier *q* now consists of four elements *a* and four elements *b*. In element *a* the wall next to be formed is transverse: the outer daughter cell is an epidermal initial, which thereafter multiplies in the ordinary fashion; the inner daughter cell later divides transversely (as a general rule), thus producing two superposed subepidermal elements, *e* and *i*, which later undergo anticlinal cytokinesis and give rise to two layers which are prominent in the embryo at the moment of origin of the cotyledonary protuberances (FIG. 53D, E). In other instances, the internal cell arising from *a* is segmented vertically several times before transverse walls appear. Finally, the wall in *a* may be oblique, separating two very unequal cells: one is smaller and triangular in outline and the other is pentagonal. The latter, situated adjacent to the axis, is segmented horizontally only to give two superposed elements. The triangular cell may be considered the mother cell of the row from which is derived the internal cortical meristems of the cotyledon; it may be designated as *mc′* (FIG. 53F). In effect, it is the derivatives of *mc′* which develop, by means of anticlinal divisions, the entire upper or internal

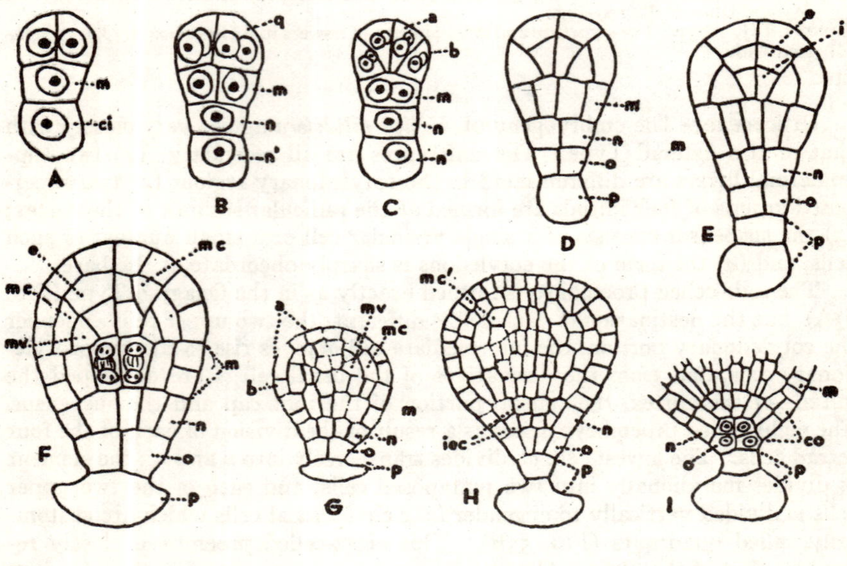

FIG. 53. — *Urtica pilulifera*. Description in text. A–C × 670; D–F × 420; G–I × 180. (*Redrawn from* SOUÈGES 1921).

cortical layers of the young cotyledon; while the derivatives of the two superposed cells adjoining the axis, solely by means of anticlinal divisions, differentiate the two subepidermal layers which occupy the intercotyledonary depression before theformation of the first leaves or of the stem apex (FIG. 53G, H).

In the element *b* the first wall is radially vertical, and the wall in each daughter cell is also vertical (FIG. 53D, E). The outer cell thus formed is an epidermal initial, while the inner cell has the same triangular outline of the original cell *b* and becomes divided trangentially (FIG. 53F). The inner daughter cell represents a plerome initial and gives rise to the vascular meristems of the future cotyledon; its sister cell is a periblem initial and contributes to the formation of the external cortical meristems of the same organ. These cells may be respectively designated as *mv* and *mc* (FIG. 53F–H). It is very difficult to describe with precision the order of divisions in these two cells.

In the sixteen-celled proembryo tier *m* consists of four cells exactly alike

(FIG. 53D). Each divides by a tangential wall; the resulting external daughter cell is an epidermal initial and in each of the internal daughter cells a wall which is usually tangential differentiates periblem and plerome initials (FIG. 53F, G). Sometimes the wall in the internal daughter cell is parallel to the meridional wall and the separation of the two histogens is not accomplished until after the erection of a new wall perpendicular to the preceding one in the larger of the two cells thus formed. The epidermal initials subsequently multiply in the ordinary manner by means of radial, vertical and horizontal segmentations. Shortly after the cotyledons have originated the elements of the epidermis adjoining tier *n* are divided tangentially and contribute to the formation of the most elongated portion of the root cap (FIG. 53I).

The tier *n* is composed of two elements in the sixteen-celled proembryo: both divide longitudinally to give four circumaxial cells, and the latter in turn are divided either tangentially or parallel to one of the two meridional walls (FIG. 53D–G). In transverse section these cells give the appearance of being in two concentric layers. The four elements of the peripheral layer are divided by vertically radial walls; finally there is established, either by horizontal walls which give rise to two superposed cells in which tangential walls are presently erected, or directly by means of tangential walls, two elements of the root cap. The four internal elements are again divided in the vertical plane, either tangentially, perpendicularly to the meridional walls, or, in an exceptional manner, obliquely. In short, there is differentiated in this region two series of cells: the one internal and circumaxial, and the others placed between the preceding ones and the peripheral elements (FIG. 53H). In these two series of cells horizontal walls differentiate two very distinct layers of four upper elements and four lower elements (FIG. 53I). These eight elements may be regarded as the initials of the root cortex (*co*).

The single cell representing tier *o* in the sixteen-celled proembryo is segmented vertically to give two juxtaposed cells; these in turn divide by similar meridional walls (FIG. 53D–H). These four cells may divide in either of two ways: as a general rule, they divide horizontally to give rise to two groups of four superposed elements (FIG. 53I); less frequently, they may be segmented longitudinally to engender a cellular plateau of four central elements and eight peripheral elements. The four central cells are next segmented horizontally. The two cell layers thus differentiated give rise to the central portion of the root cap. The elements of the lower layer multiply by anticlinal segmentations and constitute the most external row of the root cap; the elements of the upper layer may undergo one or two vertical partitionings but generally are segmented by horizontally tangential walls to form an apparent third layer of the cap. These tangential divisions characterize the mode of functioning of the calyptrogen.

It is thus evident that the root cap possesses a triple origin since three regions of the proembryo concur in its formation: region *o* in its entirety forms the central portion; the peripheral cells of tier *n* and several elements of the epidermal cells of tier *m* form the lateral portions. The origin of this structure is the same as in *Senecio vulgaris*.

Tier *p* gives rise to the suspensor (FIG. 53D *et seq.*). This single cell may become transformed directly into a large vesicle which may be either flattened or elongated, or it may divide transversely into two cells. The upper cell remains narrow and presents the aspect of a short pedicel connecting the embryo to the large vesicular lower cell.

The embryogeny of *Urtica cannabina*, *U. dioica*, *Laportea moroides* and *Fleurya aestuans* is exactly like that of *Urtica pilulifera*.

Reduction divisions do not occur in the megasporocyte of *Elatostemma sessile*; the egg develops into a diploid embryo without fertilization. *E. acuminata* fluctuates in its behavior: it may develop normally, in which case its embryonomy is similar to that of *Urtica pilulifera*; or it may behave much like *E. sessile*. *E. pedunculosum* is claimed to follow the Urtica Variation, although the evidence is somewhat inconclusive; in one instance two three-celled proembryos were observed in the same embryo sac but their origin was not explained.

FAGERLIND, F. 1944. Die Samenbildung und die Zytologie bei agamospermischen und sexuellen Arten von *Elatostema* und einigen nahestehenden Gattungen nebst Beleuchtung einiger damit zusammenhangender Probleme. Kungl. Svenska Vetenskap. Handl. III, Bd. 21, Nr. 4.

MODILEWSKI, J. 1908. Zur Samenentwicklung einiger Urticifloren. Flora 98: 423–470.

SOUÈGES, R. 1920. Embryogénie des *Urticacées*. Développement de l'embryon chez l'*Urtica pilulifera* L. C. R. Acad. Sci. Paris 171: 1009–1011.

SOUÈGES, R. 1921. Développement de l'embryon chez l'*Urtica pilulifera*. Bull. Soc. Bot. France 68: 172–188, 280–294.

STRASBURGER, E. 1910. Sexuelle und apogame Fortpflanzung bei *Urticaceen*. Jahrb. Wiss. Bot. 47: 245–288.

Celastrales

Only one of the eleven families (*Aquifoliaceae*, *Empetraceae*, *Corynocarpaceae*, *Cyrillaceae*, *Cneoraceae*, *Pandaceae*, *Hippocrataceae*, *Icacinaceae*, *Salvadoraceae*, *Stackhousiaceae*) has been studied embryologically.

Celastraceae:- In *Evonymus europaeus*, the terminal cell of the two-celled proembryo is segmented transversely, and each daughter cell in turn is divided transversely so that a row of four superposed cells is attached to the single basal cell. Longitudinal divisions begin first in the outermost cell and progress successively to that next to the basal cell. Apparently the basal cell contributes nothing to the structure of the embryo proper. This species appears to follow a typical Sagina Variation, Caryophyllad Type, course of development. *Celastrus scandens* seems to develop similarly.

ANDERSON, A. 1931. Studien über die Embryologie der Familien *Celastraceae*, *Oleaceae* und *Apocynaceae*. Lunds Univ. Årsskr., N. V. Avd. 2, 27: 1–110.

Olacales

Neither family of the two included in the order — *Olacaceae* and *Opiliaceae* — has ever been studied.

Santalales

The uninvestigated families include the *Octoknemataceae*, *Grubbiaceae* and *Myzodendraceae*. Conditions in the *Loranthaceae* are extremely puzzling (see the chapter on apomictic embryogeny).

Loranthaceae:- The *Loranthaceae*, on the whole, have not been investigated with sufficient thoroughness, but the main details of proembryo development are ascertainable. In general, the zygote elongates considerably before dividing in the strictly longitudinal plane. The two-celled proembryo then elongates still further, whereupon several transverse walls are erected to form from four to eight tiers of two cells each. All cells except the two terminal ones constitute the suspensor. In some species there are several megagametophytes, almost all of which may be fertilized and a number of proembryos are thus produced but only one survives to maturity. In most of the older accounts the two terminal cells are depicted as being separated, but this is probably a fixation artifact. Their further development has not been carefully described. In a few of the species the suspensor remains straight (*Scurrula atropurpurea*); in most it becomes considerably twisted.

The older accounts concern species in the genus *Loranthus* before it was segregated into other genera. To give the modern designations, the species investigated were: *Dendrophthoe falcata* (*L. bicolor*), *D. pentandra* (*L. pentandrus*), *Scurrula parasitica* (*L. scurrula*) and *Macroselen cochinchinensis* (*L. globosus* and *L. sphaerocarpus*). The genus *Loranthus* as now constituted remains uninvestigated. Other species that have been studied include *Viscum articulatum* (in which the zygote elongates considerably but segments transversely about thrice near the terminal end) and *Scurrula atropurpurea* (in which the zygote presumably divides longitudinally, but the cells adjacent to the proembryo proper segment longitudinally rather than transversely).

All species in which the zygote segments longitudinally are assigned to the Dendrophthoe Variation, Piperad Type.

The mature embryo of *Korthalsella opuntia* and *Ginalloa linearis* is said to possess two cotyledons and to lack a suspensor. Their early embryogeny was not followed out adequately.

GRIFFITH, W. 1838. Notes on the development of the ovule of *Loranthus* and *Viscum*. Trans. Linn. Soc. London 18: 71–91.

GRIFFITH, W. 1844. On the ovulum of *Santalum*, *Osyris*, *Loranthus*, and *Viscum*. Trans. Linn. Soc. London 19: 171–214.

RAUCH, K. v. 1936. Cytologisch-embryologische Untersuchungen an *Scurrula atropurpurea* Dans. und *Dendrophthoe plutandra* Miq. Ber. Schwes. Bot. Ges. 45: 1–61.

RUTISHAUSER, A. 1937. Blütenmorphologische und embryologische Untersuchungen an dem Viscoideen *Korthalsella opuntia* Merr. und *Ginalloa linearis* Dans. Ber. Schwez. Bot. Ges. 47: 5–28.

TREUB, M. 1881. Observations sur les *Loranthacées*, I et II. Ann. Jard. Bot. Buitenzorg 2: 54–73.

TREUB, M. 1882. Observations sur les *Loranthacées*, III. Ann. Jard. Bot. Buitenzorg 3: 1–12.

Santalaceae:- The first division in the zygote of *Santalum album* separates a small terminal cell and a large basal cell. The next two walls are longitudinal in both terminal and basal cells. The two daughter cells of the terminal cell are next segmented transversely. Subsequent divisions follow no regular sequence. The segmentation of the suspensor cell is delayed; it appears that the basal cell contributes at least partially to the construction of the embryo proper. Tricotyledonous embryos are not exceptional in *S. album*.

Events appear to be much the same in *Thesium wightianum* and *Osyris arborea*. In these it is claimed that a longitudinal wall in the terminal cell of the two-celled proembryo is succeeded by a transverse wall, but it is more likely, to judge from the figures, that a second longitudinal wall is formed before the transverse segmentation occurs, as was described above for *Santalum album*.

The proembryo of *Thesium montanum* is depicted as consisting of a single basal cell and a tier of four circumaxial terminal cells.

The evidence for assignment of the *Santalaceae* is none too satisfactory.

IYENGAR, G. S. 1937. Life-history of *Santalum album* L. Jour. Indian Bot. Soc. 16: 175–195.

RAO, L. N. 1942. Studies in the *Santalaceae*. Ann. Bot., n. s. 6: 151–175.

SCHULLE, H. 1933. Zur Entwicklungsgeschichte von *Thesium montanum* Ehrh. Flora 127: 140–184.

Balanophoraceae:- Most of the species in the *Balanophoraceae* are probably apomictic, but there exists a certain degree of uncertainty (see further the chapter on apomictic embryogeny).

In *Blanophora dioica* syngamy was not actually observed, but there is no reason to doubt that it occurs. The egg is at first spheroidal but soon becomes oviform. The first division is longitudinal. No suspensor is formed and both cells go into the formation of the proembryo and embryo. Both cells are next segmented longitudinally to form four quadrants, following which each quadrant is divided periclinally. The formation of these peculiar walls is a most· unusual phenomenon, known for no other species. The points of insertion of the opposite ends of each wall are on the meridional wall, about one-fourth of the distance in from each end, then the wall curves outwardly to conform to the curvature of the peripheral wall. A large external cell and a small lenticular internal cell are thus formed in each quadrant cell. At the next division each of the larger external cells is divided transversely into two approximately equal sections. No further growth beyond this twelve-celled stage seems to occur until after germination commences. There is no perceptible differentiation of histogens, although the peripheral cells might be considered as being epidermal initials.

Assignment is to the Balanophora Variation, Piperad Type.

The parasitic *Cynomorium coccineum* has a four-celled proembryo in two tiers of two cells each, but the derivation has not been described. Apparently all the products of both cells go into the construction of the embryo.

EKAMBARAM, T., and R. R. PANJE. 1935. Contributions to our knowledge of *Balanophora*. II. Life-history of *B. dioica*. Proc. Indian Acad. Sci., B, 1: 522–543.
JUEL, H. O. 1903. Zur Entwicklungsgeschichte des Samens von *Cynomorium*. Beih. Bot. Centralbl. 13: 194–202.

Rhamnales

Only the *Rhamnaceae* have been studied thoroughly. The other families include the *Elaeagnaceae*, *Heteropyxidaceae* and *Ampelidaceae* (*Vitaceae*).

Rhamnaceae:- The embryonomy of *Ceanothus azureus* is very similar to that of the Geum Variation, Asterad Type. The differences are secondary and reside in the destinations of the two elements *n* and *n'* of the eight-celled proembryo. In *C. azureus*, cell *n* engenders the hypophysis cell and *n'* a very much reduced suspensor sensu stricto. In *Geum urbanum*, on the other hand, the hypophysis cell is generally produced by a daughter or granddaughter cell of *n* and all the other elements derived from both *n* and *n'* enter into the construction of a well developed suspensor.

Rhamnus frangula agrees very closely with *Ceanothus azureus*, but there are certain variations in the mode of partitioning of the basal cell.

Zizyphus jujuba and *Z. oenophia* appear to be identical with *Ceanothus azureus* embryologically.

SOUÈGES, R. 1939. Embryogénie des *Rhamnacées*. Développement de l'embryon chez le *Ceanothus azureus* Desf. C. R. Acad. Sci. Paris 208: 1673–1675.
SOUÈGES, R. 1941. Embryogénie des Rhamnacées. Développement de l'embryon chez le *Rhamnus frangula* L. C. R. Acad. Sci. Paris 213: 39–41.
SRINIVASACHAR, D. 1940. Embryological studies of some members of *Rhammaceae*. Proc. Indian Acad. Sci., B, 11:107–115.

Rutales

Information concerning the *Burseraceae* is too scanty to permit the formulation of conclusions.

Rutaceae:- The embryogeny of *Ruta graveolens* most closely resembles that of the *Ranunculaceae* (Onagrad Type), but there are several differences: the most significant reside in the inconstancy in the first segmentations, in the vertical disposition of the quadrants and in the clearer organization of the embryo at the radicular end.

The terminal cell is frequently divided by a wall that is more oblique than strictly vertical, and may also be divided by a transverse wall into two superposed elements. The basal cell often divides considerably in advance of the terminal cell. As a general rule, the middle cell *m* divides transversely into *d* and *f*, but it may divide meridionally into two juxtaposed cells, or even obliquely. As for the two derivatives of the original terminal cell: they are divided transversely, if they are juxtaposed, into four quadrant cells clearly arranged in a vertical plane; if they are superposed, they divide longitudinally to engender quadrants similarly arranged in a vertical plane; if they are separated by an oblique wall, one cell divides transversely and the other vertically. The four quadrants segment meridionally to give rise to four upper octants which represent the cotyledonary region and to four lower octants which correspond to the hypocotyledonary region. When the quadrants are arranged tetrahedrally, those which are superposed divide vertically and the others horizontally, in such a manner as to give two tiers of four octants, *l* and *l'*.

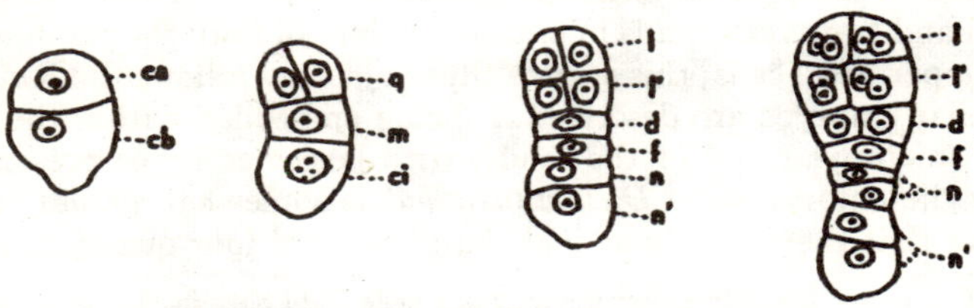

Recapitulatory Table for Ruta graveolens
(from left to right)

I. FIRST CELL GENERATION
Proembryo of two cells disposed in two tiers:

$ca = pco + pvt + phy + icc$
$cb = iec + ico + co + s$

II. SECOND CELL GENERATION
Proembryo of four cells disposed in three tiers:

$q = pco + pvt + phy + icc$
$m = iec + co + s$ (in part)
$ci = s$ (in part)

III. THIRD CELL GENERATION
Proembryo of eight cells disposed in six tiers:

$l = pco + pvt$
$l' = phy + icc$
$d = iec + co$
$f + n + n' = s$

IV. FOURTH CELL GENERATION
Proembryo of fourteen to sixteen cells disposed in six to eight tiers:

$l = pco + pvt$
$l' = phy + icc$
$d = iec + co$
$f + 2n + 2n' = s$

The hypophysis tissues originate from cell *m* when it divides vertically, or from the upper daughter cell, *d*, when it divides transversely.

The processes of embryogeny for *Ruta graveolens* may be recapitulated as shown in the above table.

Ruta patavina forms a proembryo more or less in the same fashion as in

R. graveolens, but it is destroyed through infection by a species of *Penicillium*.

In the genus *Citrus* fertilization is definitely known only for *C. trifoliata*. This occurs relatively late after pollination and the zygote remains dormant for about a month. The zygote, after growth commences, forms a linear proembryo of about six cells. To judge from the illustrations accompanying the original article, it is highly probable that the basal cell contributes to the construction of the embryo proper. It is therefore evident that this species does not conform to the Ruta Variation, but belongs under the Solanad Type and very likely to the Linum Variation. Nucellar embryony is common in this species. In *C. aurantium* and *C. nobilis* reproduction is solely by nucellar embryony.

Empleurum serrulatum is claimed to lack a suspensor but this is not borne out by the figures presented. Free-nucleate proembryos are claimed to occur in *Phellodendron sachalinense* and *Ruta patavina*; this is extremely doubtful. The proembryo of *Triphasia aurantiola* is depicted as consisting of a linear row of perhaps six cells and two terminal juxtaposed cells. A later stage shows longitudinal and transverse walls in opposite halves of both the terminal region and in the upper basal cells; this species thus evidently follows the Ruta Variation. Nucellar embryos are described as being embedded within the nucellar cells at the micropylar end of the ovule, with larger embryos presently projecting into the embryo sac. *Dictamnus albus* is somewhat similar to *T. aurantiola*, save that there is only a single basal cell and four quadrant cells.

CAPPELETTI, C. 1929. Sterilita di origine micotica nelle *Ruta patavina* L. Ann. di Bot. 18: 145-166.

MAURITZON, J. 1935. Über die Embryologie der Familie *Rutaceae*. Svensk Bot. Tidskr. 29: 319–347.

OSAWA, I. 1912. Cytological and experimental studies in *Citrus*. Jour. Coll. Agr. Imp. Univ. Tokyo 4: 83–116.

SOUÈGES, R. 1925. Embryogénie des *Rutacées*. Développement de l'embryon chez le *Ruta graveolens* L. C. R. Acad. Sci. Paris 180: 1957–1959.

SOUÈGES, R. 1926. Développement de l'embryon chez le *Ruta graveolens* L. Bull. Soc. Bot. France 73: 245–260.

Simarubaceae:- The terminal cell of the two-celled proembryo of *Suriana maritima* divides longitudinally, then the basal cell segments transversely. The next division is vertical in the upper daughter of the basal cell. Thereafter the divisions are irregular and presently a large spherical embryo is produced. A suspensor apparently is formed. It appears that the middle cell of the four-celled stage is a hypophysis initial. It is reasonably clear that this species belongs under the Onagrad Type, but the Variation to which it might be assigned can not be determined from the scanty evidence.

RAU, M. A. 1940. An embryological study of *Suriana maritima* Linn. Proc. Indian Acad. Sci., B, 11: 100–106.

Meliales

The information available on the *Meliaceae* is insufficient to permit conclusions.

Sapindales

Nothing is known for the following families: *Sapindaceae*, *Akaniaceae*, *Sabiaceae*, *Melianthaceae*, *Didieraceae*, *Picrodendraceae* and *Connaraceae*.

Staphyleaceae:- The assignment of *Staphylea trifoliata* to the Chenopodiad Type appears to be highly probable.

RIDDLE, L. C. 1905. Development of the embryosac and embryo of *Staphylea trifoliata*. Ohio Nat. 5: 320–325.

Aceraceae:- In *Acer* the first wall in the zygote is transverse, as is the second, then the third is longitudinal or oblique in the upper terminal cell and this is followed by a similar wall in the middle cell of the three-celled proembryo. The basal cell apparently divides once transversely and later may divide a few times to give rise to an irregular group of cells which form the suspensor. The pyriform developing embryo has its component cells arranged in an extremely irregular fashion. Bud embryos have occasionally been found close to the base of nearly mature embryos; their exact origin is unknown.

Disposition of this species appears to be best under the Trifolium Variation, Onagrad Type.

TAYLOR, W. R. 1920. A morphological and cytological study of reproduction in the genus *Acer*. Contrib. Bot. Lab. Univ. Penna. 5: 1–30.

Anacardiaceae:- Nothing which is of any service in determining the type of embryonomy in *Toxicodendron* (*Rhus*) *diversiloba* can be made from the sketchy account.

Anacardium occidentale is clearly referable to the Penaea Variation, Asterad Type. *Semecarpus anacardium*, on the other hand, closely conforms to the Euphorbia Variation, Onagrad Type, but there appears to be some evidence that an epiphysis initial is present (there are gaps in the series of illustrations). If an epiphysis actually occurs, the species should be transferred to the Trifolium Variation of the same Type.

COPELAND, H. F., and B. E. DOYEL. 1940. Some features of the structure of *Toxicodendron diversiloba*. Amer. Jour. Bot. 27: 932–939.
SRINIVASACHAR, D. 1940. Morphological studies in the family *Anacardiaceae*. Jour. Mysore Univ. 1, B: 83–91.

Juglandales

Nothing has been described for the *Julianiaceae*.

Juglandaceae:- In general, the embryonomy of *Carya glabra* and *Juglans mandschurica* conforms to the Geum Variation, Asterad Type, but there is some irregularity, particularly in the latter species. In *C. glabra* the terminal

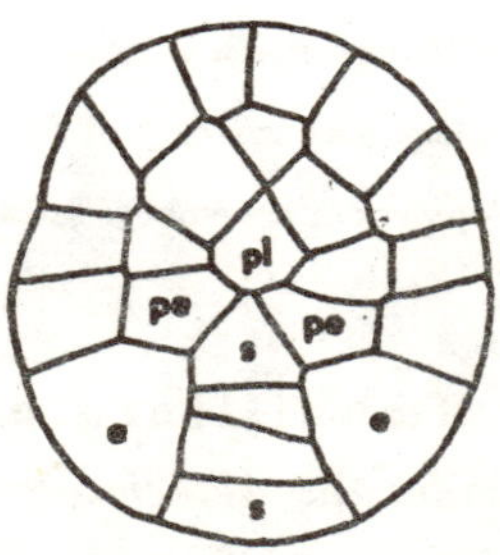

FIG. 54. — *Juglans regia*. Symbols: *s*, stalace; *pe*, periblem; *pl*, plerome; *e*, epidermal initials. Further description in text. (*Redrawn from* NEUMAN 1939).

cell of the two-celled proembryo divides diagonally into two unequal cells, *a* and *b*, exactly as in *Geum*, then the larger daughter cell divides diagonally to form an epiphysis initial. The destination of the proembryonic cells is the same as in *Geum*.

The earliest proembryonic development of *Juglans regia* has not been followed out; the youngest stage figured and described is that of a nearly globular, multicellular young embryo (FIG. 54). There is no suspensor, from which it may be concluded that all the products of both terminal and basal cells contribute to the construction of the embryo, but the precise contribution of each of these cells can not be determined from the figures or description. The distinguishing features of this embryo are the erection of a central column of cells in the lower (radicular) portion of the embryo, and the mode of partitioning of the epidermal initials in the lateral portions of the same region. The term "columella" has been applied to the central column, but this term is inappropriate; the use of the term "stalace" is to be preferred. At each side of the top cell of the stalace are located the periblem initials and above both is a single plerome initial (but whether these pertain to the root or to the hypocotyl is problematical). The epidermal cells divide first by anticlinal walls, then by a succession of periclinal walls and engender a thick layer of cells constituting the root cap.

Two types of proembryos develop in *Juglans regia*, determined by the plane of the first zygotic wall. In one type the primary wall is transverse but may be slightly oblique, and divides the zygote into either two equal or two very unequal cells. In the other type the wall is oriented in various oblique positions; this type occurs most frequently. In the first variation the terminal cell always segments longitudinally, although the resulting daughter cells may be of rather unequal size. The basal cell divides either longitudinally or obliquely; it eventually produces a short, stocky suspensor. In many respects this species resembles *Scabiosa succisa* (Scabiosa Variation, Piperad Type), in others, it is somewhat related to the Trifolium Variation, Onagrad Type.

In *Hicoria pecan* the zygote remains inactive for about two months. The large basal cell apparently does not divide again. Vertical divisions next occur in the terminal cell to form a tier of four cells. Accounts differ as to the orientation of the cotyledons with reference to the septum of the fruit; the probability is that they are oriented transversely to the septum, and they are bifurcated at the end of the septum.

LANGDON, L. M. 1934. Embryogeny of *Carya* and *Juglans*, a comparative study. Bot. Gaz. 96: 93–117.

NAST, C. G. 1941. The embryogeny and seedling morphology of *Juglans regia* L. Lilloa 6: 163–205.

NEUMANN, O. 1939. Über die Bildung der Wurzelhaube bei *Junglans*, *Mimosa* und *Lupinus*. Planta 30: 1–20.

SHUHART, D. V. 1932. Morphology and anatomy of the fruit of *Hicoria pecan*. Bot. Gaz. 93: 1–20.

WOODROOF, N. C. 1928. Development of the embryo sac and young embryo of *Hicoria pecan*. Amer. Jour. Bot. 15: 416–421.

Umbelliferales

No reports are available for the *Alangiaceae* and *Nyssaceae*.

Cornaceae:- The description and figures presented for *Cornus florida* do not permit any conclusions to be reached.

MORSE, W. C. 1907. Contribution to the life history of *Cornus florida*. Ohio Nat. 8: 197–204.

Araliaceae:- The embryology of the *Araliaceae* appears to present a certain degree of uniformity. The species which have been investigated include *Aralia racemosa*, *A. cachemiria*, *A. cordata*, *Fatsia japonica* and *Hedera helix*.

The latter species has been most thoroughly studied. In general, the embryogeny apparently conforms to the Onagrad Type, but it is difficult to determine the point with certainty from the original descriptions. It is apparent from many of the figures presented that they are not median longitudinal sections; consequently, the disposition of the cell walls may have been inaccurately described.

The description of the embryogeny of *Panax fruticosa* and *Hedera australiana* is superficial and the illustrations inadequate, but assignment of both is probably to the Chenopodiad Type.

DUCAMP, L. 1901. Développement de l'embryon chez le Lierre (*Hedera helix*). C. R. Acad. Sci. Paris 133: 651–653.
DUCAMP, L. 1902. Recherches sur l'embryogénie des *Araliacées*. Ann. Sci. Nat., Bot., VIII, 15: 311–402.
GOPINATH, D. M. 1944. Gametogenesis and embryogeny in a few members of the *Araliaceae*. Proc. Indian Acad. Sci., B, 20: 175–186.

Apiaceae (Umbelliferae):- In *Pimpinella saxifraga* both terminal and basal cells of the two-celled proembryo generally segment transversely to give rise to a tetrad composed of four superposed elements, *l*, *l'*, *m* and *ci*; but the two cells may also divide obliquely or longitudinally. The destinations of these four elements of the tetrad appears to be the same as in *Carum carvi*. The eight-celled proembryo arises by bipartitioning of each tetrad cell; normally it appears to be composed of two juxtaposed cells in *l* and *l'* and two superposed cells in *m* and *ci*. Those proembryonic forms developed subsequent to the eight-celled stage present the most varied aspects; they do not appear to be constructed according to any definite rule. The separation of the histogens and the organization of the tissues at the radicular end is produced at a very late period, after the origin of the cotyledonary protuberances.

Carum carvi follows the Sherardia Variation, Solanad Type. In this species, the cotyledonary region originates from cell *l* of the proembryonic tetrad, the hypocotyledonary region from cell *l'* and the primordium of the cortex from the cell or cells nearest to the hypocotyl derived from cell *m*; the other proembryonic elements which make up region *m*, together with all those derived from *ci*, go into the construction of a suspensor which is generally massive but occasionally may be filamentous, and of irregular form and very variable dimensions.

Anethum graveolens, *Foeniculum vulgare*, *Eryngium yuccifolium*, *Sesli montana* and *S. gracile* all appear also to follow the Sherardia Variation.

The development of the embryo in *Daucus carota* is quite comparable to that of *Carum carvi*. The principal difference between the two species resides in the destination of the four superposed elements of the tetrad: the terminal cell gives rise to the cotyledons and the upper portion of the hypocotyl, and the subterminal cell originates the lower portion of the hypocotyl, the central region of the root cap and a portion of the suspensor. There is an aberrant type of embryo in *D. carota* in which all of the embryo is derived from the terminal cell of the four-celled proembryo, the other three cells going into the formation of the suspensor. This type is referable to the Physalis III Variation, Solanad Type. The same developmental scheme prevails in *Apium inundatum*.

BORTHWICK, H. A. 1931. Development of the macrogamete and embryo of *Daucus carota*. Bot. Gaz. 92: 23–44.
HAKANSSON, A. 1923. Studien über die Entwicklungsgeschichte der *Umbelliferen*. Lunds Univ. Årsskr., N. F. Avd. 2, 18: 1–118.
JURICA, H. S. 1922. A morphological study of the *Umbelliferae*. Bot. Gaz. 74: 292–309.
SOUÈGES, R. 1926. Embryogénie des *Ombellifères*. Développement chez le *Carum carvi* L. C. R. Acad. Sci. Paris 182: 339–341.
SOUÈGES, R. 1930. Recherches sur l'embryogénie des *Ombellifères*. Bull. Soc. Bot. France 77: 494–511.
SOUÈGES, R. 1945. Embryogénie des Ombelliféres. Développement de l'embryon chez l'*Apium inundatum* Reichb. C. R. Acad. Sci. Paris 220: 128–131.

Ericales

For the *Clethraceae* and *Vacciniaceae* there are no reports on their embryology.

Ericaceae:- *Pyrola rotundifolia* agrees precisely with the Myriophyllum Variation, Caryophyllad Type. The basal cell of the two-celled proembryo never gives rise to more than two cells, which disappear early in the development of the embryo.

SOUÈGES, R. 1939. Embryogénie des *Éricacées*. Développement de l'embryon chez le *Pyrola rotundifolia* L. C. R. Acad. Sci. Paris 209: 635–637.

Epacridaceae:- The youngest described proembryo of *Styphelia longifolia* bears a remarkable resemblance to that of *Vaccaria vaccaria* at the same stage, so that this species may be assigned to the Vaccaria Variation, Caryophyllad

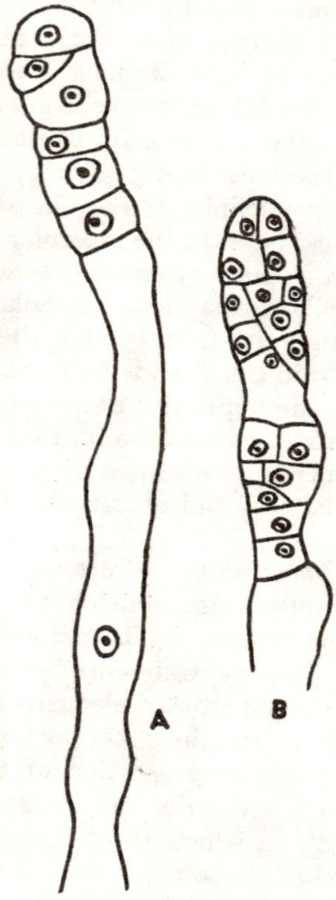

FIG. 55. — *Styphelia longifolia*. A. Young proembryo. ×600. B. Developing proembryo. ×270. (*Redrawn from* BROUGH 1924).

Type (FIG. 55A, B). The basal suspensor cell is exceedingly elongated. Exactly what portions of the filamentous proembryo go into the constitution of the embryo proper is not at all clear. The oldest stage depicted is that of a well developed embryo in which the epidermis is clearly delineated and the periblem and plerome can be observed.

BROUGH, P. 1924. Studies in the *Epacridaceae*. I. The life-history of *Styphelia longifolia* R. Br. Proc. Linn. Soc. N. S. Wales 49: 162–178.

Monotropaceae:- It is difficult to determine the accuracy of all of the published observations on *Monotropa hypopitys*, most of which are rather old. In some of the first reports it is apparent that the ellipsoidal mass of six to ten endosperm cells was mistaken for the embryo. The mass of endosperm cells is broader in one plane than in the opposite view: in the broader aspect there appear to be two superposed endosperm cells between which the embryo is embedded, while in the narrower aspect there appear to be two tiers of two cells each with the embryo located in the center. The terminal cell of the two-celled proembryo clearly divides vertically at its first segmentation. The earlier accounts state that a hypophysis initial is formed, but the latest investigation states that none could be recognized and this is borne out by the figures. When the seed is mature the embryo consists of about nine cells, but the mode of formation is obscure. At the time of germination, the basal cell has developed either three or four tiers of large cells, some tiers being two-celled and others unicellular. It is clear that the basal cell contributes nothing to the construction of the embryo proper; therefore, assignment may be to the Onagrad Type. Outside of a fairly distinct epidermal layer, no histogens are recognizable and no cotyledons are present at any period of development.

Sarcodes sanguinea may possibly conform to the Caryophyllad Type, but not enough details have been presented to allow a precise conclusion. The description states that the basal cell of the two-celled proembryo does not divide, all cells of the linear suspensor being derived from the terminal cell. This species may represent merely a slight divergence from the Myriophyllum Variation, in that the typical longitudinal segmentation of the basal cell is omitted.

DOYEL, B. E., and L. M. GOSS. 1941. Some details of the structure of *Sarcodes*. Madroño 6: 1–7.

FRANCKE, H. L. 1934. Beiträge zur Kenntnis der Mykorrhiza von *Monotropa hypopitys* L. Flora 129: 1–52.

Diapensiaceae:- *Diapensia lapponica* probably belongs under the Solanad Type.

SAMUELSSON, G. 1913. Studien über die Entwicklungsgeschichte der Blüten einiger *Bicornes*-Typen. Svensk Bot. Tidskr. 7: 97–188.

Lennoaceae:- In the parasitic *Lennoa madreporoides* the first two divisions in the zygote are transverse. The third wall is vertical in the middle cell of the three-celled stage; the fourth wall appears to be vertical in the terminal cell. When nearly mature the embryo, as in most parasitic and saprophytic plants, is more or less cylindrical, consisting in median longitudinal section of about six tiers of two cells each. A transverse section of the midportion, however, reveals that several irregularly placed longitudinal walls are present. Assignment to the Caryophyllad Type is indicated, and probably to the Sagina Variation.

SUESSENGUTH, K. 1927. Über die Gattung *Lennoa*. Flora 122: 264–305.

Ebenales

The *Sapotaceae* constitute the only uninvestigated family.

Ebenaceae:- It appears evident from the very brief description that the terminal cell of the two-celled proembryo of *Diospyros virginiana* divides transversely. A large suspensor of variable dimensions is produced and the embryo is poorly organized. Neither the figures nor the description permit reference to a specific Type.

HAGUE, S. M. 1911. A morphological study of *Diospyros virginiana*. Bot. Gaz. 52: 34–44.

Myrsinales

Myrsinaceae:- *Aegiceras majus*, *A. corniculatum* and *Maesa indica* have been very superficially described. A young embryo of the last species appears to conform to the Onagrad Type.

CAREY, G., and L. FRASER. 1932. The embryology and seedling development of *Aegiceras majus* Gaertn. Proc. Linn. Soc. N. S. Wales 57: 341–360.
DAHLGREN, K. V. O. 1916. Zytologische und embryologische Studien über die Reihen *Primulales* und *Plumbaginales*. Kungl. Svenska Vetensk. Handl. 56, Nr. 4.

Styracales

There is a report for one genus in the *Styracaceae*, but the other families (*Diclidantheraceae*, *Symplocaceae* and *Lissocarpaceae*) are still uninvestigated.

Styracaceae:- The embryonomy of *Styrax californica* conforms essentially to the Myosotis Variation, Chenopodiad Type.

The diagnosis was made by the writer from an examination of the slides described in: COPELAND, H. F. 1938. The *Styrax* of northern California and the relationships of the *Styracaceae*. Amer. Jour. Bot. 25: 771–780.

Loganiales

Loganiaceae:- No essential difference separates the embryogeny of *Buddleia variabilis* from that of the Veronica Variation, Onagrad Type. *Spigelia splendens* also appears to conform to the same Variation.

DAHLGREN, K. V. O. 1922. Die Embryologie der *Loganiazeen*gattung *Spigelia*. Svensk Bot. Tidskr. 16: 77–87.
SOUÈGES, R. 1940. Embryogénie des *Loganiacées*. Développement de l'embryon chez le *Buddleia variabilis* Hemsley. C. R. Acad. Sci. Paris 211: 139–140.

Oleaceae:- In *Fraxinus*, *Syringa* and *Ligustrum* the zygote becomes more or less elongated before it divides. The length is greatest in *Fraxinus ornus*. When the zygote divides, the wall separates a short terminal and a very long basal cell. The latter thereupon divides rapidly to form a filamentous suspensor. It was not definitely determined whether the terminal cell divides transversely or longitudinally. From the appearance of the figures accompanying the description, there is a strong presumption that development in all three genera follows the Alyssum Variation, Onagrad Type.

Ligustrum ovalifolium possesses an embryonomy almost as variable as that found in *Physalis minima* (Solanaceae) and is equally as difficult to assign satisfactorily. The basal cell *cb* divides transversely in advance of segmentation of the terminal cell, *ca*. Later the two daughter derivatives of *cb* may divide or not. Cell *ca* may divide transversely to engender a linear 4-celled proembryo, then each daughter cell segments transversely to produce a linear 6-celled proembryo. If each daughter of *cb* also divides transversely, the proembryo then consists of a row of eight cells. At the third generation, therefore, the proembryo may consist of either six or eight cells. In the case of the 8-celled proembryo only do the derivatives of *cb* contribute to the construction of the embryo proper. The generalized formula for this species is:

$$l = pco + pvt$$
$$l = \tfrac{1}{2}phy$$
$$m = \tfrac{1}{2}phy + icc + iec + co \text{ (in part)}$$
$$ci = co \text{ (part)} + s$$

However, cell *l* may divide obliquely and an epiphysis initial is then developed. Again, cell *ci*, which most often segments transversely, may divide vertically. In the first case this cell gives rise to the central portion of the root cap and several elements of the suspensor; in the other instance it produces the primordium of the cap alone. The best disposition of this species probably would be to assign it to the Physalis III Variation, Solanad Type.

ANDERSSON, A. 1931. Studien über die Embryologie der Familien *Celastraceae, Oleaceae* and *Apocynaceae.* Lunds Univ. Årsskr., N. F. Avd. 2, 27: 1–110.
SOUÈGES, R. 1942. Embryogénie des Oléacées. Développement de l'embryon chez le *Ligustrum ovalifolium* Hassk. C. R. Acad. Sci. Paris 215: 328–330.

Apocynales

Apocynaceae:- Several species in the family have been investigated; all are very similar to species in the *Hypericaceae* embryologically and may therefore be assigned to the Sedum Variation, Caryophyllad Type. It appears to be characteristic of the *Apocynaceae* that the zygote rests for a considerable period before undergoing cytokinesis.

The following species have been studied: *Lochnera rosea, Vinca minor, Gerbera odollam, Carissa carandas, Funtrumia elastica, Vallaris heyneii, Ichnocarpus frutescens* and *Wrightia tinctoria.*

ANDERSSON, A. 1931. Studien über die Embryologie der Familien *Celastraceae, Oleaceae* und *Apocynaceae.* Lunds Univ. Årsskr., N. F. Avd. 2, 27: 1–110.
RAU, M. A. 1940. Studies in the *Apocynaceae.* Jour. Indian Bot. Soc. 19: 33–44.

Asclepiadaceae:- *Cynanchum acutum* conforms to the Linum Variation, Solanad Type. The earliest proembryonic stage depicted for *Vincetoxicum officinale* is remarkably like the same stage of *Lilium parryi* and at a later stage the basal suspensor tissues are equally analogous. A similar development of the suspensor occurs in *Cynanchum nigrum,* but it is not found in *Asclepias incarnata* and *A. cornuti.*

BILLINGS, F. H. 1901. Beiträge zur Kenntniss der Samenentwicklung. Flora 88: 253–318.
FRANCINI, E. 1927. L'embriologia del *Cynanchum acutum* L. Nuovo Giorn. Bot. Ital. 34: 381–395.

Rubiales

Caprifoliaceae:- *Lonicera biflora* possesses an irregular embryogeny; the closest representative of the same degree of irregularity is *Convolvulus arvensis.* As with the latter species, arbitrary assignment is perhaps best with the Penaea Variation, Asterad Type.

SOUÈGES, R. 1945. Embryogénie des *Caprifoliacées.* Développement de l'embryon chez le *Lonicera biflora* Desf. C. R. Acad. Sci. Paris 221: 480–482.

Rubiaceae:- The embryonomy of *Sherardia arvensis,* by reason of its irregularities presently to be described, agrees in general with the Solanad Type but is aberrant and exceptional.

As in the Solanum Variation, a tetrad of four superposed cells is first formed

(Fig. 56A, B). Each of these elements is bipartitioned to give rise to an eight-celled proembryo, which may assume either of two different aspects, according to whether the terminal cell, *l*, is segmented vertically (Fig. 56D, F) or hori-

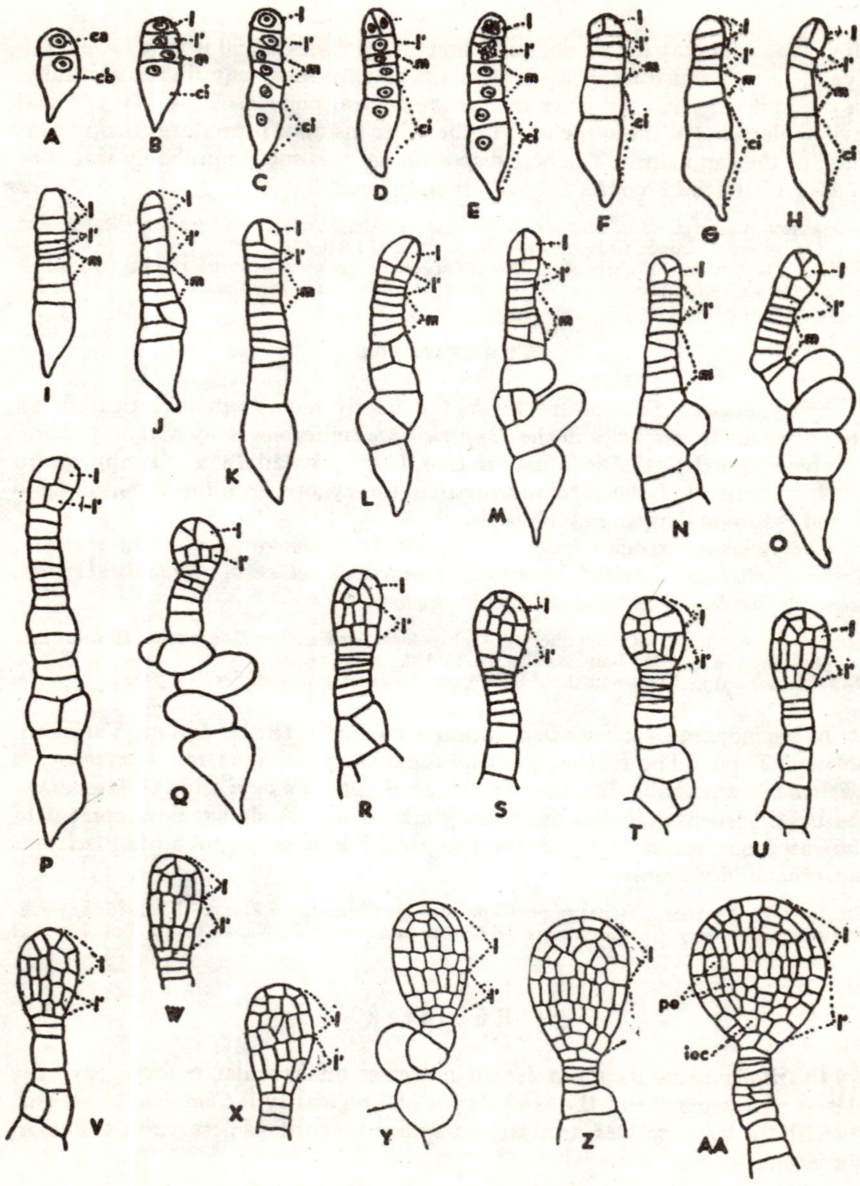

Fig. 56. — *Sherardia arvensis.* Description in text. All × 290. (*Redrawn from* Souèges 1924).

zontally (Fig. 56E, G). The mode of construction of all the observed proembryonic forms can probably be interpreted on the basis of one or the other of these two eight-celled types. For example, the proembryos delineated in

FIG. 56H, L, M and N are plainly derived from the eight-celled type of seven tiers like that in FIG. 56F, and these in turn present two important variations depending on the mode of partitioning of the two superposed cells derived from the element *l'*: generally, the upper daughter cell of this element is segmented vertically (FIG. 56L, M); in the other case, it is divided horizontally like its sister to give two new superposed elements (FIG. 56N); finally, the two superposed sister cells occasionally are partitioned by a meridional wall into two juxtaposed cells.

From the eight-celled proembryonic type disposed in eight tiers as in FIG. 56G are derived the proembryos figured in FIG. 56I and J, without much doubt. It is necessary to add certain forms, for example that represented by Fig. 56O, which is apparently derived from an eight-celled proembryo arranged in six tiers, presuming that *l'* consisted of two juxtaposed elements.

In a general way, it appears to be established that the element *l* of the tetrad, after being longitudinally or transversely partitioned, gives rise to the epidermal initials and to three or four tiers of cells from which the cotyledonary portions arise (FIG. 56Q–AA); that the element *l'*, by more or less precocious divisions, becomes disposed into two, three or four horizontal cell layers to engender the hypocotyledonary region and the initials of the root tip (FIG. 56Q–AA); that the element *m*, after partitioning by an indeterminate number of transverse walls, gives rise to a series of much flattened cells, of which the uppermost one incontestably goes into the construction of the primordium of the root cap, while all the others go into the development of the suspensor; finally, that the element *ci* produces the most important portion of the suspensor. The cells of the suspensor are swollen into the form of vesicles, are more or less dislocated and occupy the most variable positions in the endosperm (*e.g.*, FIG. 56L–Q).

Sherardia clearly ranks next to *Datura* in the Solanad Type in difficulty of interpretation of the various stages of embryo development.

The development of practically all other species, of which there is an unusually high number, agrees with that of *Sherardia*, either precisely or at least more or less closely. The species for which no further description is necessary include: *Dentella repens*, *Putoria calabrica*, *Oldenlandia alata*, *O. capensis*, *Phillis nobla*, *Leptodermis lanceolata*, *Callipeltis cucullaria*, *Crucianella* spp., *Houstonia* spp., *Diodia* spp., *Galium* spp., *Asperula* spp., *Spermacoce hispida* and *Guettarda speciosa*.

Suspensor haustoria are extensively developed in many species, such as *Crucianella aegyptica*, *Rubia tinctoria* and *Asperula setosa*. Very many species of *Galium* have been studied; there is a great deal of variation in the structure of the suspensor, depending upon the species.

The zygote of *Coffea arabica* does not divide until 60 or 70 days after opening of the flowers. A large basal and a small terminal cell are developed. The second division is transverse, but it is not clear in which cell it occurs. The third division is longitudinal in the uppermost of the three cells. Subsequent events were not followed out, but this species probably should be classified along with the other genera in the *Rubiaceae*.

FAGERLIND, F. 1936. Die Embryologie von *Putoria*. Svensk Bot. Tidskr. 30: 362–372.

FAGERLIND, F. 1936. Embryologische Beobachtungen über die Gattung *Phyllis*. Bot. Notiser 1936: 577–584.

FAGERLIND, F. 1937. Embryologische, zytologische und Bestäubungsexperimentelle Studien in der Familie *Rubiaceae* nebst Bemerkungen über einige Polyploiditätsprobleme. Acta Horti Bergiani 11: 195–470.

LLOYD, F. E. 1902. The comparative embryology of the *Rubiaceae*. Mem. Torrey Bot. Club 8: 1–112.

MENDES, A. J. T. 1941. Cytological investigations in *Coffea*. VI. Embryo and endosperm development in *Coffea arabica* L. Amer. Jour. Bot. 28: 784–789.

PIERPAOLI, I. 1917. Ricerche anatomiche, istologiche ed embriologiche sulla *Putoria calabrica* Pers. Ann. di Bot. 14: 83–100.
RAGHAVAN, T. S., and K. RANGASWAMY. 1941. Studies in the *Rubiaceae*. Part I. Development of female gametophyte and embryo formation in *Dentella repens* Forst. and *Oldenlandia alata* Koch, and some cytotaxonomical considerations. Jour. Indian Bot. Soc. 20: 341–356.
RAGHAVAN, T. S., and A. R. SRINIVASAN. 1941. Studies in *Rubiaceae*. II. *Spermococe hispida* Linn., *Guettarda speciosa* Linn. and some cytomorphological considerations. Proc. Indian Acad. Sci., B, 14: 412–426.
SOUÈGES, R. 1924. Embryogénie des *Rubiacées*. Développement de l'embryon chez le *Sherardia arvensis* L. C. R. Acad. Sci. Paris 178: 1919–1921.

Asterales

Adoxaceae:- *Adoxa moschatellina* clearly belongs under the Geum Variation, Asterad Type.

LAGERBURG, T. 1909. Studien über die Entwicklungsgeschichte und systematische Stellung von *Adoxa moschatellina* L. Kungl. Svenska Vetensk.-Akad. Handl. 44, Nr. 4.

Valerianaceae:- *Valerianella olitoria*, *Centranthus macrosiphon*, *C. ruber* and *C. angustifolius* all follow the Lamium Variation, Asterad Type, rather closely when the divisions in the proembryo and young embryo are normally oriented. In the last three species irregularly constructed embryos are frequently encountered.

ASPLUND, E. 1920. Studien über die Entwicklungsgeschichte der Blüten einiger *Valerianaceen*. K. Svenska Vetensk. Handl. 61, Nr. 3.
SOUÈGES, R. 1923. Embryogénie des *Valerianacées*. Développement de l'embryon chez le *Valerianella olitoria*. C. R. Acad. Sci. Paris 176: 1081–1083.
SOUÈGES, R. 1941. Embryogénie des *Valerianacées*. Développement de l'embryon chez les *Centranthus*. C. R. Acad. Sci. Paris 209: 718–720.

Dipsacaceae:- *Scabiosa succisa* is one of the few Angiosperms in which the first division in the zygote is nearly vertical. In most instances it is more diagonal than strictly vertical and divides the cell into two somewhat unequal cells; one, designated as *b*, is plainly the lower one and corresponds to the basal cell (FIG. 57A). The other corresponds to the terminal cell of the two-celled proembryo. The orientation of this primary wall governs all the succeeding developmental processes and thus directly influences the later organization of the embryo.

The following divisions are perpendicular to the primary wall; each cell is partitioned more or less transversely, but the wall may occasionally be vertical in one cell (FIG. 57B, C). After this, it is extremely difficult to follow the exact order of segmentation. Sometimes each of the four cells may be divided transversely to give eight cells arranged in four tiers (FIG. 57D); but most often the walls are oblique, oriented in all directions, with the external insertion on the peripheral wall and the internal insertion on one or the other of the earlier walls. The four elements of the tetrad, *c* and *d*, derivatives of *a*, and *e* and *f*, derivatives of *b*, represent the quadrants of the embryonic cell; they give rise to groups of cells which remain distinct for a long time although the processes of segmentation in each can not be followed with certainty (FIG. 57C–G).

Without doubt, however, groups *c* and *e* contribute wholly to the construction of the cotyledonary portion; groups *d* and *f* give the hypocotyledonary portion; the lowest elements of *f* produce the primordia of the root cap and the suspensor. The epidermal initials (*de*, FIG. 57H, I) are differentiated at an early stage, but the periblem and plerome are not distinguishable until after the appearance of the cotyledons.

The embryonomy of *Scabiosa succisa* is most closely comparable with that of *Valerianella olitoria*, which follows the Lamium Variation, Asterad Type. In the latter species the primary division is transverse, otherwise the divisions in the two species are oblique and both are equally difficult to interpret precisely. However, *S. succisa* may be assigned to the Scabiosa Variation, Piperad Type, because of the nearly vertical segmentation of the zygote.

SOUÈGES, R. 1937. Embryogénie des *Dipsacacées*. Développement de l'embryon chez le *Scabiosa succisa* L. C. R. Acad. Sci. Paris 204: 292–294.

Calyceraceae:- A linear tetrad is formed in *Acicarpha tribuloides*, but this is insufficient for assignment of the species.

DAHLGREN, K. V. O. 1915. Über die Embryologie von *Acicarpha tribuloides* Juss. Svensk Bot. Tidskr. 9: 184–191.

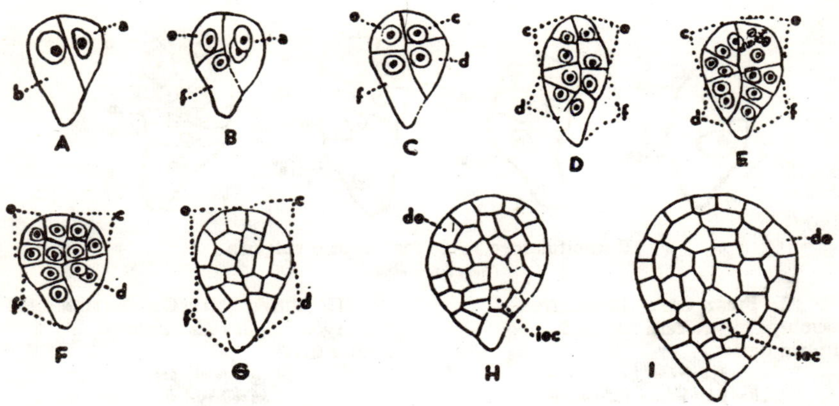

FIG. 57. — *Scabiosa succisa*. Description in text. All × 290. (*Redrawn from* SOUÈGES 1937).

Asteraceae (Compositae):- Despite a wealth of investigations on specie belonging to the *Asteraceae*, the embryonomy of the family is in none too satisfactory a condition. The family, for one thing, is too large and unwieldly. Some taxonomists have segregated tribes within the family; others have split it into a number of separate families. If the latter procedure were accepted, it is believed that the embryology of the entire group could be established on a more satisfactory basis. There is only one completely detailed investigation (that on *Senecio vulgaris*) based on modern embryological conceptions; certain of the older accounts, notably those of CARANO, indicate that there are significant minor departures which would warrant the segregation of several Variations from the accepted Senecio Variation if the splitting into several families were followed. The whole problem requires more critical consideration than can be given to it in the present connection; moreover, certain of the groups (tribes or separate families) include no carefully investigated species.

Senecio vulgaris may be taken to serve as a generalized type for the *Asteraceae, sensu latu*. Practically all other species, of which there is an unusually large number, agree with *S. vulgaris* in essential details. The designation may be either as the Asterad Type or as the Senecio Variation of that Type; the designations are identical.

The proembryonic tetrad and eight-celled stages resemble those of *Myosurus* (Onagrad Type). The sixteen-celled proembryo arises by the bipartitioning of each of the eight cells of the eight-celled stage. In the upper cells, region *q*, he wall is obliquely oriented, with one end inserted at the center of the peripheral wall and the other close to the vertical wall. Occasionally the internal end of this wall is inserted on the meridional wall, thus becoming nearly horizontal; and in still other cases the wall is nearly vertical, paralleling one of the two meridional walls. The upper cell, nearest the axis, is designated as *a*, and the other as *b*. When the wall is normally oriented, cell *b* divides first by a vertical radial wall. At this moment each of the quadrant cells has become divided into three cells, one central and two peripheral. Tangential walls are next erected in these three cells and the external derivatives become epidermal initials. New tangential walls developed in the internal cells result, in the entire cotyledonary part which this region represents, in three subepidermal layers. The cell which occupies the internal summit of the massive triangle derived from cell *b* always remains in direct connection with the innermost cells of the plerome.

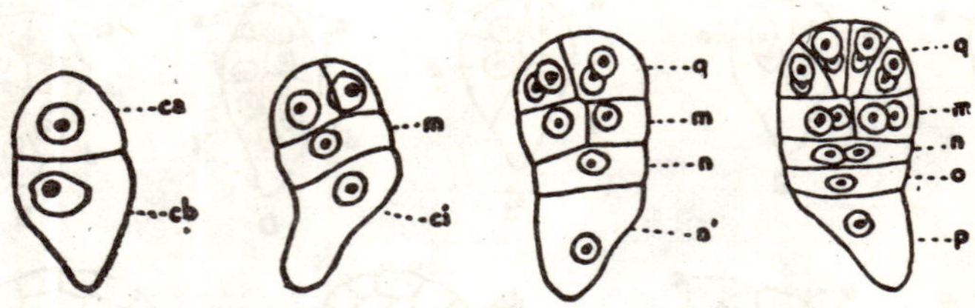

Recapitulatory Table for Senecio vulgaris
(from left to right)

I. First Cell Generation
Proembryo of two cells disposed in two tiers:

ca = *pco* + *pvt*
cb = *phy* + *icc* + *iec* + *co* + *s*

II. Second Cell Generation
Proembryo of four cells disposed in three tiers:

q = *pco* + *pvt*
m = *phy* + *ico*
ci = *iec* + *co* + *s*

III. Third Cell Generation
Proembryo of eight cells disposed in four tiers:

q = *pco* + *pvt*
m = *phy* + *icc*
n = *iec*
n' = *co* + *s*

IV. Fourth Cell Generation
Proembryo of sixteen cells disposed in five tiers:

q = *pco* + *pvt*
m = *phy* + *icc*
n = *iec*
o = *co*
p = *s*

The tier *m* gives rise to the entire hypocotyledonary region. It consists of four circumaxial cells at the sixteen-celled stage; each cell divides in the vertical tangential plane to delimit the epidermal initials, periblem and plerome. In the plerome, after one or two transverse segmentations, the first vertical walls isolate the pericycle toward the exterior.

From tier *n* are derived the initials of the root cortex. The two elements composing this region in the sixteen-celled proembryo divide by meridional walls into four circumaxial cells; then these divide again vertically, sometimes by tangential walls, sometimes by two walls perpendicular to the two meridional walls, to give rise to four internal cells and four to eight external cells. The first group of cells functions as the initials of the cortex; the others divide radially, then tangentially, and contribute to the formation of the lateral portions of the root cap.

Tier *o* engenders the central portion of the root cap. In the sixteen-celled proembryo it consists of a single cell, which becomes differentiated into four circumaxial cells. These segment tangentially to produce the four central layers of the root cap. This structure, it will be perceived, has a triple origin since three regions of the proembryo concur in its formation: all of region *o* going into the formation of the central portion; the peripheral cells of region *n* and several cells of the epidermis of region *m* forming the lateral portions.

The suspensor takes its origin from cell *p*, which is divided transversely into two superposed cells; they, in turn, by generally transverse but also longitudinal, new walls, give rise to a filamentous suspensor. The latter is eventually composed of from six to twelve cells but offers no particular differentiation.

The following species agree substantially with the Senecio Variation, but many were only superficially examined and it is entirely possible that some may conform to other Variations of the Asterad Type upon more critical study: *Eclipta erecta* (this species may possibly follow the Polygonum Variation); *Artemisia tridentata, Erechites hieracifolia, Galinsoga ciliata, Aster undulatum, A. multiflorus, A. novi-belgii, A. novae-angliae, Bellis perennis, Solidago serotina, Calendula arvensis* (which has an uncommonly long suspensor), *Helianthus annuus, Cichorium intybus, Buphthalmum salicifolium, Adenocaulon bicolor, Lactuca sativa, L. muralis, Silphium trifoliatum, S. terebinthinaceum, S. integrifolium, S. laciniatum, S. perfoliatum, Parthenium incanum, P. argentatum, Taraxacum kok-saghyz, Scorzonera tau-saghys, Mutisia candolleana, Sonchus arvensis, Ageratum mexicanum, Bidens tripartitus, Tagetes signatus, Tussilago farfara, Centaurea scabiosa, Ursinia anthemoides, Arctotis calendulacea, Dimorpotheca pluvialis, Dahlia variabilis, Hemizonia angustifolia, H. pallida, Madia madioides* and *Layia platyglossa.*

Observations on the last four species are original.

AYRES, J. A. 1915. Flower of *Adenocaulon bicolor*. Bot. Gaz. 59: 154–157.
BHARGAVA, H. R. 1935. Contribution to the morphology of *Eclipta erecta* L. Proc. Indian Acad. Sci., B, 1: 325–339.
CARANO, E. 1915. Ricerche sull'embriogenesi delle *Asteraceae*. Ann. di Bot. 13: 251–301.
CARANO, E. 1922. Nuove ricerche sulla embriologia delle *Asteraceae*. Ann. di Bot. 15: 97–196.
COOPER, G. O. 1936. Cytological investigations of *Erechites hieracifolia*. Bot. Gaz. 98: 348–355.
DAHLGREN, K. V. O. 1920. Zur Embryologie der Kompositen mit besonderer Berücksichtigung der Endospermbildung. Zeitschr. Bot. 12: 481–516.
DAHLGREN, K. V. O. 1924. Studien über die Endospermbildung der Kompositen. Svensk Bot. Tidskr. 18: 177–203.
DIANOWA, W. J., A. A. SOSNOVETZ und N. A. STESCHINA. 1935. Vergleichende zytoembryologische Untersuchungen der Varietäten von *Parthenium argentatum* Gray und *P. incanum* Gray. Beih. Bot. Centralbl. 53: 293–308.
DIETTERT, R. A. 1938. The morphology of *Artemisia tridentata*. Lloydia 1: 3–74.
JONES, H. A. 1927. Pollination and life history studies of lettuce (*Lactuca sativa* L.). Hilgardia 2: 425–497.
KIRKWOOD, J. E. 1910. The life history of *Parthenium* (guayule). Amer. Rev. Trop. Agr. 1: 193–205.
MERRELL, W. D. 1900. A contribution to the life-history of *Silphium*. Bot. Gaz. 29: 99–133.
MOTTIER, D. M. 1893. On the embryo-sac and embryo of *Senecio aureus* L. Bot. Gaz. 18: 245–253.
OPPERMAN, M. 1904. A contribution to the life history of *Aster*. Bot. Gaz. 37: 353–362.
PODDUBNAJA-ARNOLDI, W., N. STESCHINA und A. SOSNOVETZ. 1935. Der Charakter und die Ursachen der Sterilität bei *Scorzonera tau-saghys* Lipsche et Bosse. Beih. Bot. Centralbl. 53: 309–339.
POPHAM, R. A. 1938. A contribution to the life history of *Galinsoga ciliata*. Bot. Gaz. 99: 543–555.
RODOLICO, A. 1930. Embriologia del *Buphthalmum salicifolium* L. Nuovo Giorn. Bot. Ital. 37: 592–608.
SOUÈGES, R. 1920. Embryogénie des *Composées*. Les premiers stades du développement de l'embryon chez le *Senecio vulgaris* L. C. R. Acad. Sci. Paris 171: 254–256.
SOUÈGES, R. 1920. Embryogénie des *Composées*. Les derniers stades du développement de l'embryon chez le *Senecio vulgaris* L. C. R. Acad. Sci. Paris 171: 356–357.
WARMKE, H. E. 1943. Macrosporogenesis, fertilization and early embryology of *Taraxacum kok-saghyz*. Bull. Torrey Bot. Club 70: 164–173.

Gentianales

Gentianaceae:- The proembryo of *Voyria coerulea* is a linear tetrad; it is stated that both of the two upper cells are segmented longitudinally. The inference is that the terminal cell is segmented transversely and that the basal cell has little or no part in the construction of the embryo although it appears that one upper cell derived from the basal cell may constitute the hypophysis. Assignment may be to the Linum Variation, Solanad Type. *V. parviflora* appears to resemble *V. coerulea* embryologically, as does *Leiphaimos* sp.

In *Nymphoides peltatum* (*Limnanthemum nymphaeoides*) the zygote is very much elongated; it divides transversely and the basal cell becomes haustorial but presently degenerates. It would therefore appear that the basal cell does not divide again. To judge from the illustrations accompanying the original description, the terminal cell divides transversely, then the upper daughter cell is segmented vertically, but another figure shows a tier of three cells with the middle cell divided vertically. The best that can be done with this species is to assign it to the Caryophyllad Type.

A linear seven-celled proembryo is developed in *Enicostemma littorale*. The upper two cells then segment longitudinally to produce the quadrant stage; the third cell is the hypophysis, which divides longitudinally. Assignment is to the Linum Variation, Solanad Type.

A careful study of *Menyanthes trifoliata* clearly places this species under the Senecio Variation, Asterad Type.

There is a very brief description, unaccompanied by figures, for *Swertia longifolia*. Although it is stated that the basal cell takes no part in the construction of the embryo proper, nevertheless it is probable that this species agrees with the preceding one in its development.

Jacobsson-Paley, R. 1920. Etude sur la pollinisation et l'embryologie du *Swertia longifolia* Boiss. Bull. Soc. Bot. Geneve II, 12: 65–86.
Oehler, E. 1927. Entwicklungsgeschichtlicht-zytologische Untersuchungen an einigen saprophytischen *Gentianaceen*. Planta 3: 641–733.
Srinivasan, A. R. 1941. Cyto-morphological features of *Limnanthemum cristatum* Griseb. and *Enicostemma littorale* Blume. Proc. Indian Acad. Sci., B, 14: 529–542.
Souèges, R. 1943. Embryogénie des *Gentianacées*. Développement de l'embryon chez le *Menyanthes trifoliata* L. C. R. Acad. Sci. Paris 217: 488–490.
Stover, E. L. 1932. Life history of *Nymphoides peltatum*. Bot. Gaz. 93: 474–483.

Primulales

Primulaceae:- *Samolus valerandi* follows the Myriophyllum Variation, Caryophyllad Type. There is somewhat greater regularity in the sequence of divisions in *S. valerandi* than in *Myriophyllum alternifolium*, and there is also a superposition of characteristics of other Types. For instance, the four- and six-celled stages resemble those of the Solanad Type as does the process of individualization of the initials of the stele and root; the mode of formation of quadrants and octants is like that in the Onagrad Type; and the formation of the hypocotyledonary region is reminiscent of the method prevailing in the Chenopodiad Type.

Only one stage in the embryogeny of *Trientalis orientalis* has been figured; there is no description. This embryo bears a marked resemblance to the Onagrad Type. *Cyclamen* is said to have only one cotyledon.

Dahlgren, K. V. O. 1916. Zytologische und embryologische Studien über die Reihen *Primulales* und *Plumbaginales*. Kungl. Svenska Vetensk. Handl. 56, Nr. 4.
Souèges, R. 1937. Embryogénie des *Primulacées*. Développement de l'embryon chez le *Samolus valerandi* L. C. R. Acad. Sci. Paris 204: 145–147.

Plumbaginaceae:- The embryogeny of *Armeria plantaginea* conforms to that of the Solanad Type, and it may be provisionally assigned to the Linum Variation. The basal cell of the two-celled proembryo does not contribute to the formation of the embryo proper but only to the production of a very large suspensor.

Plumbagella micrantha apparently follows the Nicotiana Variation of the Solanad Type. The terminal and basal cells of the two-celled proembryo divide simultaneously in the transverse plane. The lower daughter cell of the basal cell becomes attenuated but remains capable of dividing. Further divisions in the derivatives of the basal cell are rather irregular; in *Plumbago zeylanica* the cells are even more irregularly formed and are markedly hypertrophied. Quadrant and octant cells are formed in the upper two cells derived from the original terminal cell as in *Nicotiana*, but the planes of division appear occasionally to be oblique. A hypophysis is not apparent; the epidermal initials originate as in *Nicotiana*, but nothing can be made of the origin of the other histogens. The peculiarity of the mature embryo is the formation of palisade epidermal cells at the apices of the cotyledons.

The earlier divisions in the proembryo of *Statice bahusiensis* resemble those in *Plumbagella*, but in this case a row of as many as seven cells may be formed. The first longitudinal division occurs in the derivatives of the basal cell. At the period in the development of the embryo when the cotyledons are about to be formed, the massive suspensor is as large as or even larger than the body of the embryo. The characteristic peculiarity of the suspensor is the arrangement of the cells in linear rows adjoining one another. The suspensor of *Armeria vulgaris* at the same stage is almost twice as large as the body of the embryo, but is pyriform with the broad end attached to the embryo; the cells are not organized in any semblance of regularity. The embryos of both genera are very regularly organized, but it is impossible to determine the sequence of origin of the histogens.

Plumbago zeylanica agrees with *Armeria plantaginea* embryonomically; a massive suspensor is formed.

DAHLGREN, K. V. O. 1916. Zytologische und embryologische Studien über die Reihen *Primulales* und *Plumbaginales*. Kungl. Svenska Vetensk. Handl. 56, Nr. 4.
DAHLGREN, K. V. O. 1937. Die Entwicklung des Embryosackes bei *Plumbago zeylanica*. Bot. Notiser 1937: 487–498.
SOUÈGES, R. 1937. Embryogénie des *Plombagacées*. Développement de l'embryon chez l'*Armeria plantaginea* Willd. C. R. Acad. Sci. Paris 204: 1583–1585.

Plantaginales

Plantaginaceae:- *Plantago lanceolata* and *Litorella uniflora* conform to the Veronica Variation, Onagrad Type. As in *Veronica*, the hypophysis initial is the upper daughter cell (*d*) of the intermediate cell, *m*, of the four-celled proembryo.

Development of the embryo in *Plantago major* f. *phyllostachya* is said to be "normal," whatever that means; no description or illustrations are presented.

ROSEN, W. 1940. Seed development of *Litorella uniflora* Aschers. Meddel. Göteborgs Bot. Trädg. 14: 177–184.
SHADOWSKY, A. E. 1924. Embryological researches on *Plantago major* L. f. *phyllostachya* Wallr. Bull. Soc. Nat. Moscou, Sect. Biol., n.s. 32: 234–259.
SOUÈGES, R. 1923. Embryogénie des *Plantaginacées*. Développement de l'embryon chez le *Plantago lanceolata* L. C. R. Acad. Sci. Paris 177: 964–967.

Campanulales

Campanulaceae:- *Campanula patula* follows the Solanum Variation, Solanad Type, but it appears that in the process of the divisions in the cotyledonary portion and in the formation of the initials of the root cortex and root cap, both secondary characters, the resemblance is to the Asterad Type. Characters new to *C. patula* reside in the function of cell *l'* of the proembryonic tetrad, which engenders the hypocotyledonary portion, the initials of the root cortex, the root cap and several cells of the suspensor.

Jasione montana is closely related embryonomically to *C. patula*. The linear tetrad is succeeded by an eight-celled proembryo in seven regions, with the terminal cell divided into two juxtaposed cells. The latter two cells give rise to the cotyledonary region in a manner closely following the Asterad Type. The subterminal cell has by far the greatest part in the construction of the embryo: it produces the entire hypocotyledonary region and the radicle. The initials of the root cortex and the primordium of the root cap are constituted from the derivatives of two different proembryonic regions by processes of division comparable to those prevailing in the Asterad Type.

Adenophora liliiflora also follows the Solanum Variation, as may *Specularia perfoliata* but the evidence is inconclusive.

SAFIJOVSKA, L. D. 1935. Do embriologii *Adenophora liliiflora* Led. Jour. Inst. Bot. Ac. Sc. d'Ukraine 1935: 85–98.
SOUÈGES, R. 1936. Embryogénie des *Campanulacées*. Développement de l'embryon chez le *Campanula patula* L. C. R. Acad. Sci. Paris 202: 2009–2011.
SOUÈGES, R. 1938. Embryogénie des *Campanulacées*. Développement de l'embryon chez le *Jasione montana* L. C. R. Acad. Sci. Paris 206: 278–280.
TRENT, J. A. 1942. Studies pertaining to the life history of *Specularia perfoliata* (L.) A.DC., with special reference to cleistogamy. Trans. Kansas Acad. Sci. 45: 152–164.

Lobeliaceae:- The embryonomy of *Lobelia syphilitica* has been most carefully studied; it is fundamentally identical with that of *Campanula patula* and *Jasione montana* (Linum Variation, Solanad Type. *L. trigona* is very similar to *L. syphilitica*, as are *L. cliffordiana*, and *L. excelsa*. The description of the embryonomy of *L. nicotianaefolia* is confusing and it is very probable that a misinterpretation of the sequence of divisions has been made. It is stated, for instance, that the third and fourth cells of the linear tetrad "form the hypophysis and radicle respectively," which is plainly a misconception of the function of these particular cells. From the illustrations, it is clearly apparent that this species does not differ from *L. syphilitica*.

The embryonomy of *Lobelia amoena* has been compared with those of an amazing assortment of other plants, yet it is precisely like that of other species in the genus.

BILLINGS, F. H. 1901. Beiträge zur Kenntniss der Samenentwicklung. Flora 88: 253–318.
CRÉTÉ, P. 1938. Embryogénie des *Lobéliacées*. Développement de l'embryon chez le *Lobelia syphilitica* L. C. R. Acad. Sci. Paris 207: 177–179.
HEWITT, W. C. 1939. Seed development of *Lobelia amoena*. Jour. Elisha Mitchell Sci. Soc. 55: 63–82.
KAUSIK, S. B. 1935. The life history of *Lobelia trigona* Roxb., with special reference to the nutrition of the embryo sac. Proc. Indian Acad. Sci. 2: 410–418.
KAUSIK, S. B. 1938. Gametogenesis and embryogeny in *Lobelia nicotianaefolia* Heyne. Jour. Indian Bot. Soc. 17: 161–168.

Goodeniaceae:- The second division in *Dampiera stricta* is transverse in the basal cell, after which a linear filamentous row of about ten cells is formed. The first longitudinal division is depicted for both the terminal cell of the row and in the one third from the end, with partitioning of the subterminal cell following next. This would indicate that the proper assignment of *D. stricta* is to the Linum Variation, Solanad Type.

Scaveola attenuata appears to agree with *D. stricta.*

BILLINGS, F. H. 1901. Beiträge zur Kenntniss der Samenentwicklung. Flora 88: 253–318.
BROUGH, P. 1927. Studies in the *Goodeniaceae.* I. The life-history of *Dampiera stricta* R. Br. Proc. Linn. Soc. N. S. Wales 52: 471–498.

Stylidiaceae:- Embryo development of *Stylidium adnatum* and *S. linearis* has not been described in detail, but the figures of the former species exhibit a strong conformation with the Onagrad Type. A suspensor is developed in both species, but the cotyledons are not formed until after germination commences.

BURNS, G. P. 1900. Beiträge zur Kenntniss der *Stylidiaceen.* Flora 87: 313–354.

Polemoniales

Hydrophyllaceae:- *Phacelia tanacetifolia* conforms to the Hyoscyamus Variation, Solanad Type.

CRÉTÉ, P. 1946. Embryogénie des *Hydrophyllacées.* Développement de l'embryon chez le *Phacelia tanacetifolia* Benth. C. R. Acad. Sci. Paris 223: 459–460.

Polemoniaceae:- The proembryo of *Polemonium caeruleum* consists of a tetrad of four superposed elements, *l*, *l'*, *m* and *ci*, the two latter being derived from the basal cell (FIG. 58A). The uppermost cell, *l*, divides exactly as in

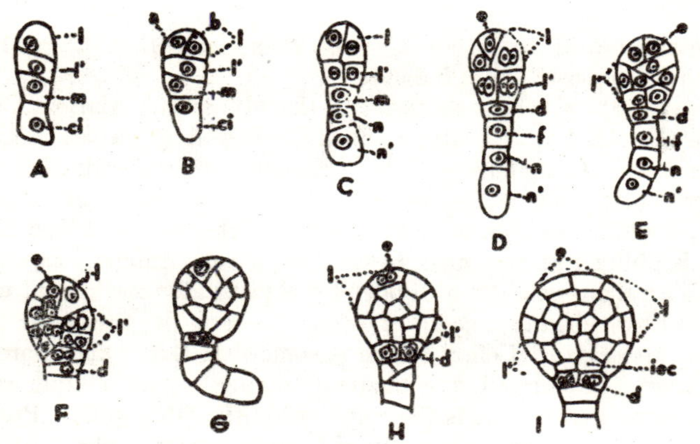

FIG. 58. — *Polemonium caeruleum.* E, F and G represent irregular embryos. Further description in text. All × 280. (*Redrawn from* SOUÈGES 1939).

Myosotis hispida and an epiphysis initial (*e*) is delimited. The epiphysis region, however, is indistinguishable by the time the cotyledons have originated.

The element *l'* of the proembryonic tetrad, by means of vertical walls, produces four circumaxial cells (FIG. 58D). These next become segmented vertically; the external derivatives become epidermal initials. The inner daughter cells are again divided vertically to give the periblem initials, while the inner cells become divided transversely and the lower derivatives become the initials of the root cortex (FIG. 58F, I). Occasionally these initials are differentiated through the formation of oblique walls in the internal cells which appear after the delimitation of the epidermal initials. Segmentation often proceeds in an irregular manner in element *l'*: in lieu of vertical meridional walls, oblique walls almost in the horizontal plane are produced. In the same way, the more or less inclined walls prevent the formation of the four circumaxial walls normally produced by cell *l'* (FIG. 58F–H).

The element *m* divides into two superposed cells, *d* and *f* (FIG. 58D); *d* is converted into four circumaxial cells and eventually engenders the primordium of the root cap, while *f*, together with all the elements originating from cell *ci* of the proembryonic tetrad, enters into the construction of the small suspensor.

Polemonium caeruleum constitutes the Polemonium Variation, Chenopodiad Type. In *P. pauciflorum*, on the other hand, cell *m* segments vertically; consequently, this species belongs to the Myosotis Variation of the same Type.

Phlox drummondii and *Leptosiphon androsace* apparently both have the same type of embryonomy as *P. caeruleum*, as does *Gilia tricolor*.

BILLINGS, F. H. 1901. Beiträge zur Kenntniss der Samenentwicklung. Flora 88: 253–318.

MILLER, H. A., and R. H. WETMORE. 1945. Studies in the developmental anatomy of *Phlox drummondii* Hook. I. The embryo. Amer. Jour. Bot. 32: 588–599.

SOUÈGES, R. 1939. Embryogénie des *Polemoniacées*. Développement de l'embryon chez le *Polemonium caeruleum* L. C. R. Acad. Sci. Paris 208: 1338–1340.

SOUÈGES, R. 1939. Les lois du développement chez le *Polemonium caeruleum* L. Bull. Soc. Bot. France 86: 289–297.

SOUÈGES, R. 1942. Embryogénie des Polémoniacées. Développement de l'embryon chez le *Gilia tricolor* Benth. C. R. Acad. Sci. Paris. 215: 543–545.

SOUÈGES, R. 1945. Embryogénie des Polémoniacées. Développement de l'embryon chez le *Polemonium pauciflorum* Wats. C. R. Acad. Sci. Paris. 220: 897–900.

Boraginales

Boraginaceae:- There is considerable difference within the family as to the embryonomic types that are followed.

Myosotis hispida, which conforms to the Myosotis Variation, Chenopodiad Type, may be taken as an example which is followed by species in other families as well as in the *Boraginaceae*. The terminal and basal cells of the two-celled proembryo both divide transversely to give rise to a tetrad of four superposed cells, *l*, *l'*, *m* and *ci* (FIG. 59A). The element *l* is the one which first divides next; the wall is oblique and separates two cells of very different aspects, *a* and *b* (FIG. 59B). Element *l'* then divides by a vertically meridional wall to give two juxtaposed cells (FIG. 59C). During this time the two lower cells, *m* and *ci*, remain undivided. Cell *a*, which is somewhat larger and more externally situated than its sister cell, *b*, is medianly divided by an oblique wall; the triangular upper daughter cell is the epiphysis initial (FIG. 59C). Presently each cell in *l'* divides, thus giving a group of four circumaxial cells (FIG. 59D); then the element *m* divides vertically, and the basal cell, *ci*, becomes segmented transversely into *n* and *n'* (FIG. 59D). The future destination of each cell group is as follows: the layer *l* gives rise to the cotyledonary region, layer *l'* gives rise to the upper portion of the hypocotyl and layer *m* to the lower portion; from *n* is differentiated the hypophysis and the derivatives of *n'* constitute a suspensor.

In the further development of the proembryo, it is the sister cell of the epiphysis initial which is the first of the three cells composing tier *l* to divide. The partition is in the vertical meridional plane, thus resulting in two juxtaposed cells (FIG. 59E). The four cells which now comprise tier *l* then divide by walls which are usually vertical, but may occasionally be tangential or even perpendicular to the meridional walls, until the layer is finally converted into a plateau of approximately twelve cells (FIG. 59E). From these cells are derived the two cotyledons and the histogens which enter into their construction.

The epiphysis initial, *e*, becomes separated into two juxtaposed cells by a meridional wall (FIG. 59E); these next segment tangentially and the four resulting cells are in turn divided vertically: there are thus two groups of four superposed cells (FIG. 59F). There are numerous variations in the order of the segmentations which eventualize in the formation of these two groups of cells. During the development of the cotyledons the cells of these two tetrads multiply by means of vertical walls and soon cease to be distinguishable from the neighboring cells (FIG. 59G). They then merely occupy the narrow cotyledonary commissure and at a rather late time engender the various regions of the stem apex.

The four cells of tier *l'* divide vertically to give rise to a layer of cells which correspond to the upper portion of the hypocotyledonary axis (FIG. 59E, F). At the time of the origin of the cotyledons the first transverse walls appear

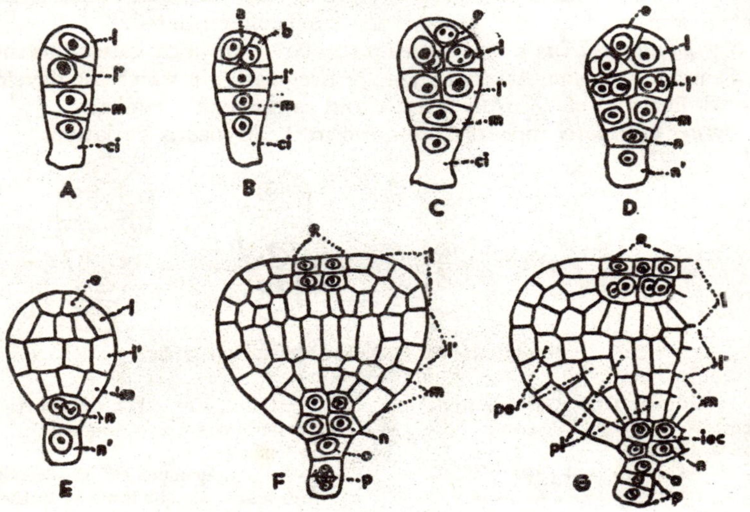

FIG. 59. — *Myosotis hispida.* Description in text. A–D × 420; E–G × 360. (*Redrawn from* SOUÈGES 1921).

and the line of demarcation between tiers *l'* and *m* become effaced (FIG. 59F, G). At this time the periblem and plerome of the hypocotyl become differentiated.

The two elements of tier *m* become segmented into four circumaxial cells by means of meridional walls (FIG. 59C–E). They continue to multiply by vertical walls and then constitute a layer which corresponds to the lower portion of the hypocotyl. As in the adjoining layer *l'*, the first transverse walls appear at the time of emergence of the cotyledons (FIG. 59G). The plerome initials of the radicular end of the hypocotyl result from the first transverse segmentations of the cells adjacent to the axis of the embryo.

The element *n*, by means of vertical meridional segmentations and then horizontal walls — or, inversely, by divisions which at first are horizontal, then vertical — gives rise to two groups of four superposed cells (FIG. 59E–G). The cells of the upper group are the initials of the root cortex; those of the lower group constitute the central rudiments of the root cap. The latter group of cells eventually undergo tangential divisions and the two upper daughter cells are the initials of the calyptrogen.

The element n' divides transversely; the two daughter cells, *o* and *p*, in turn are segmented similarly (FIG. 59F, G). The row of four cells thus engendered offer no particular differentiation and represents the suspensor.

The recapitulatory table for *Myosotis hispida* will be found below.

Echium vulgare conforms to the Polemonium Variation, Chenopodiad Type. There is little regularity in the sequence of the first segmentations. The linear tetrad is followed by three different types of eight-celled proembryos: (*1*) two superposed cells in the upper region, followed by two juxtaposed cells and four superposed cells; (*2*) two terminal cells separated by an oblique wall, then two juxtaposed cells, with the remainder in linear order, and (3) four superposed elements at the summit. The first type is the more prevalent one.

Heliotropium peruvianum conforms to the Chenopodiad Type.

Lycopsis arvensis follows the Lamium Variation, Asterad Type, closely. The behavior of the terminal cell of the two-celled proembryo is almost exactly consonant with that of *Lamium purpureum*, and that of the basal cell is likewise almost identical. A suspensor is not differentiated.

Symphytum officinale may develop regularly, in which case adherence is to the Senecio Variation, Asterad Type; or irregularly, in which the development is much like that of *Lycopsis arvensis* and assignment is similar.

Borago officinalis appears to conform to the Myosotis Variation.

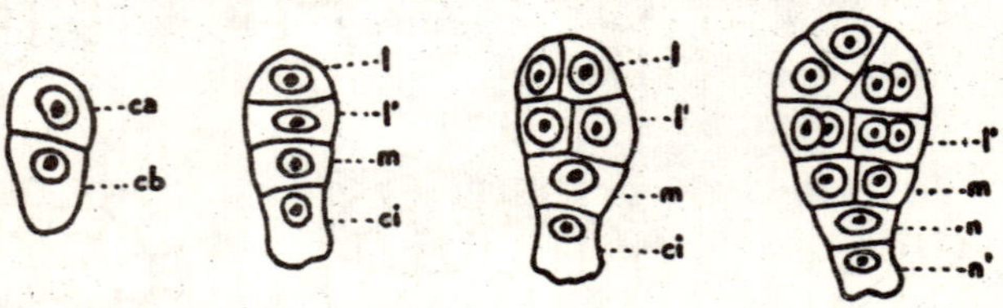

Recapitulatory Table for Myosotis hispida
(*from left to right*)

I. FIRST CELL GENERATION
Proembryo of two cells disposed in two tiers:

$ca = pco + pvt + \frac{1}{2}phy$
$cb = \frac{1}{2}phy + icc + iec + co + s$

II. SECOND CELL GENERATION
Proembryo of four cells disposed in four tiers:

$l = pco + pvt$
$l' = \frac{1}{2}phy$
$m = \frac{1}{2}phy + icc$
$ci = iec + co + s$

III. THIRD CELL GENERATION
Proembryo of six cells disposed in four tiers:

The arrangement and destinations of the four stages are the same as in the preceding generation.

IV. FOURTH CELL GENERATION
Proembryo of twelve cells disposed in five tiers:

$l = pco + pvt$
$l' = \frac{1}{2}phy$
$m = \frac{1}{2}phy + icc$
$n = iec + co$
$n' = s$

SOUÈGES, R. 1921. Embryogénie des *Boragacées*. Les premiers termes du développement de l'embryon chez le *Myosotis hispida* Schlecht. C. R. Acad. Sci. Paris 173: 726–728.
SOUÈGES, R. 1921. Embryogénie des *Boragacées*. Les derniers stades du développement de l'embryon chez le *Myosotis hispida* Schlecht. C. R. Acad. Sci. Paris 173: 848–850.
SOUÈGES, R. 1923. Développement de l'embryon chez le *Myosotis hispida* Schlecht. Bull. Soc. Bot. France 70: 385–401.
SOUÈGES, R. 1938. Embryogénie des *Boragacées*. Développement de l'embryon chez le *Lycopsis arvensis* L. C. R. Acad. Sci. Paris 207: 640–642.
SOUÈGES, R. 1938. Embryogénie des *Boragacées*. Développement de l'embryon chez l'*Echium vulgare* L. C. R. Acad. Sci. Paris 207: 871–872.
SOUÈGES, R. 1941. Embryogénie des *Boragacées*. Developpement de l'embryon chez le *Symphytum officinale* L. C. R. Acad. Sci. Paris 212: 245–246.
SOUÈGES, R. 1943. Embryogénie des *Boragacées*. Développement de l'embryon chez l'*Heliotropium pervuianum* L. C. R. Acad. Sci. Paris 217: 551–553.
SVENSSON, H. G. 1925. Zur Embryologie der *Hydrophyllaceen, Boraginaceen* und *Heliotropiaceen*. Uppsala Univ. Årsskr., Math. och Naturvetensk. 2.

Solanales

Solanaceae:- The *Solanaceae* are an exceedingly difficult group to diagnose embryonomically, mainly because of the excessive variation in the location and planes of the divisions marking the transition from the second to the third cell generation. In certain species it is also difficult to determine the exact identity of the third cell generation because of the almost endless variety of proembryonic forms which may be encountered. The various Variations which may be typified are distinguishable first by the destination of the elements of the proembryonic tetrad always formed at the second cell generation, and second by the construction of the proembryo during the third cell generation. There appears to be no other possible basis on which to erect a classificatory scheme for solanaceous embryos.

Certain of the Variations possess the identical destination of the tetrad elements at the second cell generation. The following general formula may be employed to denote the destination of the four cells at this generation:

$$\begin{aligned} l &= pco + pvt \\ l' &= phy + icc + iec \\ m &= co + s \text{ (in part)} \\ ci &= s \text{ (in part)} \end{aligned}$$

I. Nicotiana Variation. — Destination of tetrad elements: As in the typical formula.

Construction of proembryo at third cell generation: eight cells arranged in six tiers (or occasionally six cells in four tiers or seven cells in five tiers, if *l* or *l'* should divide before *m* and *ci*, or if *l*, *l'* and *m* segment before *ci*, respectively).

Species included: probably all species of *Nicotiana; Petunia nyctaginiflora;* some proembryos of *Withania somnifera, Datura stramonium* and *Physalis minima; Schizanthus pinnatus; Atropha belladonna; Solanun nigrum, S. melogena, S. villosum* and generally in *S. dulcamara.*

II. Physalis I Variation. — Destination of tetrad elements: As in the typical formula.

Construction of proembryo at third cell generation: eight cells disposed in eight tiers.

Species included: common in *Physalis minima* and *Solanum sisymbrifolium.*

III. Hyoscyamus Variation. — Destination of tetrad elements: As in the typical formula.

Construction of proembryo at third cell generation: six cells in five tiers (eight cells in six tiers at fourth cell generation).

Species included: *Hyoscyamus niger, H. albus.*

IV. Datura Variation. — Destination of tetrad elements: As in the typical formula.

Construction of proembryo at third cell generation: *Datura* is extremely irregular; it is contended that the proembryo is "always" eight-celled at the third cell generation, but this statement is contestable since, as a matter of fact, the following series of stages at what is apparently the third cell generation have been figured: five cells in four tiers, five cells in five tiers, six cells in four tiers, seven cells in five tiers, eight cells in five tiers and eight cells in six tiers.

Species included: *Datura stramonium, D. laevis, D. tatula (D. stramonium?)*; occasional in *Solanum dulcamara* and *S. nigrum.*

V. Physalis II Variation. — Destination of tetrad elements:

$$\begin{aligned} l &= pco + pvt + phy \\ l' &= icc + iec + co + s \text{ (in part)} \\ m + ci &= s \text{ (in part)} \end{aligned}$$

Construction of proembryo at third cell generation: eight cells arranged in seven tiers.

Species included: occasional in *Physalis minima*.

VI. Physalis III Variation. — Destination of tetrad elements: variable; any one of the three following formulae may prevail:

$l = pco + pvt + phy + icc + iec$
$l' = co + s$ (in part)
$m + ci = s$ (in part)

or

$l = pco + pvt + phy$
$l' = icc + iec + co + s$ (in part)
$m + ci = s$ (in part)

or

$l = pco + pvt + \frac{1}{2}phy$
$l' = \frac{1}{2}phy + icc + iec + co + s$ (in part)
$m + ci = s$ (in part)

Construction of proembryo at third cell generation: eight cells in seven tiers.

Species included: rare in *Physalis minima; Daucus carota* (*Apiaceae*) follows the third formula.

BHADURI, P. N. 1936. Studies on the embryology of the *Solanaceae* I. Bot. Gaz. 98: 283–295.

MACTANG, M. V. 1936. Floral biology and morphology of the eggplant. Philippine Agr. 25: 30–53.

PERSIDSKY, D., and J. MODILEWSKI. 1935. Citologitchnoembryologitchne doslidjennia osnovnikh sortiv *Nicotiana rustica* L. Jour. Inst. Bot. Ac. Sci. d'Ukraine 1935: 3–15.

SATINA, S., and A. F. BLAKESLEE. 1935. Fertilization in the incompatible cross *Datura stramonium* × *D. metel*. Bull. Torrey Bot. Club 62: 301–312.

SOUÈGES, R. 1920. Embryogénie des *Solanacées*. Développement de l'embryon chez les *Nicotiana*. C. R. Acad. Sci. Paris 170: 1125–1127.

SOUÈGES, R. 1920. Embryogénie des *Solanacées*. Développement de l'embryon chez les *Hyoscyamus* et *Atropha*. C. R. Acad. Sci. Paris 170: 1279–1281.

SOUÈGES, R. 1922. Recherches sur l'embryogénie des *Solanacées*. Bull. Soc. Bot. France 69: 163–178; 236–241; 352–365; 555–585.

SOUÈGES, R. 1936. Développement de l'embryon chez le *Schizanthus* et les *Petunias*. Bull. Soc. Bot. France 83: 570–577.

TOGNINI, F. 1900. Sull'embriologia di alcune *Solanaceae* di appunti lasciati. Atti Inst. Bot. Pavia II, 6: 109–112.

Convolvulaceae:- It is impossible to determine precisely the rules by which the embryo of *Convolvulus arvensis* is constructed. The first zygotic division is transverse, but thereafter the succeeding walls are without definite order and assume the most variable directions. The terminal cell appears invariably to be partitioned first by a diagonal or oblique wall, but it is not possible to determine precisely whether this wall is a deviation from an essentially vertical or an essentially horizontal wall. However, because of the marked resemblance of the embryology of *C. arvensis* to the Lotus Variation, Onagrad Type, the wall may be arbitrarily taken to be essentially vertical and therefore *C. arvensis* is assigned to that Variation. The basal cell contributes nothing to the construction of the embryo proper but merely becomes a massive suspensor.

The zygote of the parasitic *Cuscuta gronovii* divides transversely; the basal cell being much larger than the terminal one. Subsequent divisions occur in no fixed order, so that the developing embryo assumes irregular forms. The most usual form is an elongated type with a swollen base, with a one-celled or no suspensor. A spherical type is much less common; it lacks a suspensor. In neither type is there any clear indication of histogen differentiation and in mature embryos there is nothing which can be considered as cotyledons. The mature embryo is large and well developed, consisting of about two spirals. Assignment of this species is difficult; it may conform to the Penaea Variation, Asterad Type.

In *Cuscuta monogyna* and *C. epithymum* the zygote likewise divides into a small terminal and a large basal cell; the latter does not divide again but may later contain as many as three nuclei. The terminal cell divides twice to form a row of three cells, of which the lower two do not again segment but subsequently become plurinuclear. Although it is not clear whether a hypophysis is developed, it is evident that these two species are assignable to the Corydalis Variation, Caryophyllad Type.

Cuscuta europaea seems to develop more like *C. gronovii* than the two preceding species.

The early embryogeny of *Convolvulus sepium* is said to be much like that of *Cuscuta gronovii*, but there is a possibility that it may more closely resemble that of *Convolvulus arvensis*. The embryo develops an enormous suspensor. Polyembryony is said to be the rule in this species, but the exact origin of the extra embryos is not clear.

Evolvulus alsinoides develops a proembryo of seven single-celled tiers. The basal cell is said to remain undivided, which would mean that all cells save the lowest one are derived from the terminal cell. The latter then divides longitudinally. The sequence of events after this stage is not at all clear. It is claimed that this species follows the Myosotis Variation, Chenopodiad Type, but this appears to be questionable.

Ipomoea learii, *I. hederacea* and *Argyreia speciosa* follow the same type of embryonomy, which at least in the earlier proembryonic stages bears a very marked resemblance to the corresponding stages in the Myriophyllum Variation, Caryophyllad Type, save that the vertically partitioned basal cell eventually develops into a more or less massive suspensor.

FEDORTSCHUK, W. 1931. Embryologische Untersuchung von *Cuscuta monogyna* Vahl und *Cuscuta epithymum* L. Planta 14: 94–111.
MACPHERSON, G. E. 1921. Comparison of development in dodder and morning glory. Bot. Gaz. 71: 392–398.
RAO, K. V. R. 1940. Gametogenesis and embryogeny in five species of the *Convolvulaceae*. Jour. Indian Bot. Soc. 19: 53–69.
SOUÈGES, R. 1937. Embryogénie des *Convolvulacées*. Développement de l'embryon chez le *Convolvulus arvensis* L. C. R. Acad. Sci. Paris 205: 813–815.

Personales

The *Columelliaceae* represent the only uninvestigated family.

Scrophulariaceae:- In *Veronica arvensis* the quadrant and octant stages are differentiated precisely as in the Onagrad Type. The hypophysis, however, originates in a different manner: at the time of formation of the quadrant stage, the basal cell is segmented transversely; the upper daughter cell becomes the middle cell, *m*, then at the time the octants are being formed, this cell again divides transversely into two cells, *d* and *f*. Consequently, the octant proembryo in *V. arvensis* consists of four cells which have arisen from the original basal cell and eight which arose from the terminal cell. In other words, there are twelve cells in place of the six of the typical Onagrad Type. Shortly afterward, cell *d* is segmented transversely into two superposed cells, *p* and *q*; it is the upper daughter cell, *p*, which, as a general rule, becomes the hypophysis cell. A typical curved wall is then developed in this cell. The suspensor becomes long and filamentous.

The following species all appear to conform to the Veronica Variation: *Scrophularia marylandica*, *Ilysanthes hyssopioides*, *I. parviflora*, *Bonnaya tenuifolia*, *Penstemon secundiflorus*, *Striga lutea* (in this semiparasitic species the derivatives of the basal cell may produce haustoria), *Corytoloma cyclophyllum*, *Limosella aquatica*, *Paulownia tomentosa*, *Angelonia grandiflora*, *Dopatrium lobelioides*, *Stemodia viscosa*, *Sopubia trifida*. On the other hand, *Verbascum blattaria* follows exactly the Mentha Variation, Onagrad Type.

EVANS, A. T. 1919. Embryo sac and embryo of *Pentstemon secundiflorus*. Bot. Gaz. 67: 427–437.
IYENGAR, C. V. K. 1940. Development of embryo-sac and endosperm-haustoria in some members of *Scrophulariaceae*. Jour. Indian Bot. Soc. 19: 5–17.
IYENGAR, C. V. K. 1940. Structure and development of seed in *Sopubia trifida* Ham. Jour. Indian Bot. Soc. 19: 251–261.
LAURENT, V. 1923. Zur Entwicklungsgeschichte von *Corytoloma cyclophyllum* Dus., n. sp. ined. Svensk Bot. Tidskr. 17: 165–174.
MILLSAPS, V. 1936. The structure and development of the seeds of *Paulownia tomentosa*. Jour. Elisha Mitchell Sci. Soc. 52: 56–75.
MITCHELL, M. R. 1915. The embryo sac and embryo of *Striga lutea*. Bot. Gaz. 59: 124–135.
RAGHAVAN, T. S., and V. K. SRINIVASAN. 1941. Morphological and cytological studies in the *Scrophulariaceae*. III. A contribution to the life-history of *Ilysanthes parviflora* Bench. Proc. Indian Acad. Sci., B, 13: 24–32.
SCHERTZ, F. M. 1919. Early development of floral organs and embryonic structures of *Scrophularia marylandica*. Bot. Gaz. 68: 441–450.
SOUÈGES, R. 1921. Embryogénie des *Scrofulariacées*. Développement de l'embryon chez le *Veronica arvensis* L. C. R. Acad. Sci. Paris 172: 703–705.
SOUÈGES, R. 1935. Recherches embryologiques sur les *Verbascées*. Les types secondaires du développement embryonnaire. Bull. Soc. Bot. France 82: 328–334.
SRINIVASAN, V. K. 1940. Morphological and cytological studies in the *Scrophulariaceae*. Jour. Indian Bot. Soc. 19: 197–222.
SVENSSON, H. G. 1928. Zur Entwicklungsgeschichte der Blüten und Samen von *Limosella aquatica* L. Svensk Bot. Tidskr. 22: 465–476.

Orobanchaceae:- In *Orobanche hederae* (*O. hederacea* of older authors) and *O. eryngii* the basal cell of the two-celled proembryo never segments. Development is essentially that of the Lythrum Variation, Onagrad Type, but there are irregularities in the partitioning of tiers *l* and *l'*. The epidermal initials are delimited at a rather early stage and the proembryonic condition is retained for an unusually long time. *O. cumana* appears to develop like the other two species, although only a single stage is depicted.

The zygote of *Aeginetia indica* remains quiescent until long after the endosperm formation has been in progress, whereupon it elongates considerably by pushing itself between the cells of the endosperm. Two or three transverse walls are soon erected, giving a proembryo of three or four cells linearally arranged. The basal cell is the suspensor initial, which usually does not divide further but promptly degenerates. Division of the proembryonic cells appears to be rather variable; this phase was not studied extensively. A longitudinal wall may be erected in either the terminal or subterminal cell or in both to give either a five- or a six-celled proembryo. The mature embryo is a globose body entirely devoid of even the rudiments of cotyledons, root tip, stem tip or other customary embryonal regions. The best that can be deduced from this brief description is that *A. indica* belongs under the Caryophyllad Type.

Events in *Christisonia subacaulis* appear to agree with those in *Aeginetia*. The basal cell does not divide again but becomes empty by the time the terminal cell has divided transversely once or twice. A linear row of about six cells is produced, then longitudinal walls appear in the cells which are third and fourth from the top. The mature embryo is subglobular to obovoid in shape and no histogens save an epidermis are present.

CRÉTÉ, P. 1942. Recherches histologiques et physiologiques sur l'embryologie des Labiatiflores. Thesis, Univ. of Paris.
JULIANO, J. B. 1935. Anatomy and morphology of the Bunga, *Aeginetia indica* Linnaeus. Philippine Jour. Sci. 56: 405–451.
KOCH, L. 1875. Ueber die Entwicklung des Samens der *Orobanchen*. Jahrb. Wiss. Bot. 11: 218–261.
PERSIDSKY, D. 1926. Zur Embryologie der *Orobanche cumana* Wallr. und der *O. ramosa* L. Bull. Jardin Bot. Kieff 4: 6–10. (Russian with German summary)
WORSDELL, W. C. 1896. On the development of the ovule of *Christisonia*, a genus of the *Orobanchaceae*. Jour. Linn. Soc. London, Bot., 31: 576–584.

Lentibulariaceae:- The megagametophyte of *Utricularia coerulea* has been described as beginning to grow out of the micropyle at the four-nucleate stage so that the egg and synergids lie outside of the ovule. The zygote forms a bud-like outgrowth through the micropyle down into the compact endosperm which develops within the ovule. The zygote nucleus remains near the terminal end of the tube, where it divides; the resulting wall separates the zygote into a small hemispherical terminal cell and a long basal cell. Each of these cells next divides transversely. Intermediate stages were not followed out in detail, but it appears that most of the embryo proper is derived from the terminal cell alone. The mature embryo is undifferentiated except for the epidermal layer and a group of smaller, actively meristematic cells at the apical end.

It is very probable that *Utricularia vulgaris* conforms to the Polemonium Variation, Chenopodiad Type. The description and figures, although very old, are clear and precise. *U. neglecta*, *U. intermedia*, *U. bremii* and *U. stricta* are said to agree substantially with *U. vulgaris*. The embryonomy of *U. inflata* is very similar to that of *U. coerulea*. A linear proembryo of about five cells is formed; the lowermost of the basal cells is very much elongated and may divide transversely near the lower end. Longitudinal walls are erected in the first five cells of a seven-tiered proembryo, followed by periclinal walls in the top tier which delimit the epidermal initials. The earlier proembryonic stages of *U. purpurea* resemble those of *U. inflata*. The mature embryos of *U. exoleta* and *U. bifida* in general are elongated and much like those of *U. coerulea*.

The embryo of *Pinguicula alpina*, like that of the *Utricularias*, is very much elongated; quadrants and octants are said to be regularly formed. *P. vulgaris* resembles the *Utricularias*.

The sequence of events in *Polypompholyx* sp. seems to have been misinterpreted; it appears to follow the Polemonium Variation almost exactly.

Kamiensky, F. 1877. Vergleichende Untersuchungen über die Entwicklungsgeschichte der *Utricularien*. Bot. Zeit. 35: 761–775.

Kausik, S. B. 1935. The life history of *Utricularia coerulea* L. Current Sci. 3: 357–359.

Kausik, S. B. 1938. Pollen development and seed formation in *Utricularia coerulea* L. Beih. Bot. Centralbl. 58: 365–378.

Lang, F. X. 1901. Untersuchungen über Morphologie, Anatomie und Samenentwicklung von *Polypompholyx* und *Byblis gigantea*. Flora 88: 149–206.

Merz, M. 1897. Untersuchungen über die Samenentwicklung der *Utricularieen*. Flora 84: 69–87.

Stolt, K. A. H. 1936. Beiträge zur Embryologie der *Lentibulariaceen*. Svensk Bot. Tidskr. 30: 690–696.

Gesneriaceae:- The proembryo of *Rhytidophyllum crenulatum* (with which *R. tomentosum* agrees in all details) consists of a linear row of four to six cells, of which the terminal one divides longitudinally. Assignment of this species is very problematical, although it appears best to place it under the Onagrad Type without inclusion in a definite Variation.

Abnormal embryos have been observed. In two instances the embryos had elongated without developing cotyledons and lacked differentiation of plerome and periblem. In two other cases, each embryo had three well developed cotyledons. In each of these four instances, the resemblance to corresponding phenomena in *Zauschneria latifolia* (*Onagraceae*) is suggestive.

Ramondia nathaliae (and inferentially also *R. serbica*) and a hybrid between *Streptocarpus achimenoeflorus* and *S. kewensis* conform closely to the Lythrum Variation, Onagrad Type. *Ramondia pyrenaica*, although agreeing with the other species during the earlier developmental stages, later often forms a long and thick suspensor, many of whose cells appear to be plurinucleate. This, however, is not a constant feature and its significance is doubtful.

COOK, M. T. 1907. The embryology of *Rhytidophyllum*. Bull. Torrey Bot. Club 34: 179–184.
CRÉTÉ, P. 1942. Recherches histologiques et physiologiques sur l'embryologie des Labiatiflores. Thesis, Univ. of Paris.
GLISIC, L. 1924. (Development of the X-generation and embryo in *Ramondia*.) Diss., Univ. of Belgrad. (Bulgarian, with English summary)

Bignoniaceae:- The four-celled proembryo of *Catalpa kaempferi* is erected as in *Capsella* or *Veronica* (FIG. 60A, B), but instead of next dividing vertically as in those genera, each of the two upper juxtaposed cells divides, in a very constant manner, transversely as in the Solanad Type to engender four quadrants disposed in a vertical plane (FIG. 60D). These quadrants next segment

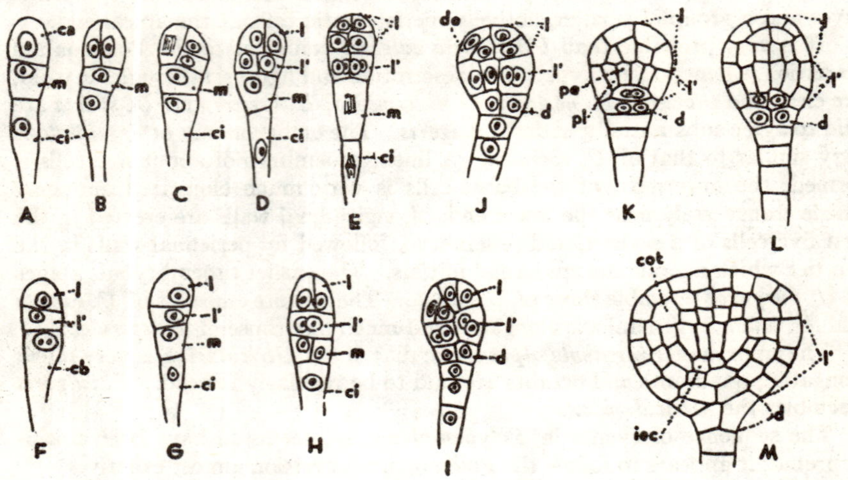

FIG. 60. — *Catalpa kaempferi*. Description in text. All × 350. (*Redrawn from* SOUÈGES 1940).

vertically to produce four upper octants and four lower octants (FIG. 60E). The upper octants engender the cotyledonary portion of the embryo; they first undergo periclinal divisions then anticlinal divisions in either a horizontal or a vertical direction to separate the epidermal initials (FIG. 60J–L). Later two cell layers are differentiated beneath the epidermis. The lower octants behave as in *Capsella* or *Veronica* and produce the hypocotyledonary portion (FIG. 60D, E).

Occasionally a linear tetrad is produced by transverse division of the terminal cell (FIG. 60F, G). The two upper cells of this tetrad, *l*, and *l'*, by means of longitudinal divisions engender four quadrants in two layers (FIG. 60D). In several other cases, the two upper quadrants, instead of dividing vertically to give four upper octants disposed in a plane around the axis, are segmented transversely and produce four cells homologous to the four upper octants and always disposed in two tiers (FIG. 60H). Tangential walls are next formed in these cells, then, in the central cells, there appears the disposition which ordinarily gives rise to the cotyledonary portion (FIG. 60I).

The hypophysis takes its origin from cell *d*, a daughter cell of *m* (FIG. 60J–L). The tissues which it produces become rather deeply embedded in the body of the embryo and produce the initials of the root cortex (FIG. 60M). Two superposed tetrads are produced by either of two procedures: either by horizontal walls followed by two vertical cruciate walls, or inversely by meridional walls followed by horizontal walls.

The long and filamentous suspensor is developed as in *Capsella*.

A second and earlier investigation on *C. kaempferi* offers no description and only a few inconclusive figures.

The embryogeny of *Bigonnia megapatomica* is said to be of the Capsella type; it probably also conforms to the Catalpa Variation.

MAURITZON, J. 1935. Etwas über die Embryologie der *Bignoniaceen*. Bot. Notiser 1935: 60–77.

SOUÈGES, R. 1940. Embryogénie des *Bignoniacées*. Développement de l'embryon chez le *Catalpa kaempferi* Sieb. et Zucc. C. R. Acad. Sci. Paris 210: 116–118.

SWAMY, B. G. L. 1941. Contribution to the life-history of *Bignonia megapatomica*. Jour. Indian Bot. Soc. 20: 299–305.

Pedaliaceae:- To judge from the inadequate illustrations accompanying the rather meager accounts of the embryogeny of *Sesamum orientale*, *Martynia louisiana* and *Pedalium murex*, the species probably follow the Onagrad Type.

ANDERSON, F. 1922. The development of the flower and embryogeny of *Martynia louisiana*. Bull. Torrey Bot. Club 49: 141–157.

MAURITZON, J. 1936. Die Endospermentwicklung von *Sesamum indicum* und *orientale*. Arkiv f. Bot. 28B, No. 5: 1–6.

SRINIVASAN, A. R. 1942. Contribution to the morphology of *Pedalium murex* Linn. and *Sesamum indicum* DC. Proc. Indian Acad. Sci., B, 16: 155–164.

Acanthaceae:- Many species of the *Acanthaceae* have been investigated but the reports on the embryogeny of all save one species are in an extremely incomplete and confusing condition. The fragmentary drawings are mostly so crude that nothing can be determined from them. Many of these figures plainly appear to have been based upon sections microtomed in an oblique plane. The one partially complete account concerns *Asteracantha longifolia*. From the figures and description there is no doubt that this species conforms precisely to the Lythrum Variation, Onagrad Type.

GIGANTE, R. 1929. Embriologia dell'*Acanthus mollis* L. Nuovo Giorn. Bot. Ital. 36: 1–33.

MAURITZON, J. 1934. Die Endosperm- und Embryoentwicklung einiger *Acanthaceen*. Lunds Univ. Årsskr., N. F. Avd. 2, 30: 1–41.

RANGASWAMY, K. 1941. Cyto-morphological studies in *Asteracantha longifolia* Nees (*Hygrophila speciosa* T. And.). Proc. Indian Acad. Sci., B, 14: 149–165.

Lamiales (Ammiales)

The *Selaginaceae* have never been studied embryologically.

Globulariaceae:- *Globularia vulgaris* conforms precisely with the Lythrum Variation, Onagrad Type. Although imperfectly illustrated, *G. cordifolia* may also follow the same type of development.

BILLINGS, F. H. 1901. Beiträge zur Kenntniss der Samenentwicklung. Flora 88: 253–318.

CRÉTÉ, P. 1943. Recherches histologiques et physiologiques sur l'embryologie des Labiatiflores. Embryologie du *Globularia vulgaris* L. Bull. Soc. Bot. France 90: 36–39.

Myoporaceae:- For *Myoporum serratum* there is depicted a two-celled proembryo with a somewhat conical small terminal cell and an elongated, irregularly inflated basal cell. At a later stage the embryo has the typical appearance of either an Onagrad or a Solanad Type embryo, but since nothing is said of the plane of the first division in the terminal cell, proper assignment is impossible.

BILLINGS, F. H. 1901. Beiträge zur Kenntniss der Samenentwicklung. Flora 88: 253–318.

Verbenaceae:- *Verbena officinalis* conforms closely to the Mentha Variation

Onagrad Type. The filamentous suspensor is rather long and somewhat haustorial, being closely surrounded with endosperm cells and deeply embedded in the latter.

Lantana indica, *Stachytarpheta indica* and *Phryma leptostachya* (the latter species is sometimes included in a separate family, the *Phrymaceae*) also appear to conform to the same Variation, as does *Tectona grandis*, which possesses an excessively long zygote and suspensor.

In *Lantana camara*, which also follows the Mentha Variation, the first wall in the hypophysis cell may be typically curved, or it may be diagonal as frequently happens in certain Onagraceae.

COOPER, D. C. 1941. Macrosporogenesis and the development of the seed of *Phryma leptostachya*. Amer. Jour. Bot. 28: 755–761.

CRÉTÉ, P. 1942. Recherches histologiques et physiologiques sur l'embryologie des Labiatiflores. Thesis, Univ. of Paris.

KOORDERS, S. H. 1896. Morphologische und physiologische Embryologie von *Tectona grandis* Linn. f. (Djati- oder Teak-Baum). Engler's Bot. Jahrb. 21: 458–498.

SOUÈGES, R. 1935 Embryogénie des *Verbénacées*. Développement de l'embryon chez le *Verbena officinalis* L. C. R. Acad. Sci. Paris 201: 234–236.

TATACHAR, T. 1940. The development of the embryo-sac and formation of haustoria in *Lantana indica* Roxb., and *Stachytarpheta indica* Vahl. Jour. Indian Bot. Soc. 19: 45–52.

Menthaceae (Labiatae, Ammiaceae, Lamiaceae):- The embryogeny of *Mentha viridis* and *Glechoma hederacea* is essentially similar, but the former is being taken as the basis for the following description. *M. viridis* follows the Onagrad Type, but differs in the following respects: (*1*) in the origin of the hypophysis cell, which is a descendant, generally in the second degree, of the intermediate cell (*m*) of the tetrad; (*2*) by the form of the suspensor, which is represented by a simple slender filament; and (*3*) by differences in the rapidity of the segmentation in the terminal and basal cells of the proembryo.

Recapitulatory Table for Mentha viridis

I. FIRST CELL GENERATION

Proembryo of two cells disposed in two tiers:

$ca = pco + pvt + phy + icc$

$cb = iec + co + s$

II. SECOND CELL GENERATION

Proembryo of four cells disposed in three tiers:

$q = pco + pvt + phy + icc$

$m = iec + co + s$ (in part)

$ci = s$ (in part)

III. THIRD CELL GENERATION

Proembryo of six cells disposed in three tiers:

Disposition and destinations are the same as in the second cell generation.

IV. FOURTH CELL GENERATION

Proembryo of twelve cells disposed in six tiers:

$l = pco + pvt$

$l' = phy + icc$

$d = iec + co$

$f + n + n' = s$

The terminal cell of the two-celled proembryo divides first, in the longitudinal plane; and the two daughter cells divide, again in the vertical plane, at the same time that the basal cell is segmented transversely. The characteristic tetrad stage of the Onagrad Type is thus elided. At the next stage the four quadrants are segmented transversely into octants: the four upper octants compose region *l* and the four lower octants represent region *l'*. Shortly after this the two cells *m* and *ci*, derived from the basal cell, are both divided transversely to give rise to a series of four superposed elements, *d*, *f*, *n* and *n'*. The uppermost of these, *d*, becomes the hypophysis cell and the others go into the formation of the suspensor. Since region *l* gives rise to the cotyledonary region and region *l'* to the hypocotyledonary axis, the eventual destination of each cell region in the twelve-celled proembryo is approximately the same as in the generalized Onagrad Type. The further development of the embryo corresponds very closely to that of the Type. The first division in the hypophysis cell is characteristic, but as in many other species which follow the Onagrad Type, it may be oblique, with one end attached to the peripheral wall. In *Glechoma hederacea* this wall may rarely even be vertical.

The table of recapitulation for *M. viridis* is given on the previous page.

The embryonomy of *Lamium purpureum* in many respects resembles that of *Urtica pilulifera*, but differs at the following points: (*1*) by the frequently oblique position of the wall separating the two lower cells of the tetrad; (*2*) by the confusion of the two regions *m* and *n* and the participation in indeterminate proportion of the region *n* in the creation of the hypocotyl; (*3*) by the origin of the "hypophysis" region and its mode of segmentation; (*4*) by the filamentous form of the suspensor, and (*5*) by the irregularity of the partitionings which do not permit description of the differentiation of the histogens with certainty.

The proembryonic tetrad assumes very irregular segmentations; they are unquestionably not accidental happenings, but are very commonly found and appear to be hereditary. The two upper cells are juxtaposed and the two lower cells are superposed in diverse fashion. The wall in the basal cell is often oblique, with the internal end attached to the wall which originally separated the terminal from the basal cell in the two-celled proembryo. The position of this wall may lead to significant modifications in the ultimate disposition of the proembryonic cells and in the constitution of the different regions of the proembryo. In the majority of the forms presently to be described, the direction of this wall is clearly visible; even at the time of origin of the cotyledons it is possible to recognize the aspect of an inclined plane traversing the hypocotyledonary region and the division into two parts which are most often unequal.

The eight-celled proembryo is created by bipartition of the four elements of the tetrad: the two upper cells divide longitudinally into four quadrant cells; the intermediate cell *m* is divided into two juxtaposed cells and the basal cell, *ci*, is segmented transversely.

The sixteen-celled proembryo arises by division of each cell of the eight-celled proembryo. The divisions in the quadrant cells are precisely like the same divisions in *Urtica pilulifera*, and the various proembryonic regions have the identical destinations. That is to say, the upper region *q* gives rise to the cotyledonary portion; the region *m* engenders the hypocotyledonary axis in its entirety or its upper part in variable proportions; region *n* contributes to the lower portion of this axis or at least the initials of the cortex; the region *o* contributes to the formation of the central region of the root cap and *p* is converted into a filamentous suspensor.

Further development in region *q* resembles that in *Urtica pilulifera* in all details. In the sixteen-celled proembryo region *m* consists of four circumaxial cells. They divide vertically by walls which are often tangential and separate the epidermal initials but the orientation of the wall is often very indefinite at first. Later transverse walls are erected, so that at the time of origin of the cotyledons region *m* most often consists of two cell tiers. The sequence of divisions just described is that prevailing in the most regular of the embryos observed. On the other hand, it is quite impossible to present an exact description of the sequence of divisions when region *m* is separated from region *n* by a more or less oblique wall. The best that can be said is that region *m* contributes by far the larger part of the hypocotyl, whose histogens, however, are not differentiated until rather late in the ontogeny of the embryo.

Region *n* of the sixteen-celled proembryo is usually composed of two juxtaposed cells, each of which divides in the most regularly constituted embryos by a meridional wall. The resulting four cells in such instances are circumaxial. In other cases two of the cells may be divided by horizontal walls. Shortly thereafter, the four cells are segmented by vertical but generally tangential walls to engender a layer of about twelve cells; the outermost are placed as a prolongation of the epidermis of region *m*; the central elements may now constitute the initials of the root cortex, or they become segmented transversely to engender several cell layers and thus contribute to the construction of a more or less extended portion of the hypocotyl. When the plane of separation of the two regions *m* and *n* is oblique, the portion of the hypocotyl contributed by region *n* is necessarily greater and the segmentations in this region are thus profoundly modified.

Region *o*, consisting of a single cell in the sixteen-celled proembryo, is generally segmented by two cruciate walls. The four circumaxial cells thus engendered are next partitioned by horizontal walls to give rise to two groups of four superposed cells: the upper group represents the first cells of the calyptrogen, while from the lower group is derived the peripheral layer of the root cap. In many cases the first wall in cell *o* is horizontal and separates two cells placed one above the other. It is very difficult to determine the future evolution of these two elements; it appears that the lower cell enters into the construction of the suspensor, while the upper cell presumably assumes the role of the mother cell *o*.

The single cell in region *p* of the sixteen-celled proembryo becomes very much elongated, then segments transversely and finally gives rise to a filamentous suspensor.

Lycopus europaeus most closely resembles *Glechoma hederacea* in development; consequently, it may be assigned to the Mentha Variation. *Physostegia virginiana* conforms similarly.

All the following species, studied by the same investigator, conform more or less to the Mentha Variation: *Ajuga reptans*, *Teucrium botrys*, *Amethystea coerulea*, *Prasium majus*, *Scutellaria minor*, *Marrubium pannonicum* (but with irregularities during the early stages), *Sideritis hirsuta*, *S. scordioides*, *Nepeta cataria*, *N. macrantha*, *Dracocephalum ruyschiana*, *D. mairei*, *D. thymifolium*, *Lophanthus chinensis*, *Lallemantia peltata*, *L. ilurica*, *Brunella vulgaris* (cotyledonary region develops rather precociously), *Melittis melissophyllum* (divisions in hypophysis cell are remarkably precocious and may be either transverse or vertical, producing an extensive hypophyseal region), *Phlomis alpina* (development of hypophyseal region also irregular), *Leonurus cardiaca*, *Salvia sclarea*, *S. pratensis*, *Ziziphora tenuior*, *Melissa officinalis*, *Calamintha grandiflora*, *Thymus serphyllum*, *Origanum vulgare*, *Perilla arguta*, *Ocimum basilicum*. On the other hand, *Galeopsis tetrahit*, *G. pyrenaica* and *Ballota foetida* all conform to the Lamium Variation of the Asterad Type.

Crété, P. 1942. Recherches histologiques et physiologiques sur l'embryologie des Labiatiflores. Thesis, Univ. of Paris.

Ruttle, M. L. 1932. Chromosome number, embryology and inheritance in the genus *Lycopus*. Die Gartenbauwiss. 7: 154–177.

Sharp, L. W. 1911. The embryo sac of *Physostegia*. Bot. Gaz. 52: 218–225.

Souèges, R. 1921. Embryogénie des *Labiées*. Développement de l'embryon chez le *Mentha viridis* L. C. R. Acad. Sci. Paris 172: 1057–1058.

Souèges, R. 1921. Embryogénie des *Labiées*. Développement de l'embryon chez le *Glechoma hederacea* L. et le *Lamium purpureum* L. C. R. Acad. Sci. Paris 173: 48–50.

Souèges, R. 1921. Recherches sur l'embryogénie des *Labiées*. Bull. Soc. Bot. France 68: 441–464.

Butomales

Butomaceae:- *Limnocharis emarginata* conforms closely to the Sagittaria

Variation, Caryophyllad Type. The basal cell increases greatly in size but never divides subsequently unless polyembryony occurs. In this case it divides repeatedly to form an embryonic mass from which several potential embryos arise. To the suspensor of the older embryo, however, are added as many as five cells derived from the terminal cell.

Butomus umbellatus and *Butomopsis lanceolata* agree likewise with the Sagittaria Variation.

HALL, J. G. 1902. An embryological study of *Limnocharis emarginata*. Bot. Gaz. 33: 214–219.
HOLMGREN, I. 1913. Zur Entwicklungsgeschichte von *Butomus umbellatus* L. Svensk Bot. Tidskr. 7: 58–77.
JOHRI, B. M. 1937. The life-history of *Butomopsis lanceolata* Kunth. Proc. Indian Acad. Sci., B, 4: 137–162.

Hydrocharitaceae:- The *Hydrocharitaceae*, altogether, all appear to conform to the Sagittaria Variation, Caryophyllad Type. The following species have been studied: *Ottelia alismoides*, *Blyxa echinosperma*, *Elodea canadensis*, *Enalus acoroides*, *Valisneria spiralis* and *Halophila* sp.

BURR, H. G. 1903. The embryology of *Valisneria spiralis*. Ohio Nat. 3: 439–443.
KAUSIK, S. B. 1940. A contribution to the embryology of *Enalus acoroides* (L. fils) Steud. Proc. Indian Acad. Sci., B, 11: 83–99.
MURTHY, S. K. N. 1935. The life-history of *Ottelia alsimoides* Pers. Proc. Indian Acad. Sci., B, 2: 59–66.
RANGASAMY, K. 1941. A morphological study of the flower of *Blyxa echinosperma* Hook. f. Jour. Indian Bot. Soc. 20: 123–133.
RIDDLE, L. C. 1905. Notes on the morphology of *Philotria*. Ohio Nat. 5: 304–305.
ROSENBERG, O. 1901. Die Embryologie von *Zostera marina* L. K. Svenska Vet.-Akad. Handl. 27, Afd. III, No. 6.
WYLIE, R. B. 1904. The morphology of *Elodea canadensis*. Bot. Gaz. 37: 1–22.

Alismatales

No species in the *Petrosaviaceae* has ever been studied.

Alismataceae:- To summarize, the embryonomy of the *Alismataceae*, from the concept of modern interpretations of the various stages, possesses a remarkable homogeneity among all species that have been investigated. These include: *Alisma plantago*, *Sagittaria variabilis*, *S. lancifolia*, *S. sagittaefolia*, *S. guyanensis*, *S. latifolia*, *Damasonium stellatum*, *D. alisma*, *Limnophyton obtusifolium*, *Machaerocarpus californicus* and *Elisma natans*.

Sagittaria sagittaefolia, which typifies the Sagittaria Variation, Caryophyllad Type, may be discussed in detail. The zygote is divided transversely into two markedly unequal cells (FIG. 61A). The basal cell, which is the larger of the two, does not divide again but becomes transformed directly into a large vesicle (FIG. 61B–G). The terminal cell becomes divided into three superposed cells, of which the upper one is next partitioned longitudinally (FIG. 61B, C). The two juxtaposed cells are destined to give rise to the cotyledon; the intermediate cell engenders the upper portion of the hypocotyl and the "stem apex"; while the derivatives of the lower cell differentiate the lower portion of the hypocotyl, the initials of the cortex, the root cap and the suspensor proper.

The eight-celled proembryo is formed by the bipartitioning of the four cells of the tetrad (the basal cell is henceforth left out of consideration): the two superior elements are partitioned vertically to give four quadrants; the two lower elements give rise to a new tetrad constituted like the original tetrad.

The sixteen-celled proembryo in turn is formed by the division of each cell of the eight-celled proembryo: the quadrants are transformed into octants by means of transverse divisions; the two juxtaposed cells derived from cell *m* are partitioned vertically to give rise to four circumaxial cells; the element *n*, derived from *ci*, is divided vertically, and its sister cell, *n′*, like the mother cell from which it arose, is segmented transversely into two new superposed cells, *o* and *p* (FIG. 61D).

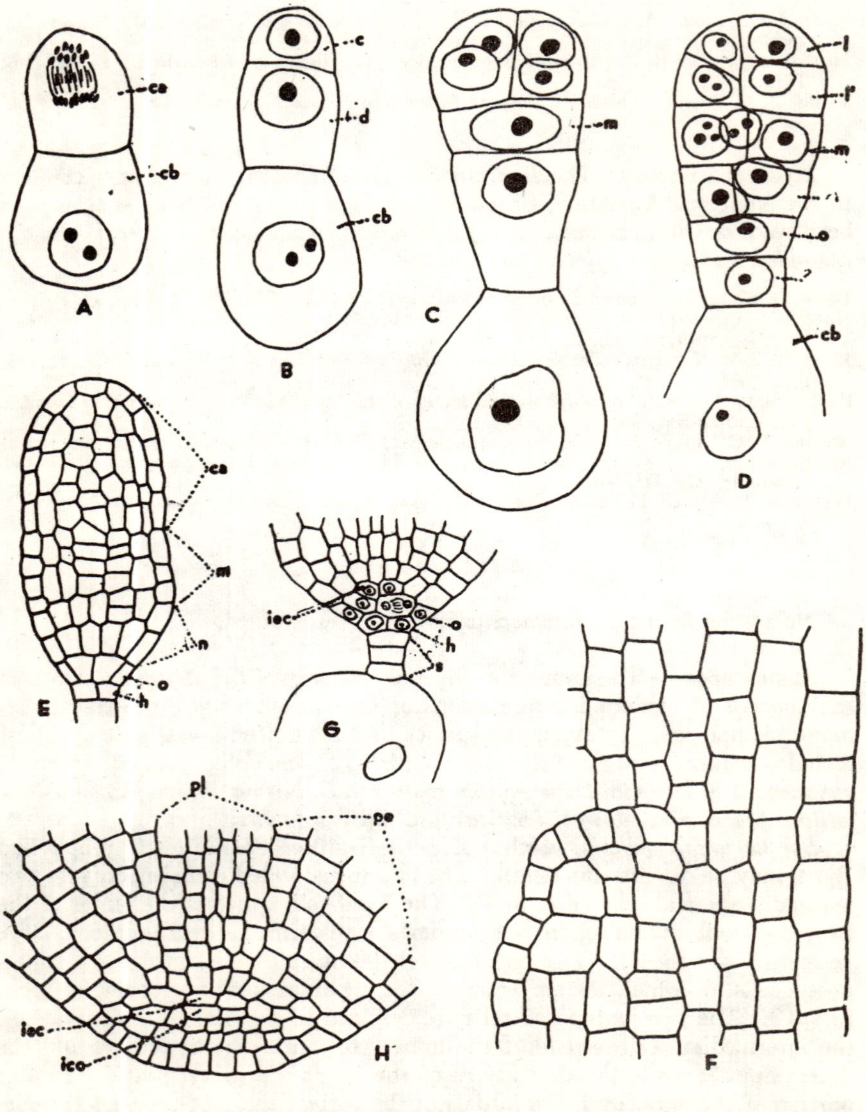

FIG. 61. — *Sagittaria sagittaefolia.* Description in text. A–D × 540; E × 280; F × 280; G–H × 240. (*Redrawn from* SOUÈGES 1931).

In the sixteen-celled proembryo, the cotyledonary part, *l* and *l′*, which become transformed in their entirety into the cotyledon, is represented by the eight octants; the hypocotyledonary portion by the two regions *m* and *n* situated below. From the derivatives of *m* are differentiated the stem apex; from those of *n*, that portion oi the hypocotyledonary axis corresponding to the radicle (FIG. 61E). The cell *o* gives rise alone to the initials of the cortex and to those of the root cap (FIG. 61E, G). After cell *p* is divided once transversely, the upper daughter cell, *h*, engenders the culminating layer of the root cap and the other cell, *r*, is converted into a true suspensor consisting of a series of from three to six superposed cells (FIG. 61E, G, H). There is thus no single hypophysis initial in *Sagittaria*.

It appears to be unnecessary to describe the sequence of divisions in each of the six regions up to the time that a new period of development appears, characterized by the appearance on one side of the embryo of a depression which indicates the point of origin of the "stem apex." This phenomenon involves a sharp change in the symmetry of the embryo; up to now it has been axial, but becomes bilateral (FIG. 61E). During this new period the following changes occur successively: (*1*) the growth of the cotyledon from the derivatives of the upper and lower cotyledonary octants; (*2*) the differentiation of the "stem apex" from the derivatives of region *m*, representing the upper part of the hypocotyledonary axis; (*3*) the construction of the radicle from the derivatives of *o* and *p* and from all the derivatives of region *n*, corresponding to the lower part of the hypocotyledonary axis.

The development of the cotyledon offers nothing of particular significance. Growth in length is accomplished by transverse segmentation of all cells, beginning with the lower octants; growth in width is exclusively at the expense of the periblem cells, comparable to a similar mode which is readily apparent in the hypocotyl. The epidermal cells, in conformity with the general rule, do not divide radially. The cells of the plerome are smaller than, but not otherwise readily distinguishable from, the neighboring periblem cells. The pericycle and endodermis are generally not well differentiated in the cotyledonary region. In the neighborhood of the hypocotyl the cotyledon develops throughout in two opposite directions, which determines the median plane of symmetry. The two extreme points continue to grow and describe a movement of curvature around the foliar bud which at this moment is clearly apparent. The two slender margins, uniquely constituted of two layers of epidermal cells applied one against the other, form a protective cover.

What is generally called the stem apex is erected at the expense of the two cell layers of region *m*. Two juxtaposed cells become distinguished by their much larger dimensions and by their mode of division. The cells divide transversely and not longitudinally; this denotes an arrest of development in the region where the two cells are located. The special mode of segmentation marks the location of the lateral depression which simultaneously separates the cotyledon from the foliar bud (FIG. 61F). The latter grows by multiplication of the periblem cells. On its surface it produces a new depression which differentiates the first leaf, homologous to the cotyledon, and a new rudimentary foliar bud, homologous to the first one. To discuss this more fully: the massive cellular protuberance appears to become bilobed, presenting two obtuse summits separated by a slight depression (FIG. 61F). One of these summits projects further than the other which appears clearly lateral. The former gives rise to the first leaf and develops in a manner preponderately in two directions, horizontal and vertical; the latter remains for a time as a very rudimentary projection, then presently it differentiates the second leaf and a small central cellular mass from which are later derived all the future leaves.

There is, in effect, no structure which can properly be called the stem apex. There is simply a massive central cellular protuberance, assymetrically placed on the embryo, which is shaped like a triangular pyramid. The three faces correspond to one complete turn of a single Archimedean spiral; from each face, in circumnutation, a leaf is cut off. Each leaf thus differentiated is represented by a segment limited exteriorly by a face of the pyramid and interiorly by a wall parallel to that face; each face of the pyramid successively gives rise to an analogous segment. In that manner, the summit of the pyramid passes successively into each foliar rudiment as it is developed; the central part of the base of the pyramid alone remains fixed and situated in the median plane. It is this point which is the seat of the constant regeneration of the protuberance since it is in direct relation with the plerome of the hypocotyl.

The development of the radicular end of the proembryo may now be described. Region *n*, at the time the cotyledon commences growth, is composed of the same three histogens (epidermis, periblem and plerome) as in region *m* and the lower cotyledonary octants (FIG. 61E). Transverse walls are erected

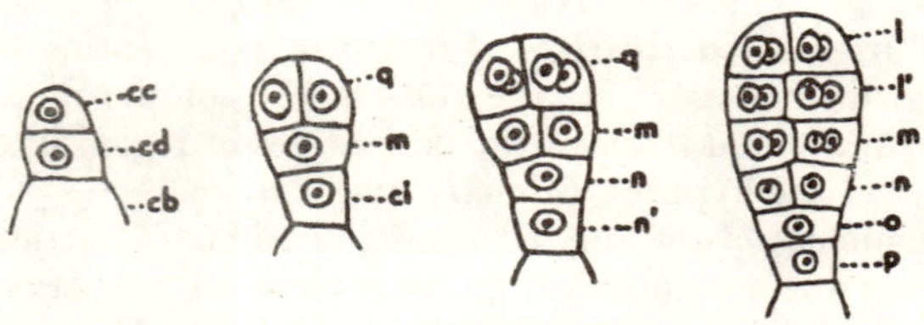

Recapitulatory Table for Sagittaria sagittaefolia
(from left to right)

I. FIRST CELL GENERATION
Proembryo of two cells disposed in two tiers (the basal cell is disregarded):
cc = *pco*
cd = *phy* + *pvt* + *icc* + *iec* + *co* + *s*

II. SECOND CELL GENERATION
Proembryo of four cells disposed in three tiers:
q = *pco*
m = *phy* (in part) + *pvt*
ci = *phy* (in part) + *icc* + *iec* + *co* + *s*

III. THIRD CELL GENERATION
Proembryo of eight cells disposed in four tiers:
q = *pco*
m = *phy* (in part) + *pvt*
n = *phy* (in part) + *icc*
n' = *iec* + *co* + *s*

IV. FOURTH CELL GENERATION
Proembryo of sixteen cells disposed in six tiers:
l + *l'* = *pco*
m = *phy* (in part) + *pvt*
n = *phy* (in part) + *icc*
o = *iec* + *co*
p = *s*

one by one until the region consists of two cell layers, then other transverse walls are formed in each of these layers so that the region is composed of four cell tiers at the time the lateral depression makes its appearance. Subsequent divisions have no definite sequence. The epidermal cells multiply by means of radial walls; they never form tangential walls as in the neighboring cells of the root cap. The periblem cells presently divide tangentially, then radial walls arise in the outer of these cells thus erected. The plerome cells multiply generally by means of vertical walls, but the planes are occasionally irregular. The traces of the primary meridional walls soon disappear and presently it becomes impossible to follow out the sequence of cell formation with certainty. In any event, the four innermost cells of the plerome constitute the initials of the central cylinder (FIG. 61H).

The elements of region *o* after three cytokineses in the horizontal plane are separated by two cruciate walls (FIG. 61E, G). The four cells thus constituted appear to divide by walls perpendicular to one another and parallel to the two meridional planes, thus giving rise to three elements in each of the four cells —

two external and one internal adjacent to the axis. Most often, however, tangential walls are formed in these four primary cells, being inserted at the upper extremity on the later walls of the epidermal cells of region *n*. After the formation of these tangential walls, each of the four primary cells of region *o* is found divided into two elements, one internal and the other external. All of the external cells, numbering from four to eight, located in a prolongation of the epidermis, thereafter multiply by radial, longitudinal or tangential walls to engender the most elongated elements at the summit of the root cap (FIG. 61G, H). They are the first to become detached when the cap commences to function in the germinating seed. The internal group of four cells becomes segmented to give rise to two groups of four superposed cells: the upper group represents the initials of the root cortex (*iec*) and the lower group constitutes the initials of the root cap or epidermis of the root (*ico*). This lower group is neither a calyptrogen nor a dermatocalyptrogen, as some authors have contended, since the cells of the epidermis never divide tangentially.

The recapitulatory table for *Sagittaria sagittaefolia* is shown on p. 236.

COOK, M. T. 1907. The embryology of *Sagittaria lancifolia*. Ohio Nat. 7: 97–101.

DAHLGREN, K. V. O. 1928. Die Embryologie einiger *Alismatazeen*. Svensk Bot. Tidskr. 22: 1–17.

JOHRI, B. M. 1935. Studies in the family *Alismaceae*. I. *Limnophyton obtusifolium* Miq. Jour. Indian Bot Soc. 14: 49–66.

JOHRI, B. M. 1935. Studies in the family *Alismaceae*. III. *Sagittaria guayanensis* H. B. K. and *S. latifolia* Willd. Proc. Indian Acad. Sci., B, 2: 33–48.

MAHESHWARI, P., and B. SINGH. 1943. Studies in the family *Alismaceae*. V. The embryology of *Machaerocarpus californicus* (Torr.) Small. Proc. Nat. Inst. Sciences of India 9: 311–322.

SOUÈGES, R. 1931. L'embryon chez le *Sagittaria sagittaefolia* L. Le cône végétatif de la tige et l'extrémité radiculaire chez les Monocotylédones. Ann. Sci. Nat., Bot., X, 13: 353–402.

Scheuchzeriaceae :- In *Scheuchzeria palustris* the basal cell presumably does not divide again once it has been formed; it becomes swollen and strongly vacuolized. Whether the terminal cell divides vertically or transversely is not stated, but it probably divides longitudinally since the four-celled proembryo depicted shows two tiers of two cells each. Proper assignment is impossible on the basis of the meager information.

STENAR, H. 1935. Embryologische Beobachtungen über *Scheuchzeria palustris* L. Bot. Notiser 1935: 78–86.

Triuridales

Triuridaceae :- Very scanty information on the embryogeny of the *Triuridaceae* is available; it is said to resemble that of the *Burmanniaceae* and the embryo is rudimentary and poorly differentiated, with a short suspensor, when mature.

WIRZ, H. 1910. Beiträge zur Entwicklungsgeschichte von *Sciaphila* spec. und von *Epirrhizanthes elongata* Bl. Flora 101: 395–446.

Juncaginales

The *Posidoniaceae* have never been investigated.

Juncaginaceae :- *Triglochin maritimum* is assignable to the Sagittaria Variation, Caryophyllad Type.

HOFMEISTER, W. 1861. Neue Beiträge zur Kenntnis der Embryobildung der Phanerogamen. II. Monokotyledonen. Abh. Kön. Sächs. Ges. Wiss. 7: 629–760.

SCHNARF, K. 1925. Kleine Beiträge zur Kenntnis der Entwicklungsgeschichte der Angio-

spermen. V. Über zwei kritische Fälle der Endospermentwicklung (*Verbena* und *Triglochin*). Oesterr. Bot. Zeitschr. 74: 40–50.
SOUÈGES, R. 1943. Embryogénie des *Scheuchzériacées*. Développement de l'embryon chez le *Triglochin maritimum* L. C. R. Acad. Sci. Paris 216: 746–748.

Lilaeaceae:- *Lilaea subulata* appears to conform to the Sagittaria Variation, Caryophyllad Type. It is much like *Zannichellia* (*Zannichelliaceae*) embryonomically, but it was not determined whether the "stem apex" is derived from tier *l* or tier *l'*.

CAMPBELL, D. H. 1898. Development of flower and embryo of *Lilaea subulata*. Ann. Bot. 12: 1–28.

Aponogetonales

Aponogetonaceae:- *Aponogeton monostachyon* and *A. crispum* are similar embryonomically and conform to the Sagittaria Variation, Caryophyllad Type. The mature embryo of the first species shows the rudiment of a foliage leaf besides the cotyledon, while *A. crispum* has three to four foliage leaves in addition to the cotyledon.

SANÉ, Y. K. 1939. A contribution to the embryology of the *Aponogetonaceae*. Jour. Indian Bot. Soc. 18: 79–91.

Zosteraceae:- In *Zostera marina* the basal cell does not divide further. The terminal cell becomes transformed into quadrants by means of longitudinally cruciate walls, from which the octant stage is formed by transverse segmentation of each quadrant cell. However, it is possible that the first wall in the terminal cell is actually transverse; in other words, the embryonomy appears to be essentially similar to that of species of *Ruppia* (Ruppia Variation, Caryophyllad Type).

HOFMEISTER, W. 1852. Zur Entwicklungsgeschichte der *Zostera*. Bot. Zeit. 10: 121–131, 137–149.
ROSENBERG, O. 1901. Die Embryologie von *Zostera marina*. K. Svensk. Vetensk. Akad. Handl. 27, Afd. III, Nr. 6.

Potamogetonales

Potamogetonaceae:- It is not possible to separate the embryonomy of *Potamogeton natans*, *P. lucens* and *P. foliosus* from that of the Sagittaria Variation, Caryophyllad Type. There are, however, a very large number of irregularly constructed embryos in *P. natans*, which indicates a certain degree of instability in this species.

COOK, M. T. 1908. The development of the embryo-sac and embryo of *Potamogeton lucens*. Bull. Torrey Bot. Club 35: 209–218.
HOLFERTY, G. M. 1901. Ovule and embryo of *Potamogeton natans*. Bot. Gaz. 31: 339–346.
SOUÈGES, R. 1940. Embryogénie des *Potamogétonacées*. Développement de l'embryon chez le *Potamogeton natans* L. C. R. Acad. Sci. Paris 211: 232–233.
WIEGAND, K. M. 1900. The development of the embryo-sac in some monocotyledonous plants. Bot. Gaz. 30: 25–47.

Ruppiaceae:- In *Ruppia rostellata* and *R. maritima* the zygote divides transversely; the basal cell does not divide again but becomes transformed into a swollen vesicular cell representing the suspensor (FIG. 62A–C). The terminal cell divides transversely and presently there is formed a proembryo consisting of four tiers of two cells each (FIG. 62B). The first longitudinal walls are erected in either of the two intermediate layers and subsequent divisions are greater in number in these two regions so that the developing embryo assumes an ovoid to globular shape (FIG. 62C). The proembryo in its earlier stages thus bears a remarkable resemblance to the Sagina Variation, Caryophyllad Type, but because of minor differences and the development of a single cotyledon, it may be made into another Variation (Ruppia Variation) under the same Type.

In the sixteen-celled proembryo disposed in four tiers of four cells each, the top tier, *l*, develops into the major portion of the cotyledon; the next tier, *l'*, presumably forms the short basal portion of the cotyledon as well as the "stem apex"; the third tier, *m*, appears to develop all of the hypocotyl, to which, however, the lowest cell, *ci*, may contribute somewhat in addition to producing a distinct group of cells which have been designated as the calyptrogen. At least, these cells constitute the only visible rudiments of the root.

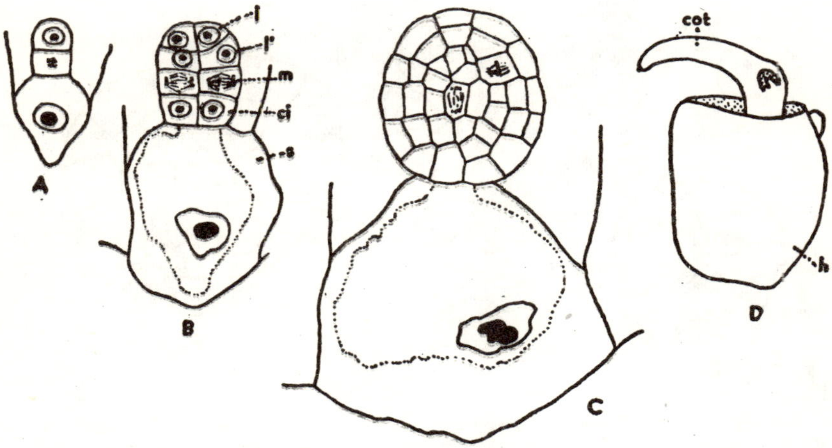

FIG. 62. — *Ruppia rostellata*. A. Three-celled proembryo. B. Ca. sixteen-celled proembryo. C. Older embryo. All × 446. D. Outline of mature embryo with cotyledon, stem apex and hypocotyl. (*Redrawn from* MURBECK 1902).

The hypocotyl develops so that its upper edges form a deep depression surrounding the cotyledon and "stem apex"; it is said to be "hollow." (FIG. 62D).

GRAVES, A. H. 1908. The morphology of *Ruppia maritima*. Trans. Conn. Acad. Arts & Sci. 14: 59–170.

HOFMEISTER, W. 1861. Neue Beiträge zur Kenntniss der Embryobildung der Phanerogamen. II. Monokotyledonen. Abh. Kön. Sächs. Ges. Wiss. 7: 629–760.

MURBECK, S. 1902. Über die Embryologie von *Ruppia rostellata* Koch. Kongl. Svenska Vetensk.-Akad. Handl. 36, Nr. 5.

WILLE, N. 1883. Om Kimens Udviklingshistorie hos *Ruppia rostellata* og *Zannichellia palustris*. Vidensk. Meddel. fra d. Naturhist. Forening i Kjøbenhavn 1882–86: 1–4.

Naiadales

Zannichelliaceae:- The embryonomy of *Zannichellia palustris* appears to be primarily like that of the Sagittaria Variation, Caryophyllad Type, but with certain distinct advances over that Variation during the later stages of development. In *Sagittaria* the two upper tiers, *l* and *l'*, of the sixteen-celled proembryo become transformed in their entirety into the cotyledon, while the stem apex, so-called, originates from tier *m*. In *Zannichellia*, on the other hand, the cotyledon and "stem apex" are both derived from the uppermost cell of the linear tetrad, *l*. In other words, these two parts are both terminal in origin and the "stem apex" is not lateral as in most monocotyledonous plants.

Zannichellia palustris is therefore referable to the Zannichellia Variation, Caryophyllad Type.

CAMPBELL, D. H. 1897. A morphological study of *Naias* and *Zannichellia*. Proc. Calif. Acad. Sci. III, 1: 1–61.

HOFMEISTER, W. 1861. Neue Beiträge zur Kenntniss der Embryobildung der Phanerogamen. II. Monokotyledonen. Abh. Kön. Sächs. Ges. Wiss. 7: 629–760.

Naiadaceae:- *Naias flexilis* and *N. major* follow the Sagittaria Variation, Caryophyllad Type, very closely.

CAMPBELL, D. H. 1897. A morphological study of *Naias* and *Zannichellia*. Proc. Calif. Acad. Sci. III, 1: 1–61.

HOFMEISTER, W. 1861. Neue Beiträge zur Kenntniss der Embryobildung der Phanerogamen. II. Monokotyledonen. Abh. Kön. Sächs. Ges. Wiss. 7: 629–760.

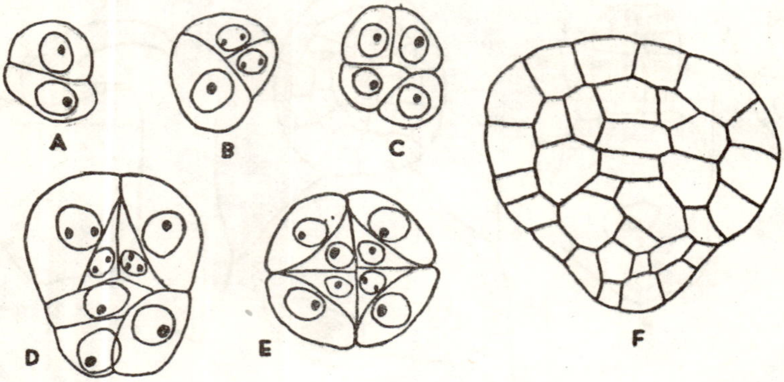

FIG. 63. — *Xyris indica*. A. Two-celled proembryo. B. Three-celled proembryo. C. Four-celled proembryo. D. Developing embryo. E. Cross section through central region of preceding. F. Fully developed embryo. All × 620. (*Redrawn from* WEINZICHER 1913).

Commelinales

Nothing has been described of the embryogeny of the *Flagellariaceae* and *Mayacaceae*.

Commelinaceae:- *Cyanotis cristata* and *Tradescantia virginiana* are both probably referable to the Penaea Variation, Asterad Type. In the first species the basal cell divides vertically in advance of a similar division in the terminal cell. Succeeding divisions take place in various planes and a spherical mass of cells is produced. There is no definite suspensor. The apical portion of the embryo presently begins to form a fairly thick circular outgrowth at the periphery. This represents the "sheath" of the cotyledon. Soon, however, the "lamina" of the cotyledon arises as a prominent lateral outgrowth of the "sheath." The lamina gradually extends its growth all around but not at a uniform rate, so that one of its lateral margins becomes more conspicuous than the other. The sheath of the cotyledon becomes indistinguishably merged with the lamina. A pore remains open toward one side of the embryo but the margins of the cotyledon presently come together and practically obliterate the opening. By this time the stem apex, which originates in the central depression of the apical portion of the embryo, has assumed the form of a mound of cells. The cotyledon forms a complete cover over the undifferentiated stem apex in the mature embryo. The basal portion of the embryo becomes very broad and disc-shaped, with a characteristic rim or shelf extending around the periphery. The radicle originates in the central portion of the basal region. The terminal end of the cotyledon which encloses the stem apex broadens and increases in thickness.

The germination of the seed essentially resembles that in the date, *Phoenix dactylifera*, but the cotyledon becomes a spherical structure retained within the seed and a long threadlike stalk connects the seedling with the ovule.

Assignment is provisionally to the Penaea Variation, Asterad Type.

HOFMEISTER, W. 1861. Neue Beiträge zur Kenntniss der Embryobildung der Phanerogamen. II. Monokotyledonen. Abh. Kön. Sächs. Ges. Wiss. 7: 629–760.

MURTHY, K. L. 1938. Gametogenesis and embryogeny in some *Commelinaceae*. Jour. Indian Bot. Soc. 17: 101–116.

Xyridales

The *Rapateaceae* are embryologically unknown.

Xyridaceae:- The first wall in the zygote of *Xyris indica* is slightly diagonal (FIG. 63A); the second wall is vertical in the terminal cell and the third wall is

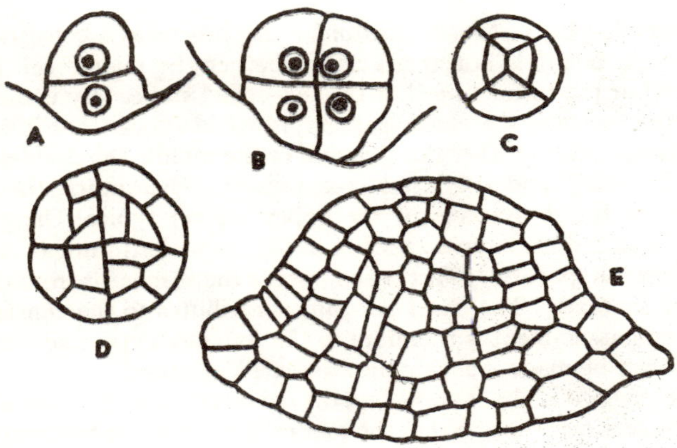

FIG. 64. — *Eriocaulon septangulare*. A. Two-celled proembryo. B. Quadrant stage. C. Periclinal walls in octant embryo cutting off epidermal initials. D. Developing embryo. E. Embryo from mature seed. All × 730. (*Redrawn from* R. W. SMITH 1910).

similarly oriented in the basal cell. The four-celled stage thus consists of two tiers of two cells each (FIG. 63B, C). The wall in the basal cell is frequently oriented at an oblique angle, but no suspensor is formed. Further divisions in both regions are very variable so that it is impossible to determine either the exact sequence of divisions or the relative respective contributions of the original basal and terminal cells. At a later stage of development there appears in the center of the embryo a complex of smaller cells delimited from the large peripheral cells by curved tangential walls (FIG. 63D, E). The nature or function of these cells remains unexplained.

The embryogeny of the *Xyridaceae* closely resembles that of the related *Eriocaulaceae* and, like that family, may be assigned to the Penaea Variation, Asterad Type.

WEINZIEHER, S. 1913. Beiträge zur Entwicklungsgeschichte von *Xyris indica* L. Flora 106: 393–432.

Eriocaulales

Eriocaulaceae :- The zygote of *Eriocaulon septangulare* divides transversely, whereupon each daughter cell segments longitudinally, the basal cell as a rule dividing first (FIG. 64A, B). The next divisions are also longitudinal and perpendicular to the preceding walls. Both cells of the two-celled proembryo thus participate equally in the formation of the embryo; there is no suspensor. The embryo is not globular in form. The octant stage is followed by the partitioning of the epidermal initials by means of periclinal walls (FIG. 64C). The process is not similar in the two halves: sometimes one half divides more rapidly, sometimes the other, but eventually those portions derived from the basal half have increased more than those derived from the terminal half (FIG. 64D). The mature embryo is bell-shaped with flaring edges, and quite frequently the edges are upturned owing to pressure exerted by the endosperm (FIG. 64E). There is no differentiation of internal histogens. Assignment is best to the Penaea Variation, Asterad Type.

SMITH, R. W. 1910. The floral development and embryogeny of *Eriocaulon septangulare*. Bot. Gaz. 49: 281–289.

Bromeliales

Bromeliaceae :- In *Tillandsia usneoides* the proembryo consists of three superposed cells, but it is not at all clear whether the middle cell is derived from the basal or the terminal cell of the two-celled stage. By comparing the drawings of the two- and three-celled stages, it would appear that it is the basal cell which divides first. Either the terminal or the middle cell divides next, the former longitudinally and the latter transversely. Quadrants, then octants, are produced in the derivatives of the upper cell, in typical Onagrad Type fashion. It is stated that the cotyledon arises from the terminal cell (*l*) and the "stem apex," hypocotyl, root tip and part of the suspensor from the middle cell (*l′* rather than *m*). The basal cell thus contributes to the construction of the embryo and assignment is therefore to the Asterad Type, and the best disposition that can be made is under the Muscari Variation.

When the embryo is about three-fourths developed, there occurs a degradation of certain cortical cells of either the root or the end of the hypocotyl nearest the root tip. It progresses until the cell contents have almost completely disappeared at the time the embryo reaches full size. This phenomenon indubitably has an intimate relation to the complete atrophy of the root in the mature plant.

BILLINGS, F. H. 1904. A study of *Tillandsia usneoides*. Bot. Gaz. 38: 99–121.

Zingiberales

For most of the families (*Musaceae*, *Strelitziaceae*, *Lowiaceae* and *Cannaceae*) there is no available description.

Zingiberaceae :- In *Costus speciosus* the first zygotic mitosis is transverse; the basal cell does not divide further. This species would appear to conform essentially with the Sagittaria Variation, Caryophyllad Type, but further study is required. Two stages are depicted for *Brachychilus horsfieldii*; a two-celled suspensor appears to be present, but otherwise there is no description of the embryonomy.

BANERJI, I. 1940. A contribution to the life-history of *Costus speciosus* Smith. Jour. Indian Bot. Soc. 19: 181–196.
MAURITZON, J. 1935 (1936). Samenbau und Embryologie einiger *Scitamineen*. Lunds Univ. Årsskr., N. F. Avd. 2, 31: Nr. 9.

Marantaceae:- *Calathea picturata* and *Stromanthe lutea* have been very superficially investigated. The proembryo appears to be a linear row of four superposed cells; the first vertical wall is erected in the cell second from the top (*l'*?) The developing embryo presently becomes more or less globular in shape. The epidermal initials are early delimited. In *S. lutea* two basal cells are slightly haustorial in nature, but degenerate at an early stage in the growth of the embryo. In this species the cells of the older embryo are depicted with two and three tiny nuclei.

It is impossible to make an accurate diagnosis of these species, other than to presume that they belong under the Chenopodiad Type, but the evidence is insufficient to allow the creation of a Variation to accommodate these monocotyledonous genera.

MAURITZON, J. 1936. Samenbau und Embryologie einiger *Scitamineen*. Lunds Univ. Årsskr., N. F. Avd. 2, 31, Nr. 9.

Liliales

The *Tecophilaeaceae, Smilacaceae* and *Ruscaceae* are still uninvestigated.

Liliaceae:- *Lilium Variation, Onagrad Type.* — The embryonomy of *Lilium* is essentially a modification of the Muscari Variation and is characterized by an excessive degree of irregularity in the location and disposition of the cell walls during both the early and later proembryonic stages. In *L. parryi* the basal cell generally divides first, into *m* and *ci*. Cell *ci* may divide or not. If not, it quickly degenerates. If it does divide, it may develop an irregular, laterally projecting embryonal mass much like that in certain species of *Erythronium* or *Tulipa* but which rarely develops distinct embryonal buds. In still other cases cell *ci* may develop a distinct suspensor with a more or less broadened base and a narrowed neck.

Cell *m* and its derivatives persist, even when the derivatives of *ci* have disappeared, and it is clear that these derivatives of the basal cell contribute at least the root cap and a portion of the suspensor if one is secondarily developed. It is equally apparent that the contribution of cell *m* is not as great in the case of *L. parryi* as it is with *Muscari conicum*; hence *Lilium* is referable to the Onagrad Type rather than to the Asterad Type.

The terminal cell, which in the case of *Lilium* may be designated as *q*, behaves in either of two ways. It always divides longitudinally. This division is rarely prompt; when this occurs, construction of the proembryo is regular and the derivatives are disposed in a constant manner: quartets and octants are formed. When division is long delayed, as appears to happen frequently, the cell assumes all the characteristics of an apical cell and this appearance is continued into the well developed embryo. However, it is hardly a true apical cell since it divides in almost any plane; the divisions, and consequently the structure of the embryo, are extremely irregular in very many instances. Since many degenerating structures are observable at late proembryonic and early embryonic stages, it is presumed that the divisions shortly become so irregular that the embryo can no longer continue development. When growth is normal, the epidermal initials are developed early, but there is no perceivable differentiation of the internal histogens.

At the second cell generation the formula for *L. parryi* is probably as follows:

$$q = pco + phy + pvt + icc$$
$$m = iec + co + s \text{ (part or all)}$$
$$ci = s \text{ (occasionally, or } ci \text{ disappears)}$$

Erythronium dens-canis and *E. americanum* develop their embryos in the identical fashion. The zygote of the latter species lies free in the endosperm some distance from the micropylar end of the embryo sac. After the first transverse partitioning, both basal and terminal cells are segmented longitudinally, but the division is occasionally transverse in the basal cell. The latter contributes little or nothing to the embryo proper during the subsequent development of the latter; consequently, these species are assignable to the Lilium Variation.

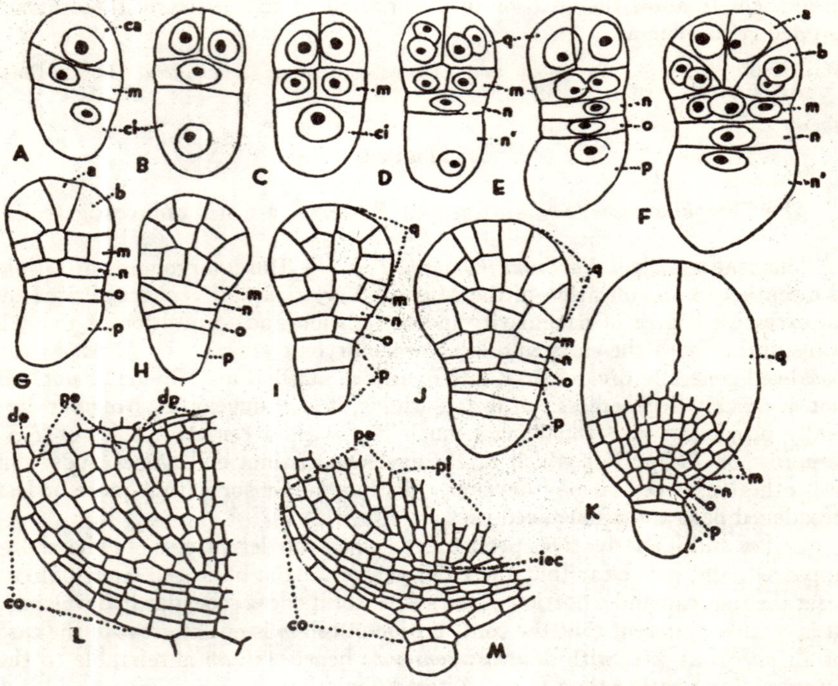

FIG. 65. — *Muscari comosum.* Description in text. A–F × 300; G–I × 210; J–L × 150. (*Redrawn from* SOUÈGES 1932).

The only stage in the embryonomy of *Fritillaria persica* which is depicted bears a strong resemblance to the same stage in *Lilium parryi*, even to the "apical cell."

The embryonomy of *Asparagus officinalis*, although very imperfectly described, seems to be like that of *Lilium parryi*.

Species of *Gasteria* (unidentified) seem to conform to the Lilium Variation.

Muscari Variation, Asterad Type. — The embryogeny of *Muscari comosum* is particularly instructive and is easier to follow out than in most other species of the *Liliaceae*. The earlier stages present a remarkable similarity to corresponding stages in certain dicotyledonous species which follow the Asterad Type, such as *Urtica pilulifera* and *Malva rotundifolia*.

The basal cell of the two-celled proembryo divides transversely in advance of the longitudinal division in the terminal cell (FIG. 65A). The eight-celled proembryo arises when each of these four derivatives is partitioned: the upper two cells (*q*) divide longitudinally; the middle cell (*m*) divides similarly and the lowest cell (*ci*) is segmented transversely (into *n* and *n'*, FIG. 65B–D). All these divisions occur exactly as in *Anthericum ramosum* (see below).

The sixteen-celled proembryo originates by the bipartitioning of each of the cells of the eight-celled proembryo. Cell *n'* divides first: it is segmented transversely to give cells *o* and *p*; then the adjoining cell *n* is segmented vertically. The two juxtaposed cells in tier *m* divide vertically, but in a somewhat irregular fashion, to produce four circumaxial cells (FIG. 65F–H); the events in *m* occur exactly as in *Myosurus minimus*, *Sagittaria sagittaefolia*, *Polygonum persicaria* and *Anthericum ramosum*.

Recapitulatory Table for Muscari comosum

I. FIRST CELL GENERATION
Proembryo of two cells disposed in two tiers:

ca = *pco*
cb = *phy* + *pvt* + *icc* + *iec* + *co* + *s*

II. SECOND CELL GENERATION
Proembryo of four cells disposed in three tiers:

q = *pco*
m = *phy* + *pvt* + *icc*
ci = *iec* + *co* + *s*

III. THIRD CELL GENERATION
Proembryo of eight cells disposed in four tiers:

q = *pco*
m = *phy* + *pvt* + *icc*
n = *iec*
n' = *co* + *s*

IV. FOURTH CELL GENERATION
Proembryo of sixteen cells disposed in five tiers:

q = *pco*
m = *phy* + *pvt* + *icc*
n = *iec*
o = *co*
p = *s*

In a very general way, the four elements of tier *q* are segmented by oblique walls, with the insertion of one end at the center of the peripheral wall and the other end near where the horizontal wall separating tiers *q* and *m* meets the original first vertical wall of the terminal cell (FIG. 65F). This wall isolates two cells which, when viewed in median longitudinal sections of the proembryo, present a form and dimensions which are quite dissimilar. The larger one adjacent to the axis appears to be clearly quadrilateral; the other, smaller, and outer one presents a triangular outline. The first may be designated in the figures by the letter *a*, the other by *b* (FIG. 65F, G). There are numerous divergences from the customary position of the wall in tier *q*. For instance, it may be nearly horizontal in one or more cells, as happens in *Anthericum ramosum*; again, it may be nearly vertical or in various tangential positions.

The eventual destinations of the various tiers is exactly the same as the corresponding tiers in an embryo which follows the Asterad Type, but differ solely in the point of origin of the stem apex. The tier *q*, briefly, engenders the cotyledon alone; *m* gives rise to the hypocotyl and the stem apex; *n* furnishes the initials of the root; *o* provides the root cap and *p* the suspensor.

The element *a* of tier *q* now divides by a horizontally tangential wall (FIG. 65H); the external cell is the first epidermal initial and next divides vertically. The internal daughter cell is the one from which the tissues of the cotyledon presently arise. The sister cell *b* divides vertically to give two juxtaposed cells; the subsequent walls are oriented very diversely — they may be vertical, more or less oblique, horizontal or tangential (FIG. 65H–J). After these first divisions the cells multiply irregularly to produce first a massive semispherical, then a subspherical, and finally a more or less elongated mass which represents the young cotyledon. At its base there is produced a lateral depression which separates it from the hypocotyl and also marks its lower limits.

In the sixteen-celled proembryo the tier *m* is composed of four circumaxial cells (FIG. 65F–H). They are partitioned by walls which are generally vertical and meridional, but may be oriented tangentially (FIG. 65J). The next series of walls appear to be transverse; sometimes they appear first in the peripheral cells, sometimes in the central cells. All the internal cells correspond to the plerome of the hypocotyl. The peripheral cells are segmented vertically; the external daughter of each cell is an epidermal initial and the inner daughter represents a periblem initial (FIG. 65K). While the epidermal initials may next divide by walls which may be either anticlinal, horizontal or vertical, the periblem initials usually first divide by anticlinal walls, but these may also be periclinal and thus contribute an increased number of cortical layers. The limits separating the three histogens are none too distinct at this stage of development so that it is not possible to follow the subsequent succession of divisions with certainty; they do not appear distinctly until the moment of differentiation of the cotyledon and are then already well advanced. The tissues derived from *m* are then composed of five to seven cell layers which correspond to the hypocotyledonary portion of the embryo; later the upper portion gives rise to the stem apex.

The tier *n* of the sixteen-celled proembryo consists of two juxtaposed elements (FIG. 65G). They divide vertically by meridional walls to give four circumaxial cells which next behave like the four circumaxial cells of tier *m* (FIG. 65J). By walls which may be vertical, tangential or parallel to the meridional walls, they give rise to four central cells and to peripheral cells varying in number from four to eight. The inner cells next divide transversely to give rise to two cell tiers; these internal cells furnish the initials of the root cortex and the peripheral cells of the lateral portions of the root cap (FIG. 65K–M). At a more advanced stage, only a single layer of initials appears to be truly functional (*iec*, FIG. 65M).

The single cell *o* of the sixteen-celled proembryo is segmented vertically into two juxtaposed cells (FIG. 65I); they then divide by similar meridionally vertical walls to form four circumaxial cells. These four cells most often divide vertically to give a cellular plateau normally composed of four central elements and eight peripheral elements. The central cells next are segmented transversely to give the appearance of two superposed cell layers (FIG. 65J–K). The peripheral cells at the same time are segmented tangentially but tardily form transverse walls. These cells go into the formation of the root cap, while the central two cell layers soon function as the initials of this tissue (*co*, FIG. 65L, M).

Cell *p* of the sixteen-celled proembryo is generally divided by a transverse wall into two superposed cells (FIG. 65I); the upper daughter cell next divides vertically and the lower daughter is divided transversely. The later divisions are few in number and rather variable; the entire mass constitutes a short conical or cylindrical suspensor.

The recapitulatory scheme for *Muscari comosum* has been presented above.

Muscari racemosum, though imperfectly described, apparently agrees with *M. comosum*.

The two-celled proembryo of *Anthericum ramosum* divides like that of *Muscari comosum*. The eight-celled stage results when each of the two upper juxtaposed cells (*l*) divides longitudinally; the middle cell (*m*) divides similarly, and the lower cell (*ci*) is partitioned transversely (into *n* and *n'*). Each of all these cells divides to constitute the sixteen-celled proembryo: the quadrants (*l*) divide transversely to engender octants (*l* and *l'*); each daughter of the middle cell, *m*, divides vertically to give four circumaxial cells; cell *n* divides vertically and cell *n'* divides transversely into *o* and *p*. The sixteen-celled proembryo thus consists of six tiers of cells.

The two upper tiers *l* and *l'* give rise to the cotyledon, as in *Sagittaria*. The tier *m* engenders the hypocotyl and the stem apex; in *Sagittaria* this layer produces the upper half of the hypocotyl as well as the stem apex. The tier *n* becomes the hypophysis; in *Sagittaria* it gives rise to the lower half of the hypocotyl. The tiers *o* and *p* become transformed into a suspensor in the form of several superposed flattened cells; in *Sagittaria* the hypophysis originates from tier *o* and the upper daughter derivatives of *p* while the other daughter engenders the suspensor.

The hypophysis develops in a manner uniquely similar to that prevailing in the *Ranunculaceae*. When the embryo approaches maturity the hypophyseal region has become thick and massive and develops the root cap and the initials of the root cortex.

Heloniopsis Variation, Onagrad Type. — The terminal cell of the two-celled proembryo of *Heloniopsis breviscapa* develops much as in the Muscari Variation, but this species does not otherwise conform to this embryonomic type. The basal cell divides transversely in advance of the vertical segmentation of the terminal cell, but a wall apparently is not always erected. The basal region remains either two-celled or two-nucleate and thus contributes nothing to the construction of the embryo proper.

This mode of development does not conform to that of any other Variation under the Onagrad Type, to which it unquestionably belongs; consequently, the Heloniopsis Variation may be erected to accommodate it.

Calochortus. — The first division in the zygote results in the formation of two unequal superposed cells; the second and third divisions are likewise transverse. The first longitudinal walls are laid down in the apical and the third cells followed by one in the second cell. The basal cell is the suspensor initial and as the embryo grows successive transverse walls produce a suspensor of five or more cells. The suspensor remains more or less intact when the embryo is mature.

This species certainly does not follow either the Onagrad or the Asterad Type but is probably referable to the Chenopodiad Type.

The early proembryonic divisions in *Asphodelus tenuifolius* are similar to those described for *Calochortus*.

Other Species. — *Gagea lutea* forms a linear six-celled proembryo whose basal cell is swollen and vesicular. *Pushkinia scilloides* has a globular embryo with a well developed filamentous suspensor. Neither species can be properly assigned to a Type.

The observations on *Lilium parryi* are original.

Bambacioni, V. 1930. Ricerche sulla ecologia e sulla embriologia di *Fritillaria persica* L. Ann. di Bot. 18: 7–37.

Coulter, J. M. 1897. Contribution to the life history of *Lilium philadelphicum*. I. The embryo sac and associated structures. Bot. Gaz. 23: 412–422.

Guérin, P. 1931. L'ovule et la graine de l'*Erythronium dens-canis* L. Le Botaniste 23: 185–195.

Guérin, P. 1937. L'ovule et la graine des *Erythronium* et des *Calochortus*. Ann. Sci. Nat., Bot. X, 19: 255–265.

Hofmeister, W. 1861. Beue Beiträge zur Kenntniss der Embryobildung der Phanerogamen. II. Monokotyledonen. Abh. Kön. Sächs. Ges. Wiss. 7: 629–760.

Maheshwari, P., and U. B. Singh. 1930. The development of the female gametophyte of *Asphodelus tenuifolius*. Jour. Indian Bot. Soc. 9: 31–39.

Ono, T. 1927. Embryologische Studien an *Heloniopsis breviscapa*. Sci. Rep. Tohoku Imp. Univ., IV, 2: 93–104.

ROBBINS, W. W., and H. A. BORTHWICK. 1925. Development of the seed of *Asparagus officinalis*. Bot. Gaz. 80: 426–438.
SCHAFFNER, J. H. 1901. A contribution to the life history and cytology of *Erythronium*. Bot. Gaz. 31: 369–387.
SCHNARF, K., und R. WUNDERLICH. 1938. Zur vergleichenden Embryologie der *Liliaceae-Asphodeloideae*. Flora 133: 297–327.
SOUÈGES, R. 1918. Embryogénie des *Liliacées*. Développement de l'embryon chez l'*Anthericum ramosum*. C. R. Acad. Sci. Paris 167: 34–36.
SOUÈGES, R. 1926. Embryogénie des *Liliacées*. Développement de l'embryon chez le *Muscari comosum* L. C. R. Acad. Sci. Paris 183: 233–235.
SOUÈGES, R. 1931. Recherches sur l'embryogénie des *Liliacées*. Bull. Soc. Bot. France 78: 662–681.
SOUÈGES, R. 1932. Recherches sur l'embryogénie des *Liliacées*. Bull. Soc. Bot. France 79: 11–23.
WUNDERLICH, R. 1937. Zur vergleichende Embryologie der *Liliaceae-Scilloideae*. Flora 132: 48–90.

Trilliaceae:- Reports of the failure of fertilization to be effected, or of the proembryo to be aborted, in certain species of *Trillium*, with the subsequent development of an endosperm cell into an embryo, have not been substantiated by other investigators and are therefore not accepted here. It is believed that the sequence of events leading up to the erection of the megagametophyte have been misinterpreted.

In *Trillium erectum* an octonucleate megagametophyte is formed following the Scilla (Allium) Type and the egg is fertilized. The first division in the zygote is transverse. The next divisions may be transverse in both the basal and terminal cells, thus resulting in a linear tetrad; or longitudinal in both cells to give two tiers of two juxtaposed cells each. These two types of proembryos occur about equally. Subsequent divisions are not simultaneous, so that the developing proembryo consists of an unequal and variable number of cells. The later development has not yet been worked out. It is not possible at the moment to assign *T. erectum* to a definite Type because of the variation in the planes of division during the second cell generation and as nothing is known concerning the contribution of the basal cell to the construction of the embryo proper.

The information presented in the preceding paragraph has been derived from the unpublished observations of AURAY BLAIN (McGill Univ.).

Pontederiaceae:- Fertilization does not occur in *Eichhornia crassipes* and no embryo is formed.

In *Pontederia cordata* a three-celled linear proembryo is produced; the middle cell appears to be derived from the basal cell. The terminal cell is segmented longitudinally. Subsequent events assign this species definitely to the Muscari Variation, Asterad Type.

The description of the embryonomy of *Monochoris hastaefolia* does not agree with the figures presented. The latter plainly show that the first division in the two-celled proembryo is longitudinal in the terminal cell, whereas the description states that this division is transverse in the basal cell, and that the terminal cell then "divides by an anticlinal wall." It is also said that "in some ovules the terminal cell divides again by a periclinal wall," although this cell has not yet undergone division. Altogether, this species appears to develop essentially as in *Pontederia cordata*.

BANERJI, I., and S. HALDAR. 1942. A contribution to the morphology and cytology of *Monochoria hastaefolia* Presl. Proc. Indian Acad. Sci., B, 16: 91–106.
SMITH, W. R. 1898. A contribution to the life history of the *Pontederiaceae*. Bot. Gaz. 25: 324–337.

Alstroemeriales

None of the three families, *Alstroemeriaceae*, *Petermanniaceae* or *Philesiaceae*, has ever been investigated embryologically.

Arales

Araceae:- It appears to be established beyond question that in most species belonging to the *Araceae* the terminal cell of the two-celled proembryo divides transversely, usually once but perhaps twice in a few species. In either case, the outer daughter of the terminal cell is next divided longitudinally. The great question is what part, if any, the basal cell plays in contributing to the development of the embryo proper. All authors are silent on this important point. That the basal cell does divide is beyond dispute, since definite suspensors are formed in a few species. The arbitrary conclusion here reached is that the basal cell does not contribute to the structure of the embryo proper. On this basis, assignment of the *Araceae* in toto to the Caryophyllad Type is indicated. There is a marked resemblance between the embryogeny of the *Araceae* and that of the Myriophyllum Variation of this Type. However, since but one cotyledon is produced, the family may be distinguished by the creation of a separate division, the Aglaonema Variation.

In *Aglaonema commutatum* the terminal cell of the two-celled proembryo is segmented transversely. In *Lysichiton camtschatensis* (= *L. americanum*), the proembryo appears to be erected in a similar fashion, whereupon the upper daughter of the terminal cell is divided obliquely, then the larger of the two derivatives is segmented in a transversely diagonal fashion. Subsequent divisions are very irregular and mostly oblique. Presently a pyriform to globular body, undifferentiated except for epidermal initials, is developed. In *Anthurium cordifolium* a single large lateral cotyledon is formed. The three-celled proembryo of *Anthurium violaceum* is erected as in *Aglaonema commutatum*, then both daughters of the terminal cell are segmented longitudinally. The basal cell may also be partitioned vertically but this does not always seem to be the case. The developing embryo becomes nearly globular, then conical; there is no differentiation of any histogens save for the epidermis. The embryogeny of *Nephthytis liberica* follows that of *Anthurium violaceum*; that of *N. gravemeuthii* appears to do so also. *Dieffenbachia daraquiniana*, *Richardia africana*, *Aglaonema nitidum*, *A. versicolor*, *Anthurium crystallinum*, *Arum maculatum*, *Philodendron wendlandii*, *P. gloriosum*, *Pistia stratiotes*, *Symplocarpus foetidus* all follow the Aglaonema Variation. In *Pistia stratiotes*, however, one old account states that both the terminal and basal cells are divided vertically, but subsequent divisions are extremely irregular. All the products of both cells go into the organization of the embryo.

The following species form a linear tetrad: *Spathynema foetida*, *Pothos longifolia*, *Peltandra virginica*, *Acorus calamus* and probably *Arisaema triphyllum*.

It is evident that a more critical survey of the somewhat variable embryological forms occurring in the *Araceae* is desirable.

Buell, M. F. 1938. Embryogeny of *Acorus calamus*. Bot. Gaz. 99: 556–568.
Campbell, D. H. 1899. Notes on the structure of the embryo sac in *Sparganium* and *Lysichiton*. Bot. Gaz. 27: 153–166.
Campbell, D. H. 1900. Studies on the *Araceae*. Ann. Bot. 14: 1–25.
Campbell, D. H. 1905. Studies on the *Araceae*, III. Ann. Bot. 19: 329–349.
Goldberg, B. 1941. Life history of *Peltandra virginica*. Bot. Gaz. 102: 641–662.
Gow, J. E. 1907. Morphology of *Spathyema foetida*. Bot. Gaz. 43: 131–136.
Gow, J. E. 1908. Embryogeny of *Arisaema triphyllum*. Bot. Gaz. 45: 38–44.
Gow, J. E. 1908. Studies in *Araceae*. Bot. Gaz. 46: 35–42.

GOW, J. E. 1913. Observations on the morphology of the aroids. Bot. Gaz. 56: 127–142.
HEGELMAIER, F. 1874. Zur Entwicklungsgeschichte monokotyledoner Keime nebst Bemerkungen über die Bildung der Samendeckel. Bot. Zeit 32: 673–686; 689–700; 705–716.
HOFMEISTER, W. 1861. Neue Beiträge zur Kenntniss der Embryobildung der Phanerogamen. II. Monokotyledonen. Abh. Kön. Sächs. Ges. Wiss. 7: 629–760.
MITCHELL, M. R. 1916. The embryo sac of *Richardia africana* Kth. Bot. Gaz. 61: 325–336.
ROSENDAHL, C. O. 1909. Embryo-sac development and embryology of *Symplocarpus foetidus*. Minn. Bot. Studies 4: 1–9.

Lemnaceae:- The young embryo of *Lemna minor* bears an astonishing resemblance to that of *Lilium parryi*, even to the persistence of an apparent "apical cell"; this species may therefore be assigned to the Lilium Variation, Onagrad Type.

In *Lemna perpusilla* development is similar to that of *L. minor*. It is claimed that a suspensor is not formed, but this is hardly correct since the illustrations reveal that one is present. No definite root tip region appears in the mature embryo.

BLODGETT, F. H. 1923. The embryo of *Lemna*. Amer. Jour. Bot. 10: 336–342.
CALDWELL, O. W. 1899. On the life history of *Lemna minor*. Bot. Gaz. 27: 37–66.

Typhales

There is no record of an embryological investigation on the *Typhaceae*.

Sparganiaceae:- *Sparganium simplex* and *S. ramosum* conform closely in embryological details to the Juncus Variation, Onagrad Type.

CAMPBELL, D. H. 1899. Studies on the flower and embryo of *Sparganium*. Proc. Calif. Acad. Sci., III, 1: 293–328.
HEGELMAIER, F. 1874. Zur Entwicklungsgeschichte monocotyledoner Keime nebst Bemerkungen über die Bildung der Samendeckel. Bot. Zeit. 32: 631–639; 648–656.
SOUÈGES, R. 1924. Embryogénie des *Typhacées*. Développement de l'embryon chez le *Sparganium simplex* L. C. R. Acad. Sci. Paris 179: 198–200.

Amaryllidales

Amaryllidaceae:- *Allium ursinum* offers one of the best examples among monocotyledonous embryos in which a suspensor is not formed; this species, moreover, has been carefully investigated. In general the embryonomy resembles that of *Anthericum ramosum*, but differs: by the disposition of the elements in the eight-celled proembryo; by the mode of segmentation of the quadrants and the constitution of the twelve-celled proembryo; by the general configuration of the proembryo and the absence of a suspensor; and by the apparently more distinct organization of the radicular extremity.

The proembryonic tetrad is erected as in *Anthericum ramosum* and the eventual destination of each element is the same, save that the basal element, *ci*, does not contribute to the formation of a suspensor. The four-celled proembryo is erected as in the Onagrad Type and resembles that of *A. ramosum*.

Proembryos of *Allium flavum*, *A. paniculatum*, *A. zebdanense*, *A. rotundum*, *A. paradoxum*, and *A. moly* have been superficially described and illustrated; all more or less agree with the Anthericum Variation, but all are figured with short suspensors. In the last two species the proembryos are early aborted. In *A. mutabile* a suspensor several cells long and two cells in transverse section is produced.

Cyrtanthus parviflorus conforms to the Anthericum Variation. The younger embryo of *C. sanguineus* has an enveloping sheath, still meristematic and with four distinct lobes at its apex. Each lobe has its own vascular strand, four separate strands arising from the four poles of the root. The lobes at this stage are approximately equal. At the base of the sheath is the region from which the stem tip will later arise. At a much older stage of growth of the embryo the sheath consists of two regions, a lower portion which still envelops the growing point, and a long upper projection which has resulted from the greater growth in length of one side of the sheath. The abortion of one side of this sheath is presumed to be due to pressure exerted by the growth of the extensive mass of endosperm cells. At this stage of development, the growing point has differentiated the first and second leaves, the first arising on the side corresponding to the aborted side of the sheath, and the second one about opposite the first. The sheath is presumed to be "fusion of two or more cotyledons."

The basal cell of the two-celled proembryo of *Cooperia drummondii* is said to divide either once or not at all. This species probably conforms to the Lilium Variation, Onagrad Type.

Hippeastrum aulicum, *Leucojum vernum*, *Habranthus chilensis* and *Crinum capense* appear also to follow the Anthericum Variation.

In *Crinum latifolium* two types of embryos are to be found: (*1*) those surrounded by endosperm, and (*2*) those which are not so surrounded — *i.e.*, naked. In both groups the embryo arises from a normally fertilized egg, but in the second group the polar nucleus is not fertilized. Otherwise, embryogenesis in both groups is similar. When the endosperm is not developed, the growing embryo absorbs all of the nucellus and comes thus to lie naked in the ovule. It appears that the terminal cell of the two-celled proembryo divides first and transversely, and the basal cell then divides in the same plane. Apparently the two daughters of the basal cell do not divide further but remain as a transitory two-celled suspensor. In the case of embryos surrounded by endosperm, the developing embryo is about thrice as long as broad, while those not surrounded by endosperm are globular-pyriform in shape. The mature embryo is differentiated into cotyledon, plumule and radicle in both instances.

Church, M. B. 1916. The development of the embryo sac and embryo of *Cooperia drummondii*. Bull. Torrey Bot. Club 43: 397–405.

Farrell, M. E. 1914. The ovary and embryo of *Cyrtanthes sanguineus*. Bot. Gaz. 57: 428–436.

Hofmeister, W. 1861. Neue Beiträge zur Kenntniss der Embryobildung der Phanerogamen. II. Monokotyledonen. Abh. Kön. Sächs. Ges. Wiss. 7: 629–760.

Porter, T. R. 1936. Development of the megagametophyte and embryo of *Allium mutabile*. Bot. Gaz. 98: 317–327.

Souèges, R. 1926. Embryogénie des *Liliacées*. Développement de l'embryon chez l'*Allium ursinum* L. C. R. Acad. Sci. Paris 182: 1344–1346.

Souèges, R. 1931. Recherches sur l'embryogénie des *Liliacées*. Bull. Soc. Bot. France 78: 662–681.

Stenar, A. H. S. 1925. Embryologische Studien. II. Die Embryologie der *Amaryllidaceen*. Akad. Abhandl. Uppsala 1: 1–197.

Taylor, W. R. 1921. The embryogeny of *Cyrtanthus parviflorus* Baker. Amer. Jour. Bot. 8: 502–506.

Tomita, K. 1931. Über die Entwicklung des nackten Embryos von *Crinum latifolium* L. Sci. Rep. Tohoku Imp. Univ., IV, Biol. 6: 163–169.

Weber, E. 1929. Entwicklungsgeschichtliche Untersuchungen über die Gattung *Allium*. Bot. Archiv. 25: 1–44.

Iridales

Iridaceae:- In *Crocus vernus* the terminal cell of the two-celled proembryo divides transversely, then the upper daughter cell is segmented longitudinally. The basal cell appears to remain undivided; it becomes fairly large but not vesicular. Subsequent divisions appear to be rather irregular. The embryonomy appears to correspond with that of the Sagittaria Variation, Caryophyllad Type.

HOFMEISTER, W. 1861. Neue Beiträge zur Kenntniss der Embryobildung der Phanerogamen. II. Monokotyledonen. Abh. Kön. Sächs. Ges. Wiss. 7: 629–760.

Dioscoreales

No reports are available for the *Stenomeridaceae*, *Trichopodaceae* and *Roxburghiaceae*.

Dioscoreaceae:- The first division in the zygote of *Dioscorea villosa* is generally very oblique, occasionally approaching the vertical. The second division is vertical in the basal cell, and the third is similarly oriented in the terminal cell. This species very likely belongs to the Lilium Variation, Onagrad Type, particularly as the developing embryo even shows the apparent "apical cell."

SMITH, P. M. 1916. The development of the embryo and seedling in *Dioscorea villosa*. Bull. Torrey Bot. Club 43: 545–558.

Agavales

The *Xanthorrhoeaceae* are still uninvestigated, while the account for one species in the *Agavaceae* is sketchy.

Agavaceae:- In *Yucca glauca* a linear proembryo consisting of from four to eight cells is formed. Apparently most of the cells are derived from the terminal cell of the two-celled proembryo as it has been stated that the basal cell is larger (though not perceptibly so, to judge from the illustrations) than the others and may occasionally divide by an oblique wall.

This species appears to be definitely assignable to the Caryophyllad Type, and to either the Sagittaria or Zannichellia Variation.

FOLSOM, D. 1916. Studies in the morphology of *Yucca glauca*. Minn. Bot. Studies 4: 427–435.

Palmales

The *Palmaceae* have never been studied embryologically.

Pandanales

Pandanaceae:- The zygote of *Pandanus coronatus* divides transversely, then both derivatives are segmented longitudinally. Development thereafter is very irregular and an undifferentiated ellipsoidal to pyriform embryo is produced. Assignment is probably best under the Penaea Variation, Asterad Type.

CAMPBELL, D. H. 1911. The embryosac of *Pandanus*. Ann. Bot. 25: 773–789.

Cyclanthales

For the *Cyclanthaceae* there is no report.

Haemodorales

Uninvestigated families: *Haemodoraceae*, *Velloziaceae*, *Apostasiaceae*, *Taccaceae* and *Philydraceae*.

Hypoxidaceae:- For *Hypoxis decumbens* there is depicted a typical "T-shaped" proembryo of four cells; to judge from the structure of a later embryo, this species probably conforms to the Asterad Type.

STENAR, A. H. S. 1925. Embryologische Studien. II. Die Embryologie der *Amaryllidaceen*. Akad. Abhandl. Uppsala 1: 1–197.

Burmanniales

The *Corsiaceae* have never been studied.

Burmanniaceae:- The first division in the zygote of *Burmannia candida* is transverse; this is followed by a longitudinal division in the terminal cell, then by a transverse division in the basal cell. Assignment is perhaps to the Asterad Type, but the evidence is too meager. *B. capitata* has a ten-celled proembryo, while *B. javanica* and *B. maburnia* resemble species of *Thismia* (*Thismiaceae*). *B. coelestis* is apogamous; from one to three embryos may be formed, depending on whether the egg alone functions or the synergids also develop into embryos.

ERNST, A., and C. BERNARD. 1912. Beiträge zur Kenntnis der Saprophyten Javas. IV. Entwicklungsgeschichte des Embryosackes und des Embryo von *Burmannia candida* Engl. und *B. Championii* Thrv. Ann. Jard. Bot. Buitenzorg II, 10: 161–188.

PFEIFFER, N. E. 1914. Morphology of *Thismia americana*. Bot. Gaz. 57: 122–135.

Thismiaceae:- In *Thismia americana*, which is not a saprophytic species as are most of the others, the embryo is inconspicuous and consists of three or four mostly linearly arranged cells.

The earliest proembryonic developmental stages of *Thismia versteegii* and *T. clandestina* were not followed out; hence, the sequence of divisions is not known. The figures for the former species show a four-celled proembryo consisting of two superposed cells in the basal region topped by two juxtaposed cells. On the presumption that the two lower cells were derived from the basal cell, the vertical partitioning of the terminal cell would place these species in either the Onagrad or the Asterad Type, depending on the contribution of the basal cell to the construction of the body of the embryo. Older embryos of *T. clandestina*, although crudely drawn, bear a strong resemblance to typical Onagrad Type embryos.

ERNST, A., and C. BERNARD. 1911. Beiträge zur Kenntnis des Saprophyten Javas. VI. Beiträge zur Embryologie von *Thismia clandestina* Miq. und *Thismia verstegii* Sm. Ann. Jard. Bot. Buitenzorg II, 9: 70–77.

PFEIFFER, N. E. 1914. Morphology of *Thismia americana*. Bot. Gaz. 57: 122–135.

Orchidales

Orchidaceae:- The embryogeny of the *Orchidaceae* is not consistent throughout this large and diverse family. It is extremely puzzling as to the Type to which the *Orchidaceae* as a whole or the individual species should be assigned, for the simple reason that the exact sequence of divisions in the young proembryo is most imperfectly known. The proembryo generally consists of three superposed cells when the first longitudinal division occurs, but in other species a linear filament of as many as twenty cells is produced before any longitudinal walls appear. Generally speaking, this would indicate adherence to the Chenopodiad Type. In *Epipactis palustris* the terminal cell is segmented longitudinally at the two-celled stage, then the basal cell may divide in either plane. Since all the products of the basal and terminal cells enter into the construction of the embryo in this species, this would indicate assignment to the Asterad Type. In *Stanhopea oculata* both basal and terminal cells are segmented longitudinally; the four-celled proembryo bears a strong resemblance to the Penaea Variation, Asterad Type. The basal cell of *Phalaenopsis grandiflora* divides longitudinally; the terminal cell first segments transversely, then the upper daughter cell is partitioned either transversely or vertically; the derivatives of the two basal cells produce embryonal tubes and do not definitely contribute cells to the body of the embryo proper. Such a procedure calls for disposition under the Solanad Type.

It is therefore possible to assign certain species to definite Types, but not always to Variations already established under these Types since too little is known of the exact sequence of divisions in the late proembryo and early embryo. Matters are greatly complicated by the fact that no histogens save the epidermis are usually identifiable at the time the embryo is mature. There is also no definite resting period for most orchid seeds. Until such time as a large number of species are carefully investigated and the exact sequence of divisions determined, it appears to be better to follow a different procedure for the classification of orchidaceous embryos. The characters which are employed include the presence or absence of a suspensor and the presence or absence of embryonal tubes and whether the latter arise directly or indirectly from the suspensor initial cell. On this basis thirteen distinct developmental procedures are recognizable and may be designated as Patterns.

There exists a definite possibility that the usual clear separation of suspensor from embryo proper in the organization of essential regions of the mature embryo breaks down more or less completely in the case of many species in the *Orchidaceae*. *Epidendrum cochleatum* is a good example: a definite suspensor is depicted at a late proembryonic stage, yet it is impossible to recognize one at the early embryo stage. This raises the question: has the suspensor developed to the extent that it has become an integral part of the mature embryo? Or, on the other hand, has it reverted to a possible ancestral condition and lost its identity?

The outgrowths from the suspensor initial or from cells of a more or less well developed suspensor, which have been given various and mostly inappropriate names in the literature, are here called embryonal tubes, from the obvious analogy with similar structures found in various species of the Coniferophyta.

The following descriptive key serves to elucidate the thirteen Patterns; to each Pattern has been given the name of the genus and species whose development has been most fully followed out.

Suspensor absent or represented by a single cell .... **1. Epipactis palustris Pattern**
Suspensor or embryonal tubes or both present.
  Embryonal tubes absent.
    Suspensor linear or slightly curved, consisting of one row of cells only.
      Suspensor cells relatively uniform .... **2. Orchis latifolia Pattern**
      Suspensor cells variously swollen or elongated.
        Terminal cell of suspensor alone affected, excessively elongated .... **3. Goodyera discolor Pattern**
        Terminal and subterminal cells of suspensor extruded from micropyle, more or less swollen.
          Terminal cell only branched .... **4. Serapias linguae Pattern**
          All extruded cells branched .... **5. Herminium monorchis Pattern**
    Suspensor massive.
      Embryo lacking "cotyledon." .... **6. Epidendrum ciliare Pattern**
      Embryo with presumptive "cotyledon" .... **7. Sobralia macrantha Pattern**
  Embryonal tubes present.
    Suspensor initial giving rise directly to embryonal tubes.

Embryonal tubes eight or less in number.
  Embryonal tubes growing in one direction only (over body of embryo) .... **8. Vanda parviflora Pattern**
  Embryonal tubes growing in all directions.
    Embryonal tubes essentially uniform in diameter.
      Tubes unbranched .... **9. Cymbidium bicolor Pattern**
      Tubes each with two protuberances .... **10. Phalaenopsis grandiflora Pattern**
    Embryonal tubes extended behind embryo becoming several times the diameter of those covering embryo . **11. Phalaenopsis schilleriana Pattern**
Embryonal tubes often more than eight in number .... **12. Stanhopea oculata Pattern**
Suspensor initial developing into a linear primary suspensor with embryonal tubes arising from terminal cell of latter .... **13. Eulophea epidendracea Pattern**

1. *Epipactis palustris Pattern.* — Several orchid embryos are known in which no suspensor definitely recognizable as such is developed: *Epipactis palustris*, *Listera ovata*, *Zeuxine sulcata* (in which compound megasporic embryogeny occurs) and *Cypripedium spectabile*; of these the development of the

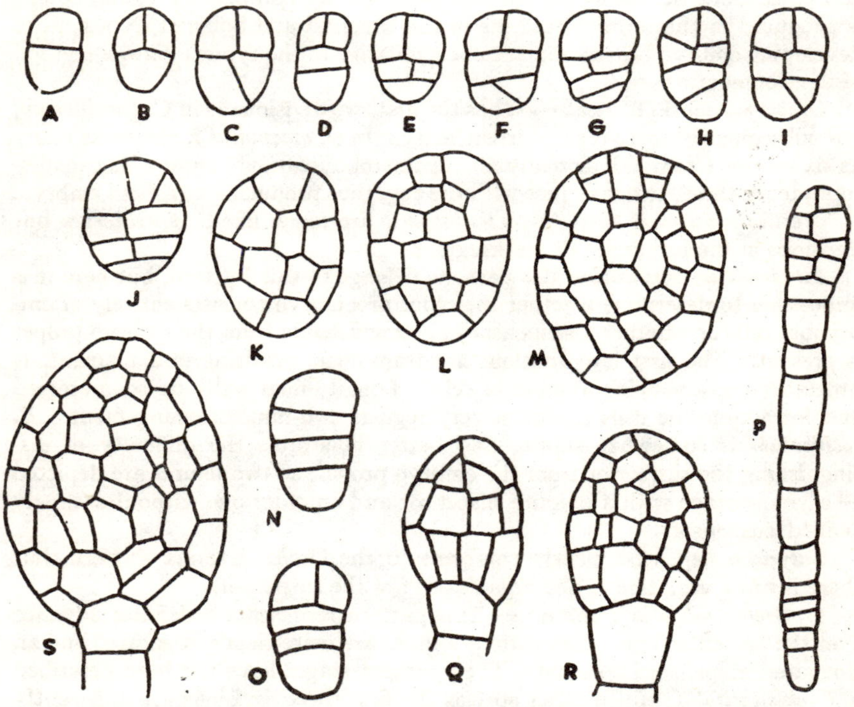

Fig. 66. — *Epipactis palustris*. A–I. Proembryos. I, J. Young embryos. K, L. Developing embryos. M. Presumably mature embryo. *Orchis latifolia*. N. Four-celled proembryo. O. First division in terminal cell. P. Four-tiered quadrant (sixteen-celled) embryo with suspensor. Q–S. Developing to mature embryos. All × 240. All these figures are optical sections. (*Redrawn from* Treub 1879).

first two has been most thoroughly investigated. The upper and lower daughter cells of the two-celled stage contribute equally to the sum total of the embryo. The second division is longitudinal in the terminal cell and the third is either transverse or oblique in the basal cell (Fig. 66A–D). The impression one gains is that the basal cell, in embryos of this pattern, was originally a suspensor initial whose normal function during the course of evolution became merged with that of the terminal cell so that the two share equally in contributing to the body of the embryo. On the other hand, however, the basal cell may still have retained its normal function of producing the suspensor but it has become impossible to distinguish between suspensor and embryo proper. This latter conception is illustrated by a survey of the series of embryos in FIG. 66A–J, in all of which the appearance of the lower half of each embryo suggests the persistence of the suspensor, but this begins to disappear in an embryo such as Fig. 66K and is completely obliterated in nearly mature embryos such as the one in Fig. 66M.

In *Cypripedium parviflorum* the terminal cell of the two-celled proembryo divides longitudinally and the basal cell transversely. However, the basal cell apparently does not always divide because some figures of the developing proembryo plainly reveal it as a single cone-shaped cell with no evidence that it has added any derivatives to the embryo proper. In still other figures there is evidence that the upper daughter cell of a transversely divided basal cell may add at least one tier of two to four cells to the embryo proper. A regular octant is produced in the terminal cell, as in the Onagrad and Solanad Types. The developing embryo bears a remarkable similarity to the typical *Lilium* method of development.

Orchis maculatus f. *ericetorium* has the first two divisions as in *C. parviflorum*, but subsequently there is some irregularity. In *O. maculatus* f. *meyeri* as many as six sperms (from half as many microgametophytes) may enter a megagametophyte; of these, two may fuse with the egg, thus producing a triploid embryo.

Neottia nidus-avis and *Epipactis latifolia* are much like *C. parviflorum* but divisions in the terminal cell are irregular.

Epidendrum variegatum also perhaps belongs to this Pattern, but here it is impossible to determine whether the older proembryo consists entirely of embryonal cells or whether a suspensor indistinguishable from the embryo proper is present. The first few divisions are transverse, resulting in a filamentous proembryo of a varying number of cells. Longitudinal walls are soon erected and for a time the cells exhibit a very regular two-ranked arrangement. *E. cochleatum* is somewhat similar and better illustrates the difficulty of distinguishing the suspensor from the embryo proper, as two figures are depicted — a young stage with a definite suspensor and an older one without a clearly defined suspensor.

Satyrium nepalense clearly conforms to the Orchis latifolia Pattern, but there is some variation in the organization of the suspensor.

2. *Orchis latifolia Pattern.* — This pattern represents a definite advance over the preceding one. The various regions are more clearly organized and an elongated suspensor is present. The youngest stages have not been described in *Orchis latifolia*, but in other species the first three divisions are apparently transverse and the fourth is longitudinal in the uppermost cell (FIG. 66N, O). It is questionable whether the basal cell divides further after this stage; most of the cells of the long suspensor appear to be derived from the lower of the two middle cells and its derivatives, all of which divide solely in the transverse plane (FIG. 66P). The body of the embryo presently consists of three tiers of four cells each, with a large cell in the position corresponding to that occupied by the hypophysis in dicotyledonous genera (FIG. 66Q). Whether this cell is actually a hypophysis cell is an extremely dubious point; at least, it has not divided, except possibly transversely, to add cells to the suspensor, even in approximately mature embryos. Periclinal walls erected first in the upper layer of quadrant cells and quickly followed by similar walls in the two lower

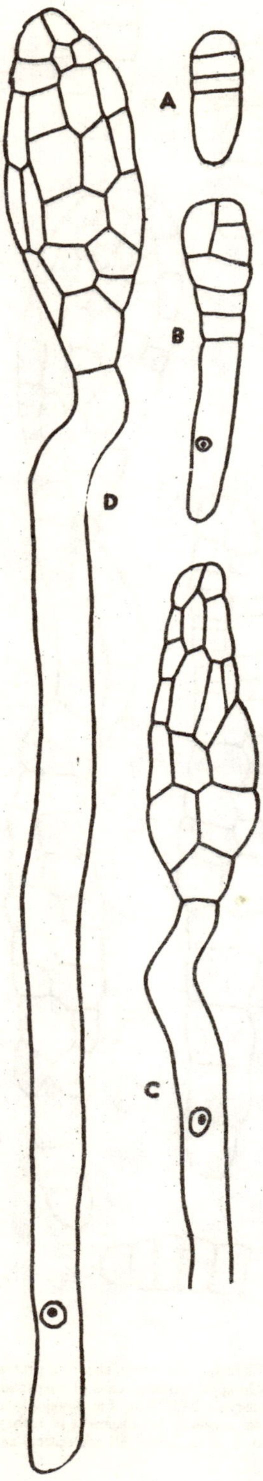

tiers delimit the epidermal initials (FIG. 66Q). The epidermis is more clearly defined in this Pattern than in the preceding one.

Species in the genera *Phajus*, *Platanthera*, *Nigritella* and *Gymnadenia*, as well as *Cypripedium venustum*, *C. barbatum* and *Epidendrum verrucosum*, have an embryonomy essentially similar to that of *Orchis latifolia*, but all have much shorter suspensors; that of *C. barbatum*, for instance, may be considerably twisted.

Events in *Habenaria ciliaris* and *H. integra* are very similar to those in *Orchis latifolia*. The second division is transverse in the basal cell and the third is longitudinal in the terminal cell. Divisions then occur rapidly in the derivatives of the basal cell until a row of about eight cells is produced. The cell immediately below the two juxtaposed terminal cells is equivalent to the large cell usurping a similar position in *O. latifolia*, but in *Habenaria* it more clearly has the functions of a hypophysis initial since it divides either transversely or vertically and contributes to the construction of the embryo. What part the derivatives might have in the development of the radicle has not been described. The two terminal cells divide transversely to form two superposed groups of two cells each, after which stage divisions are rather irregular. *H. platyphylla* is essentially similar to the other two species.

In *Himantoglossum hircinum* the basal cell divides transversely and the terminal cell longitudinally. The middle cell, derived from the basal cell, forms a portion of the embryo proper and the very suggestive claim is advanced that it is these derivatives which become infected with fungal mycelium when the seed germinates. The embryo in the mature seed consists of from 52 to 56 cells; of these, 32 are contributed by the terminal cell, 16 by the middle cell and from four to eight by the sister of the middle cell.

3. *Goodyera discolor Pattern.* — The early stages are essentially like those in the preceding Pattern, but the sequence of divisions in the developing embryo is far less regular (FIG. 67A–D). The basal cell of the four-celled proembryo never divides after its origin but commences to elongate rapidly

FIG. 67. — *Goodyera discolor*. A. Four-celled proembryo; basal cell is already beginning to elongate. B. Young embryo. C. Developing embryo. D. Nearly mature embryo with elongated suspensor cell. All × 240. All these figures are optical sections. (*Redrawn from* TREUB 1879).

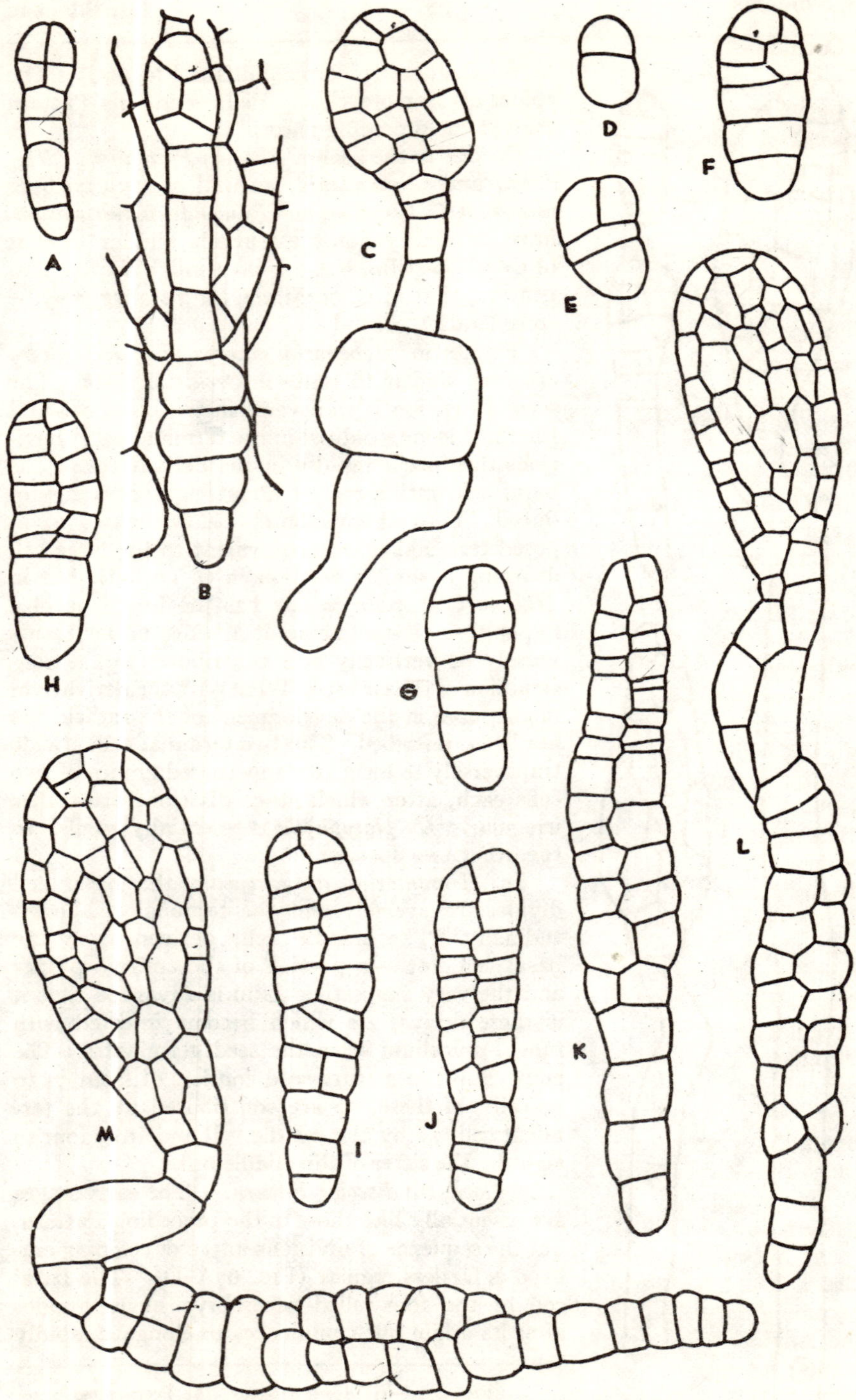

FIG. 68. — *Herminium monorchis.* A. Young embryo at octant stage. B. Slightly older stage showing protrusion of half of suspensor from micropyle. C. Beginning of swelling and branching of terminal cells of suspensor. All × 240. *Epidendrum ciliare.* D. Two-celled proembryo. E. Quadrant embryo. F–J. Stages in development of embryo proper, with early stages of suspensor growth in I and J. K, L. Later stages in development of embryo and suspensor. M. Approximately mature embryo. D–F × 730; G–I × 600; J–M × 400. All are optical sections. (*Redrawn from* TREUB 1879).

and when the embryo is approximately mature it may even, especially in *Spathoglottis plicata*, extend out of the micropyle. The development of the embryo has not been followed out very closely, but the divisions are extremely irregular. The approximately mature embryo is very narrow, with the epidermal cells poorly delimited (FIG. 67D). An apparent hypophysis initial is present.

A similar embryonomy is present in *Corallorhiza maculata*.

4. *Serapias linguae Pattern.* — In several species of *Habenaria* and in *Serapias linguae* the filamentous suspensor grows out of the micropyle and becomes variously swollen. In *Habenaria* the terminal cell of the suspensor becomes embedded in the placental tissue and acts like an aggressive haustorium, while in *Serapias* the terminal cell is both swollen and profusely branched. In *Phajus grandiflora* the terminal cell of the two-celled suspensor pushes out of the micropyle, becomes slightly branched and acts like a haustorium, but the structure later perishes.

This Pattern is apparently a transition form between the Goodyera discolor and Eulophea epidendracea Patterns, differing from the former in the branching habit of the suspensor and from the latter in that the branches are not true embryonal tubes.

5. *Herminium monorchis Pattern.* — This Pattern greatly resembles the Serapias linguae Pattern in that the filamentous suspensor projects from the micropyle for half of its length, but differs in that all cells which protrude become swollen and then branch profusely (FIG. 68A–C).

6. *Epidendrum ciliare Pattern.* — Early developmental stages correspond greatly with those in the Orchis latifolia Pattern (FIG. 68D–F), but shortly after the quadrant stage one layer after another, each consisting of four cells, is erected until the developing embryo may include from eight to twelve such tiers before the suspensor initials commence dividing (FIG. 68E–J). The latter at first consist of three or four superposed cells, but it appears that the terminal two or three cells do not divide further and that all of the suspensor is derived from one or two cells located adjacent to the embryo proper, beginning when the embryo consists of about four or six tiers of quadrant cells. Most of the divisions in the suspensor cells are transverse, but a daughter cell may divide longitudinally only once (FIG. 68L). Presently, however, the divisions become irregular and the cells are rather uneven in size but preserve the appearance of being arranged in two ranks. At intermediate stages of development it is very difficult to distinguish between embryo and suspensor, but the embryo presently becomes ovaliform, with smaller cells, and the epidermal initials are delimited. In the nearly mature embryo the suspensor may reach a length twice that of the embryonic body (FIG. 68M).

Other species of *Epidendrum* evidently also have the same embryogenic pattern.

7. *Sobralia macrantha Pattern.* — Development of the embryo is none too clear. The suspensor is either long or short at first (FIG. 69A–F), eventually becoming much like that of the preceding Pattern but remaining shorter than the length of the embryo proper. The epidermis is delimited early and is regular in development. The embryo rapidly becomes greatly elongated and cells are presently produced in greater number at one side of the apex so that an overarched protuberance is produced, with a distinct notch at the base of the curvature (FIG. 69F–H). This protuberance has every appearance of being a primitive or at least a potential cotyledon: this is the distinguishing character of the Sobralia macrantha Pattern of orchid embryogeny, and is, in fact, the only Pattern in which anything resembling a cotyledon is produced by the embryo. (FIG. 69I).

The embryo of *Platyclinis* is said also to possess a presumptive cotyledon.

8. *Vanda parviflora Pattern.* — After the first transverse wall is formed, the embryonic initial divides transversely to form the three-celled proembryo characteristic of most orchids. The suspensor initial enlarges and divides longitudinally by two walls intersecting at right angles to form a basal tier of four cells. Each of these four cells then becomes divided by a vertical wall and all eight resulting cells become enlarged and produce embryonal tubes which grow up and almost completely surround the embryo (FIG. 70A). These tubes eventually wither at the time of dispersal of the seed.

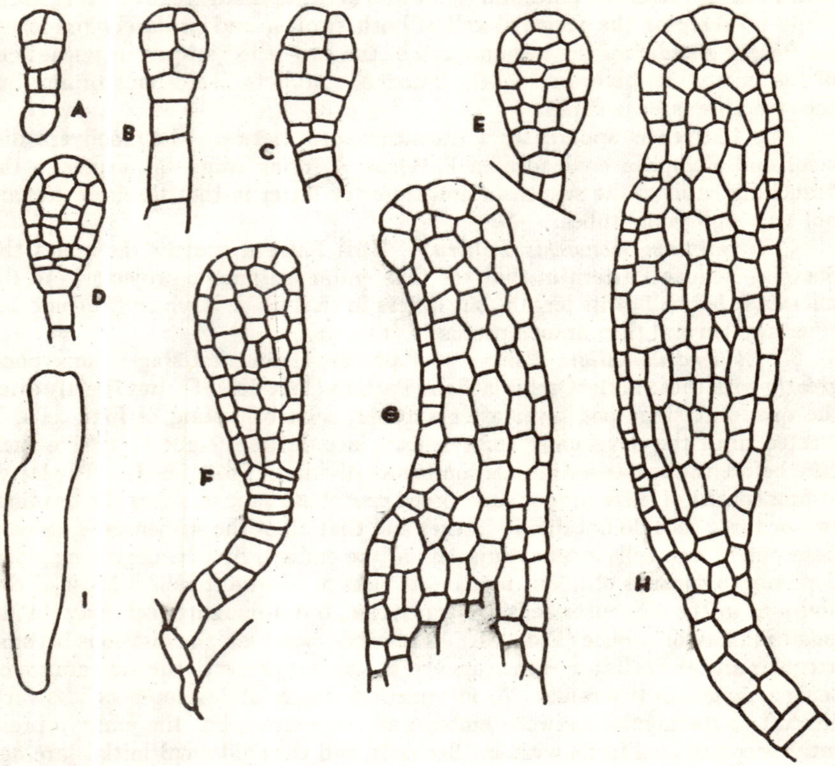

FIG. 69. — *Sobralia macrantha.* A–E. Young embryos; the suspensor is long in the first four, short in E. F. Developing embryo at commencement of "cotyledon" development. G. "Cotyledon" well developed. H. Upper half of more or less mature embryo. I. Outline of mature embryo with "cotyledon." A–H × 240; I × 90. All figures are optical sections. (*Redrawn from* TREUB 1879).

At about the time the suspensor initial has become four-celled, the terminal cell of the proembryo divides by a longitudinal wall, then a similar wall is produced in the lower cell. Subsequent divisions are irregular and a spherical or ovoid embryo is eventually formed. A cotyledon or internal histogens can not be distinguished.

Vanda spathulata and *V. roxburghii* are very similar in development to *V. parviflora.*

9. *Phalaenopsis grandiflora Pattern.* — After the first transverse division the basal cell divides longitudinally (FIG. 70B) and then the terminal cell is

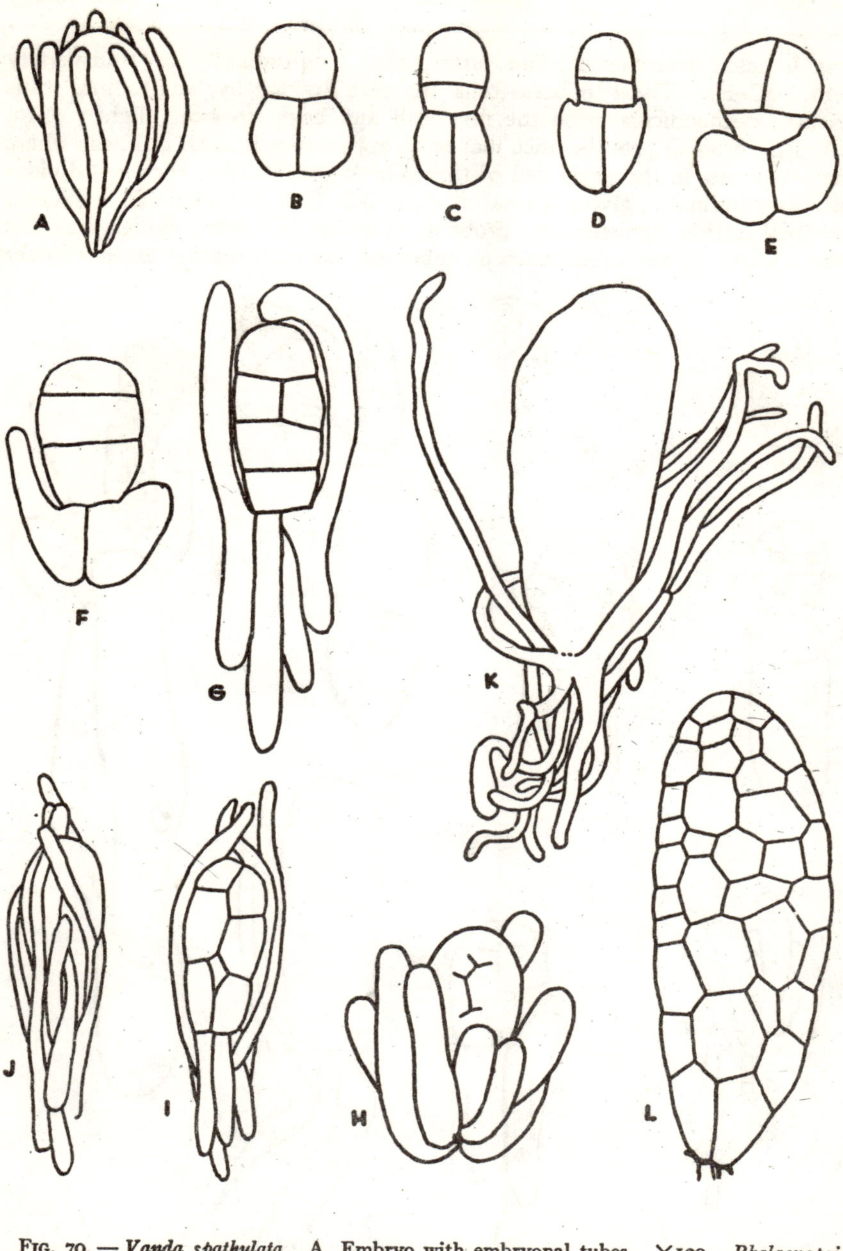

FIG. 70. — *Vanda spathulata*. A. Embryo with embryonal tubes. ×120. *Phalaenopsis grandiflora*. B. Three-celled proembryo. C. Four-celled proembryo. D. Quadrant proembryo. E–G. Early embryonal tube development, H, I. Surface views of embryonal tubes at two successive stages. J. Optical view of I. K. Embryonal tubes in surface view at optimum development. L. Mature embryo with strophiole at base, representing all that remains of embryonal tubes. B–G × 730; H, J, L × 600; K × 250. Most of the figures, except as noted otherwise, are optical sections. (A *redrawn from* SWAMY 1943; B–L *redrawn from* TREUB 1879).

first bisected transversely and later either longitudinally or transversely (Fig. 70C–E). The two basal cells are next divided by longitudinal walls oriented perpendicularly to the first wall and begin to swell slightly (Fig. 70B–F). A small protuberance makes its appearance in each cell near where the cell abuts on the lower cell of the embryo proper, then each cell divides once longitudinally, giving a total of eight cells (Fig. 70F, G). It is difficult to visualize this segmentation; probably it occurs somewhat earlier than has been stated. In any event, each protuberance becomes rapidly extended over

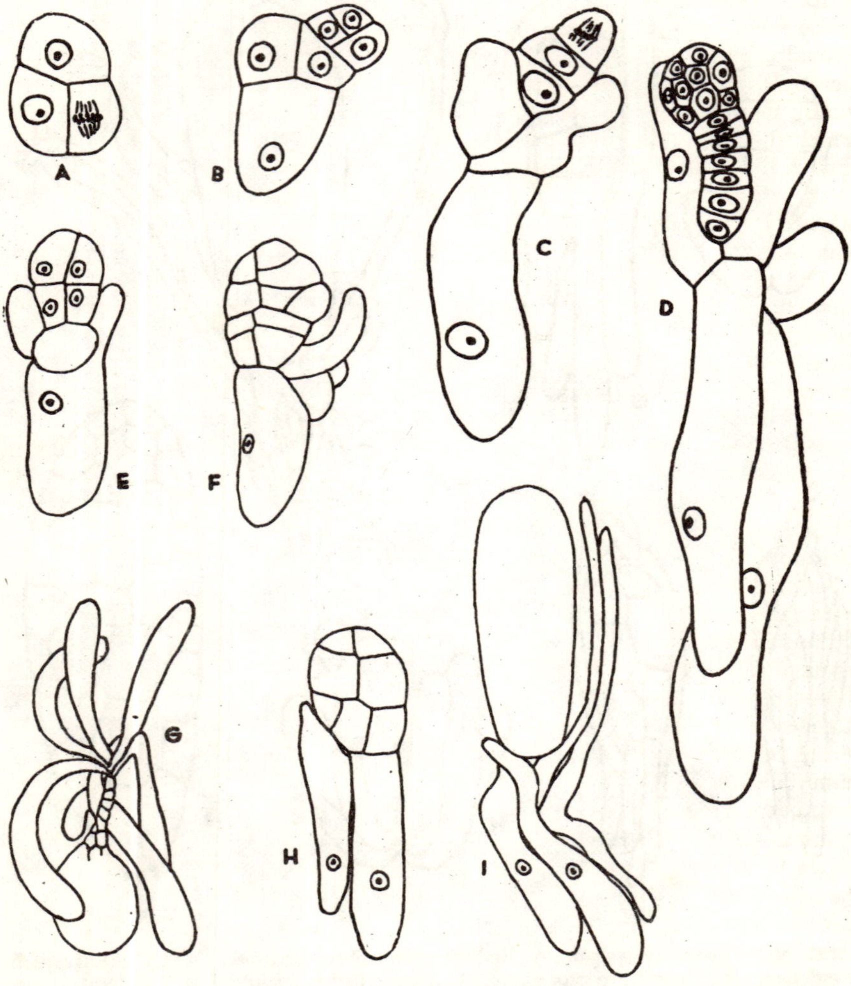

Fig. 71. — *Cymbidium bicolor*. A–D. Description in text. *Phalaenopsis schilleriana*. E, F. Early development of embryonal tubes. H. Unicellular suspensor and two embryonal tubes (only one shown). I. Nearly mature embryo with embryonal tubes. *Eulophea epidendracea*. G. Embryo with embryonal tubes at end of suspensor. A–C × 560; D × 160; E–H × ca. 320; G × 100. (A–D *redrawn from* Swamy 1942; E–H *redrawn from* Treub 1879; G *redrawn from* Swamy 1943).

both the body of the embryo and the other end of each cell produces a similar protuberance which grows backward (Fig. 70G–J). The two arms of each embryonal tube are approximately equal in diameter. At first those tubes covering the embryo are closely applied to the latter but as the embryo approaches maturity they become twisted or convoluted and those extending backward become interlaced (FIG. 70K). Each tube is a unit; it is very rare that a transverse wall may be found within a tube. The whole mass of tubes is solidly attached to the embryo. As the latter begins to attain maturity the mass of tubes slowly commences to shrivel until a small strophiole at the base of the mature embryo is all that remains of them (FIG. 70L).

There is no regularity in the sequence of divisions in the developing embryo. At maturity it is elongated, with a fairly distinct epidermis (FIG. 70L).

10. *Cymbidium bicolor Pattern.* — The basal cell divides to produce from five to seven cells which enlarge, at first somewhat unevenly, to produce embryonal tubes (FIG. 71A–B). Those tubes first developed project toward the micropyle, while the later ones grow over the embryonic body. The latter are usually more numerous than the former (FIG. 71C, D). The tubes are unbranched. Except for this unbranched character of the tubes, this Pattern is much like that of the following one.

11. *Phalaenopsis schilleriana Pattern.* — The earlier stages could not be followed out, but are said to resemble those of *Vanda tricolor*, which also belongs to the P. schilleriana Pattern but was not described or illustrated in detail.

The number of embryonal tubes produced in this Pattern is rather variable: they may be as few as two and rarely exceed four (FIG. 71E, F). In any case the region protruding backward is always from twice to four times the diameter of the arm projecting over the embryo (FIG. 71H, I). It is debatable whether the embryonal tubes in *P. schilleriana* and *V. tricolor* represent a reduced or a more primitive type than the extensively developed ones of *Phalaenopsis grandiflora*. In a few cases a unicellular suspensor is present in addition to the embryonal tubes.

12. *Stanhopea oculata Pattern.* — The first division is described as being oblique with reference to the longitudinal axis of the ovule, although the zygote itself appears to be divided transversely (FIG. 72A). Next each daughter cell is bisected by a longitudinal wall (FIG. 72B, C), after which the divisions are extremely irregular and all appear to occur in the derivatives of the basal cell of the two-celled stage, and in such planes that a group of cells surround the terminal cell in a single layer (FIG. 72D–H). Each of these derivative cells is a potential embryonal tube initial. The whole mass is composed of from eight to fifteen cells by the time the embryonal initials begin to elongate. No single suspensor cell nor definite suspensor, derived from the basal cell, is apparent. The embryonal tube initials begin to elongate in different directions (FIG. 72H, I); some cover the embryo but most of them extend backward. Division of the terminal cell does not commence until the embryonal tubes are considerably elongated (FIG. 72I). At least the first two divisions are transverse (whatever the orientation of the proembryo), then longitudinal walls are erected in all save the lowermost cell (FIG. 72J, K). The epidermal initials are formed comparatively early (FIG. 72L, M). The embryonal tubes eventually penetrate to all portions of the ovule until restrained by the epidermis of the latter.

13. *Eulophea epidendracea Pattern.* — The earlier stages of development of this Pattern have not been described in detail; it is stated that the succeeding few divisions of the zygote after its first division are most irregular, then one of the cells at the chalazal end develops a filamentous row of cells (apparently a true suspensor, although probably derived indirectly from the basal cell of the two-celled proembryo). The description is vague on this point, but it appears that the tip cell of the suspensor divides irregularly into several cells (eight?) which are potential embryonal tube initials. These elongate, become

somewhat inflated, and grow in all directions but do not cover the embryo (FIG. 71G).

Conditions appear to be decidedly similar in some species of *Cyr bidium*.

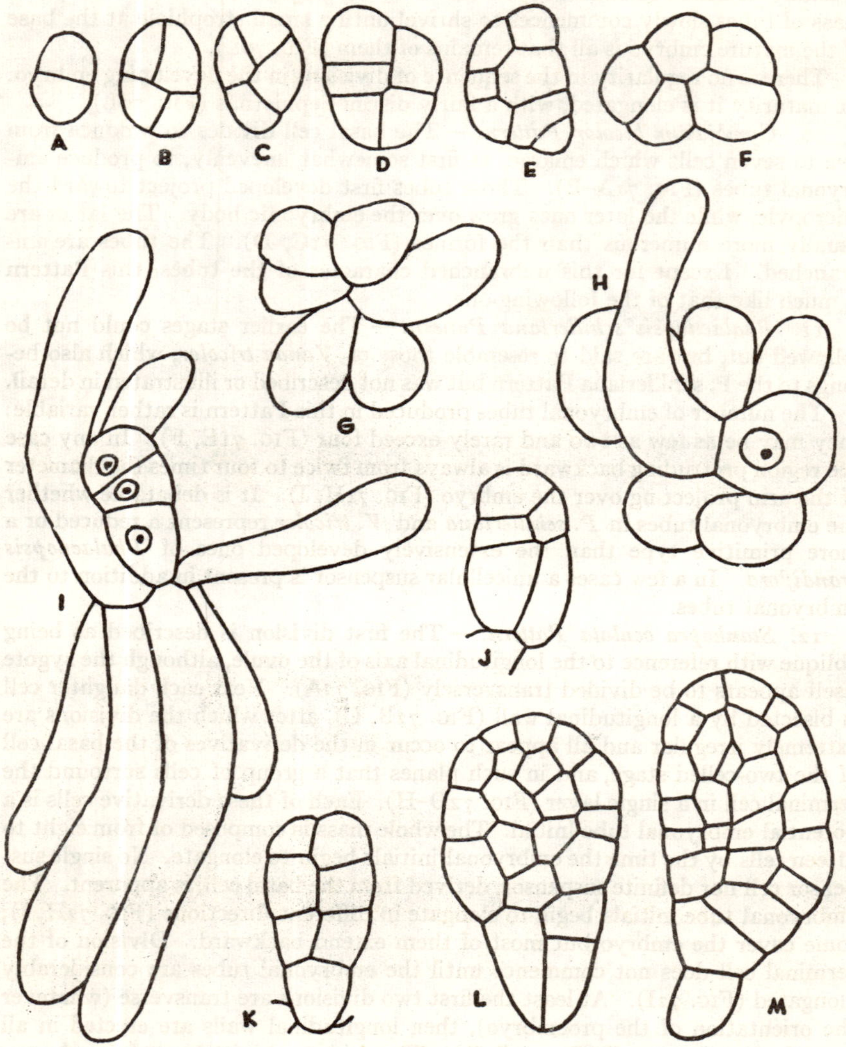

FIG. 72.—*Stanhopea oculata*. A. Two-celled proembryo. B. Four-celled proembryo. C–H. Stages in formation of embryonal tube initials (H is in surface view). I. Terminal cell of embryo (with nucleus) surrounded by embryonal tubes. J. Three-celled embryo with embryonal tubes. K. Young embryo. L. Developing embryo. M. Adult embryo. All figures are optical sections. A–I × 400; J–M × 250. (*Redrawn from* TREUB 1879).

Polyembryony:- The cleavage type of polyembryony has been reported for *Cymbidium bicolor* and *Eulophea epidendracea*. In the first species the first wall in the zygote is claimed to be vertical (which would assign the species to the Piperad Type) and each half develops independently into an embryo. In the latter species any cell of the young embryo may proliferate to produce additional embryos.

Brown, W. H. 1909. The embryo sac of *Habenaria*. Bot. Gaz. 48: 241–250.
Carlson, M. C. 1940. Formation of the seed of *Cypripedium parviflorum*. Bot. Gaz. 102: 295–301.
Dumee, P. 1910. Quelques observations sur l'embryon des *Orchidées*. Bull. Soc. Bot. France 57: 83–87.
Hagerup, O. 1944. On fertilisation, polyploidy and haploidy in *Orchis maculatus* L. sens. lat. Dansk Bot. Arkiv 11: 1–25.
Heusser, K. 1915. Die Entwicklung der generativen Organe von *Himantoglossum hircinum* Spr. (=*Loroglossum hircinum* Rich.). Beih. Bot. Centralbl. Abt. I, 12: 218–277.
Pace, L. 1907. Fertilization in *Cypripedium*. Bot. Gaz. 44: 353–374.
Sharp, L. W. 1912. The orchid embryo sac. Bot. Gaz. 54: 372–385.
Swamy, B. G. L. 1942. Female gametophyte and embryogeny in *Cymbidium bicolor* Lindl. Proc. Indian Acad. Sci., B, 15: 194–201.
Swamy, B. G. L. 1942. Morphological studies in three species of *Vanda*. Current Science 11: 285–286.
Swamy, B. G. L. 1943. Embryology of *Orchidaceae*. Current Science 12: 13–17.
Swamy, B. G. L. 1944. The embryo-sac and the embryo of *Satyrium nepalense* Don. Jour. Indian Bot. Soc. 23: 66–70.
Swamy, B. G. L. 1946. Embryology of *Habenaria*. Proc. Nat. Inst. Sci. India 12: 314–326.
Treub, M. 1879. Notes sur l'embryogénie de quelques *Orchidées*. Verhandel. Konink. Akad. Wetensch., Afdeel. Natuurkunde, 19[2] Deel. Amsterdam.
Ward, H. M. 1880. On the embryo sac and development of *Gymnadenia conopsea*. Quart. Jour. Micr. Soc. 20: 1–18.

Juncales

The *Juncaceae* have been carefully studied, but the other families — *Thurniaceae*, *Centrolepidaceae* and *Restionaceae* — have never been investigated.

Juncaceae:- The embryonomy of *Juncus effusus* and *Luzula forsteri* is essentially identical.

In *L. forsteri* the terminal cell first segments longitudinally into two juxtaposed elements (Fig. 73A, B), then the basal cell divides transversely into two superposed cells, *m* and *ci* (Fig. 73C). The two upper cells thereupon divide longitudinally to give four quadrants disposed in a horizontal plane; the middle cell *m* is similarly partitioned by a vertical wall; and later the basal cell, *ci*, is divided transversely into two superposed cells, *n* and *n'* (Fig. 73D). This is as far as the proembryo of the *Juncaceae* agrees with those of other Anthophyta; after this eight-celled stage an entirely new character appears. The proembryo, contrary to the usual course of events, does not immediately thereafter form octants.

The behavior of each stage may now be taken up in detail, first indicating the eventual designation of each: the superior tier, *q*, engenders the cotyledon and the hypocotyl; the tier *m* gives rise to the initials of the root cortex and to those of the root cap; from *n* is differentiated the culminating portion of the root cap; and finally *n'* is converted into a very short suspensor.

The four elements which compose tier *q* behave in an identical manner. In each is erected a wall that is tangential and curved outwardly to parallel the rounded external wall (Fig. 73E). These walls separate two cells in each quadrant cell; the outer one now becomes specialized as an epidermal initial (*de*). The inner daughter cell of the original quadrant cell is divided transversely. As the result of this last segmentation the entire upper portion of the proembryo corresponding to tier *q* becomes divided into two new tiers, *l* and *l'*, each composed of eight elements — four peripheral epidermal cells and four central cells (Fig. 73F–I). The peripheral or epidermal initial cells now divide; the wall is perpendicular to the curved wall, at about the center of each cell, and divides it equally (Fig. 73F, G). The four central upper cells of tier *l* are each divided by two vertical walls, giving rise to three new elements from each mother cell.

Altogether, twelve cells disposed in a layer are thus formed: four circumaxial cells and eight elements separated from the axis. Next each of the twelve cells may be partitioned transversely, or only the four circumaxial cells are so divided while the other eight cells segment tangentially (FIG. 73I). Occasionally, however, the central primitive cells of tier *l* divide horizontally rather than vertically to give two circumaxial superposed tetrads; then, in the lower tetrad and later in the upper tetrad, vertical walls are erected, thus giving rise to a cellular plateau of twelve elements, four medial and eight peripheral. The

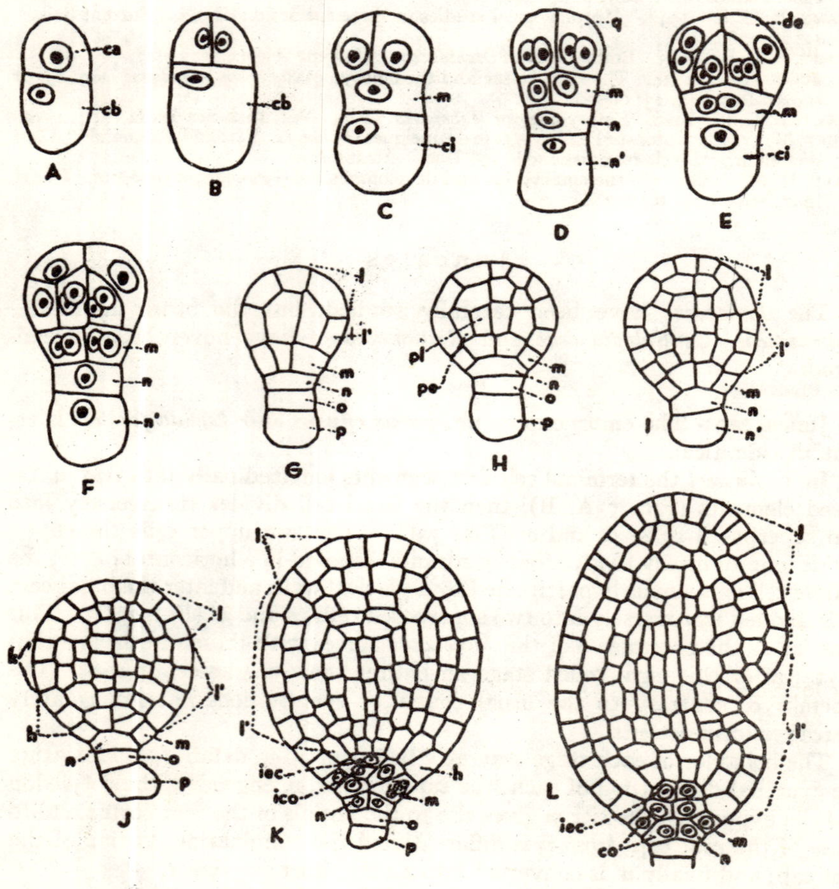

FIG. 73. — *Luzula forsteri*. Description in text. A–F × 650; G–L × 370. (*Redrawn from* SOUÈGES 1933).

divisions which follow are irregular; there finally results a massive body in which no differentiation into periblem and plerome has yet appeared (FIG. 73J, K). This mass is for a long time distinctly separated from the lower portion of the embryo.

The four lower central cells of tier *l′* are segmented, like their corresponding sister cells in tier *l*, by vertical walls (FIG. 73H), which demarcate the four primary elements of the plerome around the axis and eight to twelve periblem elements under the epidermal layer (FIG. 73I). Transverse walls next separate this region into two new tiers, *k* and *h* (FIG. 73J). Segmentations proceed more rapidly in the upper layer, *k*, than in the lower layer *h*; the periblem elements display the most activity. They divide vertically and then transversely to give two or three rows of cells, while the circumaxial plerome cells do not divide transversely. The cells of the lower tier, *h*, divide transversely at first, then longitudinally when the embryo approaches maturity. These walls separate the pericambium exteriorly.

In the eight-celled proembryo region *m* is composed of two juxtaposed cells (Fig. 73D). They divide by meridionally vertical walls to give four circumaxial cells (Fig. 73D). Next each divides either by two vertical walls to engender one central and two peripheral cells, or by a tangential wall more or less inclined and with one end attached to the lateral wall of the epidermal cell of the hypocotyl so as to separate an internal and an external cell. In the first instance, tier *m* now consists of a total of twelve cells, and of eight cells in the second instance. Subsequently all of the external cells multiply through radial,

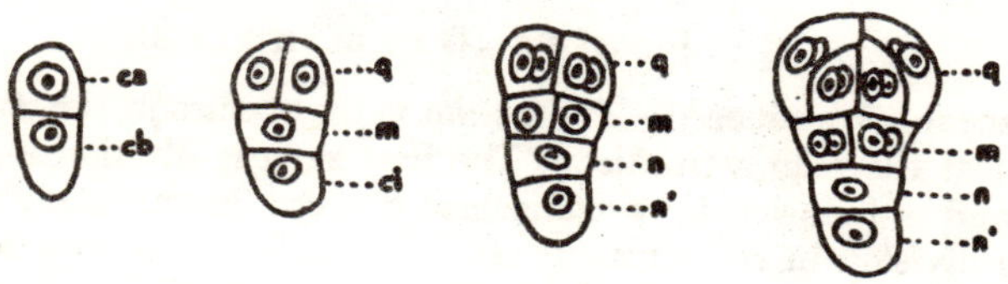

Recapitulatory Table for Luzula forsteri
(from left to right)

I. First Cell Generation
Proembryo of two cells disposed
in two tiers:
ca = pco + phy + pvt + icc
cb = iec + ico + co + s

II. Second Cell Generation
Proembryo of four cells disposed
in three tiers:
q = pco + phy + pvt + icc
m = iec + ico
n = co + s

III. Third Cell Generation
Proembryo of eight cells disposed
in four tiers:
q = pco + phy + pvt + icc
m = iec + ico
n = co
n' = s

IV. Fourth Cell Generation
Proembryo of fourteen cells disposed
in four tiers:
Arrangement and destinations as in third cell generation.

longitudinal and transverse segmentation to engender the more elongated apical portions of the root cap (Fig. 73J, K). The internal cells always divide by obliquely curved walls whose upper insertion is below the elements of the periblem of the hypocotyl and the other end on the two primary meridional walls. Each such wall separates an upper cell, *iec*, which is triangular in section, and a lower cell, *ico*, quadrilateral in section. The upper group of four circumaxial cells represents the initials of the root cortex; the tetrad of four lower cells, also circumaxial, constitutes the initials of the root cap.

Occasionally the cells in tier *m* may develop somewhat asymmetrically, exactly as frequently occurs in other species which also follow the Onagrad Type.

In the eight-celled proembryo, tier *n* is unicellular, and is still in this condition when the remainder of the embryo is quite well developed. Eventually the cell divides longitudinally and each daughter cell later divides similarly (Fig. 73F *et seq.*). These four cells complement the root cap and do not divide further to any great extent.

Tier *n'* of the eight-celled proembryo generally divides transversely into *o* and *p*, but may divide longitudinally or obliquely. The region corresponds to a very rudimentary suspensor, sensu stricto.

The recapitulatory table for *Luzula forsteri*, and hence for the Juncus Variation, Onagrad Type, is given on p. 267.

SOUÈGES, R. 1923. Embryogénie des *Joncacées*. Développement de l'embryon chez *Luzula forsteri* DC. C. R. Acad. Sci. Paris 177: 705–708.

SOUÈGES, R. 1933. Recherches sur l'embryogénie des *Joncacées*. Bull. Soc. Bot. France 80: 51–69.

Cyperales

Cyperaceae:- The embryonomy of *Carex punicea* resembles that of the Juncus Variation, Onagrad Type.

HOFMEISTER, W. 1861. Neue Beiträge zur Kenntnis der Embryobildung der Phanerogamen. II. Monokotyledonen. Abh. Kön. Sächs. Ges. Wiss. 7: 629–760.

Poales (Graminales)

Poaceae (Gramineae):- *Poa annua* is the species in the *Poaceae* which has been most thoroughly studied. The first zygotic division is transverse (FIG. 74A), and is followed by a transverse division in the basal cell and a longitudinal division in the terminal cell (FIG. 74B). The two derivatives of the basal cell, *m* and *ci*, next divide simultaneously: the former is partitioned vertically and the latter transversely into two superposed cells, *n* and *n'* (FIG. 74C, D). The two upper cells, designated as *q*, are partitioned by vertical walls oriented perpendicular to the first wall, then each of the four resultant cells is partitioned by more or less transverse or horizontally curved walls to give two distinct layers, *l* and *l'* (FIG. 74D, E). The cotyledon is derived from these two tiers. More vertical walls are erected in the two regions *m* and *n*, while *n'* is divided transversely into tiers *o* and *p* (FIG. 74E). The last cell, *p*, forms a very short suspensor (FIG. 74G *et seq.*).

At the moment when the embryo begins to exhibit laterality and the depression which marks the separation of the cotyledon and hypocotyl appears, this depression is observed to be produced at the common limit of the two regions *q* and *m* (FIG. 74I). The upper lip of the coleoptile (*cl'*), which is differentiated shortly after at the expense of the epidermal cells at the base of the cotyledon, takes its origin from the region of the proembryo corresponding, from all appearances, to the tier *l'* (FIG. 74J, K). From tier *m* are derived the hypocotyledonary axis, the stem apex, the lower lip of the coleoptile, the initials of the central cylinder and the cortex of the root.

The stem apex, the first leaf (*pv*) and the lower lip of the coleoptile (*cl''*) first appear as three rudimentary projections (FIG. 74K). The first structure develops but little, while the other two, on the contrary, grow rapidly at the top and bottom. The two lips of the coleoptile slowly grow until they meet and enclose the stem apex and the rudiment of the first leaf (FIG. 74L). The epidermis is quickly differentiated but the other two histogens are developed more slowly; eventually they produce in the region which corresponds to the plane of separation of the two regions *m* and *n* a detachment of tissue, which remains united only by the axial and marginal parts.

Tier *n* engenders the root cap, coleorhiza and epiblast (*ep*, FIG. 74J, K). The latter originates from the epidermal cells of region *n*. The central cells of this region, united above to the initials of the cortex, multiply by tangential and radial segmentations and give rise to a slightly extended tissue, of conical form, which represents the root cap (*co*, FIG. 74L). The lateral and lower elements of the same region contribute to the formation of the coleorhiza (*cr*, FIG. 74L). The regions *o* and *p* remain composed of elements which multiply but little;

from them is developed the lowermost portion of the embryo which corresponds to a suspensor and is designated generally by the name of hypoblast (*hb*, FIG. 74L). In *Poa annua* the cells of the hypoblast are distinguished by their large dimensions and the accumulation of reserve foodstuffs. The limits of the three regions, *n*, *o* and *p*, can not be clearly distinguished during the last stages of development.

In *Poa pratensis* normal fertilization may or may not occur; when a fertilized egg develops into an embryo, the latter may grow to maturity or may be inhibited in growth by nucellar embryos. The suspensor is massive and from one to several extra embryo-like buds may develop therefrom. The normal embryogeny is like that of *Poa annua*.

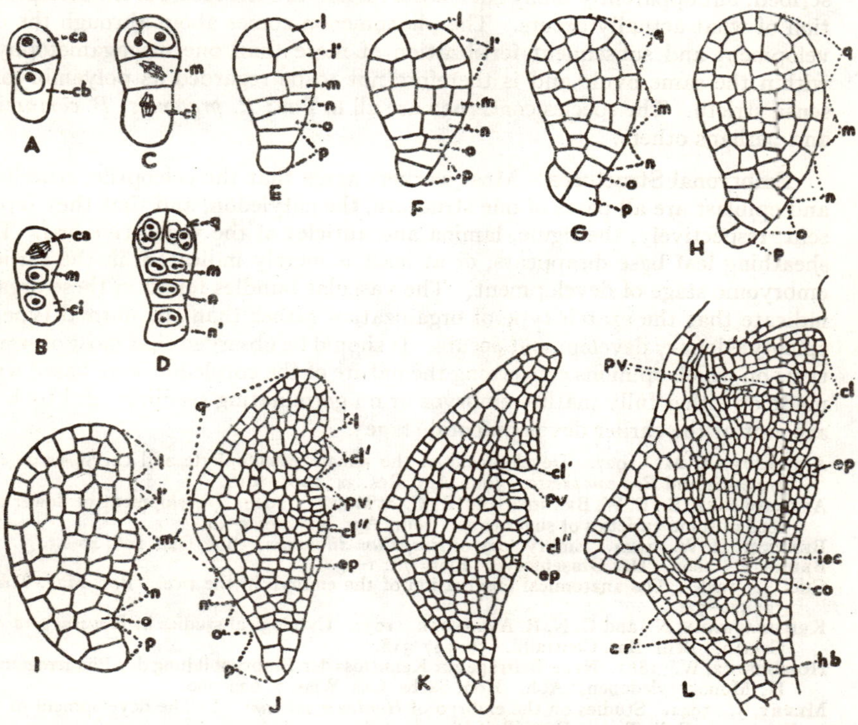

FIG. 74. — *Poa annua*. Description in text. Partial explanation of symbols: *cl*, coleoptile; *cl'* and *c''*, two lips, upper and lower; *pv*, stem apex and first leaf; *ep*, epiblast, *hb*, hypoblast; *iec*, initials of root cortex; *cr*, coleorhiza; *co*, root cap. A–I × 375; J–L × 190. (*Redrawn from* SOUÈGES 1924).

The proembryonic development of *Secale cereale*, *Triticum vulgare* and *Hordeum vulgare*, although figured more than eighty years ago, bears a remarkable resemblance to that of *Poa annua*.

The development of *Hordeum vulgare*, *H. sativum*, *H. distichum*, *Zea mays*, *Eragrostis chiliensis*, *Paspalum dilatatum*, *Triticum vulgare*, *Secale cereale* and *Avena sativa* in each case is essentially similar to that of *Poa annua*, although an early attempt was made to separate them into three different developmental types. The differences reside mainly in variations in the planes of the walls laid down in various embryonal regions and these are probably not of sufficient importance to warrant separation from the Poa Variation.

There are numerous fragmentary observations and illustrations in the literature which do not seem to be worthy of special mention. In general, the embryogeny is essentially similar to that of *Poa annua*. Among the species involved are *Poa pratensis*, *P. compressa*, *Saccharum officinarum* and *Oryza sativa*.

The terminal cell of the two-celled proembryo of *Eleusine coracana* divides transversely, then the upper daughter cell is segmented longitudinally. It is said that all cells below these two juxtaposed terminal cells go into the formation of the suspensor and only these two create the embryo proper. The embryo portion is two cells wide for a time, then increases to six cells, at which time the epidermis, plerome and periblem are more or less well differentiated. This species seems to differ considerably from *Poa annua*.

"Polyembryony" of a nucellar or suspensor type has sometimes been described, but apparently many such observations are the result of misinterpretation of what actually occurs. The phenomenon comes about through the development and subsequent fertilization of more than one megagametophyte within the same ovule and is therefore not to be regarded as polyembryony sensu strictu. The species concerned are all in Poa: *P. pratensis*, *P. compressa* and perhaps others.

Embryonal Structures:- Most workers agree that the coleoptile, scutellum and epiblast are all parts of one structure, the cotyledon, and that they represent, respectively, the ligule, lamina and auricles of the vegetative leaf. The sheathing leaf base disappears, or at least is merely indicated, in the earliest embryonic stage of development. The vascular bundles found in these organs indicate that the exarch type of organization rather than the normal type of epicotyledonary development occurs. It should be observed that most observations on which opinions concerning the nature of the cotyledon were based were made on either fully mature embryos or on germinating seedlings and took no account of the earlier developmental stages.

ANDERSON, A. M. 1927. Development of the female gametophyte and caryopsis of *Poa pratensis* and *Poa compressa*. Jour. Agr. Res. 34: 1001–1018.

ARTSCHWAGER, E., E. W. BRANDES and R. C. STARRETT. 1929. Development of flower and seed of some varieties of sugar cane. Jour. Agr. Res. 39: 1–30.

BENNETT, H. W. 1944. Embryology of *Paspalum dilatatum*. Bot. Gaz. 106: 40–45.

BRUNS, E. 1892. Das Grasembryo. Flora 76: 1–33.

CHO, J. 1938. The anatomical observation of the embryo in the rice. Bot. Mag. Tokyo 52: 520–531.

KRISHNASWAMY, N., and G. N. R. AYYANGAR. 1937. Cytological studies in *Eleusine coracana* Gaertn. Beih. Bot. Centralbl. 57: 297–318.

HOFMEISTER, W. 1861. Neue Beiträge zur Kenntniss der Embryobildung der Phanerogamen. II. Monokotyledonen. Abh. Kön. Sächs. Ges. Wiss. 7: 629–760.

MERRY, J. 1941. Studies on the embryo of *Hordeum sativum*. I. The development of the embryo. Bull. Torrey Bot. Club 68: 585–598.

MODILEVSKI, J. S., und R. A. BEILIS. 1937. Zur Embryologie und Zytologie von Weizen. I. Embryogenesis des Weizen. Vom Archesporium bis zum Embryo. Zhurn. Inst. Bot. Akad. Nauk URSR (Jour. Inst. Bot. Acad. Sci. RSS Ukraine) 21/22: 127–141. (In Ukrainian with German summary.)

MODILEVSKI, J. S., and R. A. BEILIS. 1938. On the embryology and cytology of the wheat plant. II. The stages of the maturing of the embryo and caryopsis, of their germination and of earing. Zhurn. Inst. Bot. Akad. Nauk URSR (Jour. Inst. Bot. Acad. Sci. RSS Ukraine) 26/27: 13–39. (In Ukrainian and Russian with English summary.)

NISHIMURA, M. 1922. Comparative morphology and development of *Poa pratensis*, *Phleum pratense* and *Setaria italica*. Jap. Jour. Bot. 1: 55–85.

NISHIMURA, M. 1922. On the germination and the polyembryony of *Poa pratensis* L. (Preliminary note.) Bot. Mag. Tokyo 36: 47–54.

NOGUCHI, Y. 1929. Zur Kenntnis der Befruchtung und Kornildung bei der Reispflanze. Jap. Jour. Bot. 4: 385–403.

PERSIDSKY, D. J. 1940. Embryological and cytological investigations of barley, *Hordeum distichum* L. Bot. Zhurn. URSR (Jour. Bot. Acad. Sci. RSS Ukraine) 1: 145–153. (In Ukrainian with English summary.)

RANDOLPH, L. F. 1936. Developmental morphology of the caryopsis in maize. Jour. Agr. Res. 53: 881–916.

SOUÈGES, R. 1924. Embryogénie des *Graminées*. Développement de l'embryon chez le *Poa annua* L. C. R. Acad. Sci. Paris 178: 860–862.

STOVER, E. C. 1937. The embryo sac of *Eragrostis ciliensis* (All.) Link. A new type embryo sac and a summary of grass embryo sac investigations. Ohio Jour. Sci. 37: 172–181.

Chapter XIX

APOMICTIC EMBRYOGENY

Apomixis, briefly, is a reproductive process in which no sexual fusion is involved. It is a collective term embracing several different phenomena which will be described presently. It is sometimes broadened to include various forms of vegetative reproduction, such as apocarpy and apoflory; but since embryos are not known to be concerned, these phases will not be considered further. The term "polyembryony" has frequently been employed in connection with apomictic phenomena but as it does not take into account the mode of origin of the various embryos which might be present, its use in this sense should be abandoned.

All types of apomixis exclusive of purely vegetative reproduction may be described as agamospermy, or reproduction by means of seeds. Agamospermy, in turn, consists of diplospory (diploparthenogenesis), apogamety, apospory and adventitious embryony. In all save adventitious embryony an alternation of generations occurs, at least in the morphological sense.

The term agamogony has been applied to all types of apomixis in which an embryo develops directly from the egg of a diploid gametophyte. This reproductive cycle invariably involves two processes not included in the normal sexual cycle: a substitute for meiosis and a substitute for fertilization. Several substitutes are known for each process but none is wholly independent. If an agamogonic plant, for instance, is diploparthenogenetic, it is also either aposporous or diplosporous. Similarly, if an agamogonic plant is aposporous, it is either diploparthenogenetic, pseudogamous or apogamous. (Adventitious embryony may also enter into the picture almost anywhere, but for purely practical purposes this phase of apomixis will later be treated independently.) The various substitutes for meiosis which result in gametophytes with diploid nuclei are collectively designated as apomeiosis. Three fairly distinct types have been recognized, although many transitions between them are known. In the first type, diplospory, a megagametophyte with the unreduced number of chromosomes is developed from an archesporial (or rarely from a nucellar) cell; the first division in the megasporocyte is strongly meiotic in character. The second type comprises various manifestations of apospory in which the megagametophyte develops by a series of purely mitotic divisions. If the gametophyte arises from an indubitably somatic cell, usually one in the chalazal region of the ovule but occasionally one located elsewhere in the nucellus, the process is known as somatic apospory. If the gametophyte develops directly from an archesporial cell, with meiosis omitted, the designation is generative apospory. Occasionally the distinction between archesporial and nucellar cells is not sharply defined, so that conditions intermediate between somatic and generative apospory are to be found in certain plants. When a pseudohomotypic division, in which the chromosomes are strongly contracted as in meiosis but are not at all paired, occurs, the condition is known as semiapospory.

The preceding classification is not entirely satisfactory and should not be considered as a fixed one since all sorts of intermediate conditions are known and even different types of apomeiosis may be found in different ovules of the same plant.

Apomeiosis is always followed by one of three abnormal types of embryogeny: parthenogenesis, pseudogamy or apogamety. These terms are often preceded by the adjective "unreduced" to distinguish them from the corresponding phenomena ("reduced") occurring in normal haploid megagametophytes. In unreduced (or diplo-) parthenogenesis, the commonest method of embryogeny in agamogamic plants, the embryo develops autonomously from the diploid egg. The embryogeny is very similar to that of a normal embryo, but the time at which the egg first divides is extremely variable. In all known instances of unreduced pseudogamy the origin of the megagametophyte is aposporic and ordinarily the egg is stimulated into development by the fertilization of the polar or primary endosperm nucleus. Apogamety is the development of a synergid, antipodal cell or endosperm nucleus into an embryo. The term "apogamy" was formerly employed to designate these phenomena but has been so misused that the substitution of "apogamety" is to be preferred.

When the haploid egg of a megagametophyte which originated from normal meiosis is in some way stimulated into development without the intervention of fertilization, reduced (or haplo-) parthenogenesis or pseudogamy results. When some cell other than the egg develops into an embryo, reduced apogamety is the result. All these processes usually end in sterile haploids and are not repeated from one generation to the next; they are thus described as nonrecurrent apomixis.

At one time it was considered, and it is still maintained by a few investigators, that agamogonic megagametophytes could be classified into definite types according to the presence or absence of a wall subsequent to the first division in the megasporocyte. Thus there resulted what were called the Alchemilla, Taraxacum and Antennaria Types, but such a grouping has been shown not to stand the test of critical examination. The Alchemilla Type, particularly, has practically no foundation in fact since the genus *Alchemilla* exhibits only generative apospory and is not parthenogenetic. The majority of other genera described as conforming to the Alchemilla Type are also not parthenogenetic.

Diplospory:- Diplospory designates the formation of an unreduced megagametophyte from an archesporial cell in which a division of relatively strong meiotic character occurs. The unreduced chromosome number results in either of three ways: by the formation of restitution nuclei, through pseudohomotypic divisions (equivalent to semiapospory), or by so-called somatic divisions. The first two imply meiotic processes, the third a change from meiotic to mitotic division with transition divisional types traceable. It is true that the chromosome number is not reduced, but the two processes differ from normal sexual meiosis only in that pairing of the chromosomes is omitted. Earlier investigators have regarded them simply as somatic divisions, considering that the heterotypic division was suppressed, but more modern investigations have revealed that meiosis has merely been altered.

The three methods by which the diploid chromosome number originates are illustrated by the appended diagrammatic scheme (FIG. 75). The upper row (A–G) depicts the formation of restitution nuclei. Following the failure of pairing in zygotene (A), the diploid number of univalents appears at diaphase (B); these univalents, after the dissolution of the nuclear membrane, form a semiheterotypic metaphase (C) with or without secondary association. The univalents are carried toward the poles by the stretching of the spindle (D), but a few remain near the equatorial plate. A nuclear membrane next envelops the spindle and chromosomes (E). A homotypic metaphase follows; it has a normal appearance and gives rise to a dyad each of whose nuclei is diploid. One of the dyad cells (if a wall is formed) or both nuclei (if a wall is not erected) then develop into a megagametophyte. The middle row (H–L) illustrates the pseudohomotypic metaphase as occurring in *Taraxacum* and other genera. The first three stages (H–J) are identical with the corresponding ones in the preceding process. The spindle fails to stretch and a nuclear membrane is not

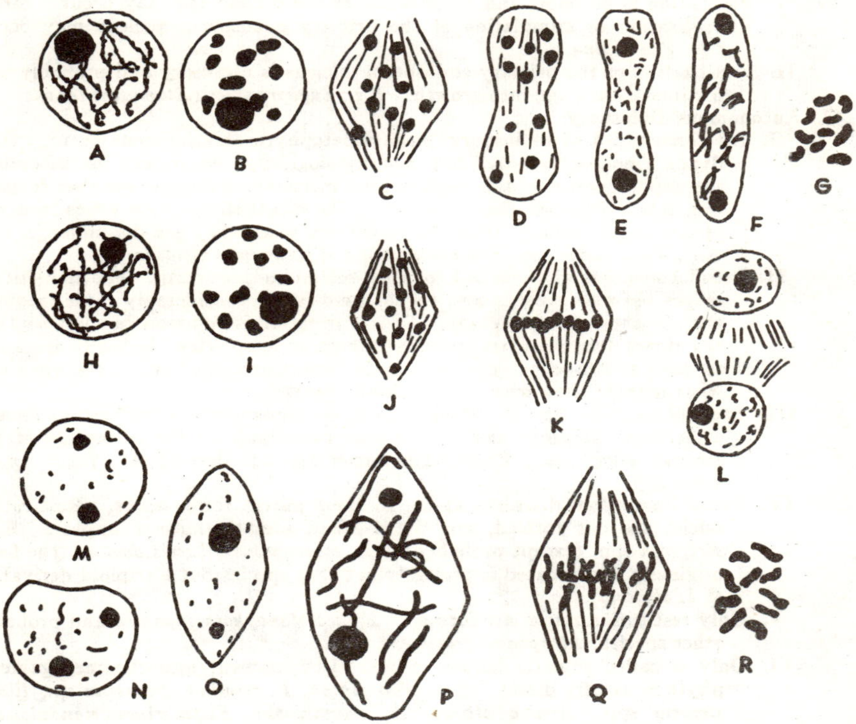

FIG. 75. — Schematic diagrams depicting the formation of restitution nuclei, pseudohomotypic metaphases and so-called somatic divisions in megasporocytes. Further description in text. (*Redrawn from* GUSTAFSON 1936).

formed. The chromosomes next move to the equatorial plate, the halves separate and then result in two daughter nuclei between which a wall may or may not be formed (L). The pseudohomotypic division differs somewhat in *Hieracium* and a few other genera. The resting stages resemble those in the so-called somatic division (M–R), and the prophases contain long but rather thick chromosomes which do not acquire a spherical shape until metaphase. The so-called somatic division (lower row, M–R) involves a retardation of the onset of prophase; it is probably nothing more than a degeneration phenomenon (an ageing of the cell). In prophase (P) the unpaired chromosomes have a mitotic appearance and at metaphase (Q, R) they are somewhat contracted. At metaphase-anaphase (Q) the halves of each chromosome move to opposite poles, resulting in two diploid nuclei between which a wall is apparently not formed. This division proceeds with extreme rapidity. In none of the preceding cases is anything other than a dyad or a binucleate megagametophyte ever produced; quartets have never been observed. One of the dyads develops directly into the megagametophyte.

Although many of the cytological details underlying diplospory and diploparthenogenesis have been carefully investigated, there is an incredible paucity of detailed embryological studies. The best that can be done is to present the following classificatory scheme to indicate the means by which diplosporic embryos may arise, together with examples of species in which each type occurs.

A. There is no regular apomictic propagation. *E.g.*: *Ochna serrulata, Artemisia nitida.*

B. Induced diplospory (equivalent to pseudogamy).

- I. Pollination is probable and fertilization of the diploid egg may occur. Either fertilization or stimulation of the primary endosperm nucleus may occur. *E.g.*: *Poa alpina.*
- II. Fertilization of the primary endosperm nucleus is necessary and probably acts to stimulate the egg into growth. *E.g.: Zephyranthes* (*Atamosco*) *texana.*

C. Autonomous diplospory.

- I. All three types of diplosporic megagametophytic development occur. Only dyads (morphologically, but not cytologically, equivalent to secondary megasporocytes in plants with normal megagametophytes) are ever formed; quartets have never been observed. The megagametophyte arises from one of the dyads, usually from the larger one when they are dissimilar in size. *E.g.: Taraxacum* spp.; *Chondrilla* spp.; *Wikstroemia viridiflora.*
- II. Pseudohomotypic divisions and so-called somatic mitoses occur. The transitional stages between meiosis and sexual mitosis are considerably more frequent than in the following group. Dyad formation is common but less so than the direct transition to binucleate megagametophytes. In other instances as many as five cells originate from the megasporocyte, but it is not known in what manner they arise. *E.g.: Hieracium* spp.
- III. Pseudohomotypic and so-called somatic divisions occur; a wall is not formed after these divisions and both nuclei participate in the formation of the megagametophyte. Restitution nuclei are not developed. *E.g.: Antennaria* spp.
- IV. Pseudohomotypic divisions alone, in their purest form, occur. Restitution nuclei are not formed, nor do so-called somatic mitoses appear. *E.g.: Erigeron* spp. (except probably for *E. mucronatus*), *Ixeris dentata* (the form originally investigated is probably not that species but a triploid derivative of *I. alpicola*).
- V. Only restitution nuclei are formed. *E.g.: Balanophora japonica* and probably other species; *Atraphaxis frutescens.*
- VI. Only so-called somatic mitoses occur. Development into the megagametophyte is usually direct. *E.g.: Poa alpina, P. serotina, Pellionia* sp., *Elatostemma* spp., *Arnica alpina, A. chamissonis, Eupatorium glandulosum, Hieracium umbellatum, Cotylanthera tenuis* (?), *Dendrophthora opuntioides* (?), *Parthenium argentatum.*

In addition to the instances mentioned above, there are numerous records of obviously diplosporic phenomena in the literature which can not be accurately classified because of paucity of essential details.

Apospory:- Aposporic megagametophytes, containing from one to several diploid nuclei, have been described for numerous genera, including such as *Godetia, Fuchsia, Crepis* and *Leontopodon.* They appear in the nucellus, usually at the chalazal end, and adjacent to the quartet or megagametophyte. The embryos arising in such structures must not be confused with nucellar embryos, which arise directly from nucellar cells. These supernumerary megagametophytes appear to occur far more commonly than is generally realized. The significant fact is that embryos have very rarely been described as originating in these abnormal megametophytes. The mere presence of such a megagametophyte is far from being conclusive evidence *per se* that apomixis prevails; the occurrence of indubitable embryos is required as proof.

In the majority of published reports it is difficult to determine precisely how the aposporous megagametophyte originated, since the authors are silent on this important point. The tendency among those who have written on the subject in a broad manner appears to be to disregard the matter, so that the following key may serve to distinguish the various manifestations of apospory. This scheme is admittedly incomplete and may even be incorrect in certain details.

A. Apomictic propagation is not a regular occurrence but aposporous megagametophytes are regularly or sporadically formed. *E.g.: Leontodon hispidus, Fuchsia riccartoni, F. magellanica, F. hybrida* and probably other species, *Godetia amoena, Picris hieracioides, Coreopsis bicolor, Ochna serrulata* (?).

B. Induced apomixis (equivalent to pseudogamy).

I. The egg possesses a marked sexual tendency in that fertilization is still possible despite the fact that its nucleus possesses the diploid number of chromosomes. When the egg does grow without fertilization, the stimulation comes from the microgametophyte or from the developing endosperm. *E.g.: Hypericum perforatum.*

II. The egg can either be fertilized or develop parthenogenetically. Fertilization of the primary endosperm nucleus is questionable, as is the stimulation or lack of such of the egg. *E.g.: Poa pratensis.*

III. The primary endosperm nucleus must be fertilized before the egg can develop parthenogenetically. Stimulation may possibly also be furnished by a microgametophyte. *E.g.:* forms of *Potentilla collina* and probably other apomictic species of *Potentilla.*

IV. The primary endosperm nucleus is fertilized but parthenogenetic growth of the egg is autonomous. *E.g.:* other forms of *Potentilla collina.*

V. Pollination is a necessary stimulation, but there is probably no actual fertilization. The endosperm and egg begin simultaneous (and probably autonomous) growth. *E.g.: Alchemilla arvensis, Parthenium argentatum.*

C. Autonomous apomixis.

VI. The egg and primary endosperm nucleus develop parthenogenetically. The polar nuclei usually fuse (to form the primary endosperm nucleus), and internal three-nuclear fusions are common. *E.g.:* many apomictic forms of *Alchemilla vulgaris.*

VII. The egg and primary endosperm nuclei develop parthenogenetically, but there are no three-nuclear fusions. *E.g.: Hieracium pilosella* (also *H. excellens, H. ramosum, H. aurantiacum* and *H. flagellare?*), *Crepis* spp. (?).

VIII. This type is similar to the preceding but the polar nuclei do not fuse. The egg commences growth rather early. *E.g.: Alchemilla alpina, Atraphaxis frutescens* (?).

There are not enough reports of the development of aposporic embryos to warrant further discussion of this subject.

Pseudogamy:- The term pseudogamy is employed to designate all instances in which an embryo is formed without fertilization subsequent to the stimulating influence exerted by the entrance of a foreign microgametophyte. Pseudogamy is of two types, reduced and unreduced. In the first type, the megagametophyte is formed in the customary fashion following meiosis; in the second type, all known examples involve megagametophytes which developed through diplospory or apospory. Instances of unreduced pseudogamy, consequently, are described under those topics.

Unreduced pseudogamy usually implies the simultaneous occurrence of haploparthenogenesis (or, as it is called by some writers, azygoparthenogenesis). One of the best examples of this type of apomixis is that of emasculated flowers of *Solanum nigrum* (n-36) pollinated with pollen from *S. luteum* (n-24). The microgametophyte of the latter species enters a megagametophyte which developed normally. Fertilization is not effected, although one or both male nuclei may enter an egg. The male nuclei eventually disintegrate slowly, as remnants of these nuclei may still be observed at even as late a time when the embryo has become ten-celled. The haploid egg is stimulated into growth by the foreign nuclei and an embryo that appears to be normal is produced. Certain other embryos appear to have approximately the haploid maternal number of chromosomes while others had the diploid maternal number. The latter condition presumably arose by the failure of a cell wall to be formed after the first proembryonic mitosis (this process is known to occur in other plants), with the result that the initial chromosome number was doubled in the succeeding mitosis when the spindles of the two free nuclei fused.

Megasporic Embryogeny:- A most unusual type of apomixis occurs in the orchid *Zeuxine sulcata*. The megasporocyte undergoes regular meiosis to form a linear quartet of four megaspores. There is no degeneration of any megaspore and megagametogenesis is entirely omitted. Each megaspore divides further and together all produce a haploid apomictic embryo. There is no organization of histogens in these embryos and no suspensor. Occasionally one of the megaspores contains a number of free nuclei but a megagametophyte is not organized.

Since this type of apomixis can not be classified under any of the other types, the designation "compound megasporic embryogeny" may be employed to include such instances. The word "compound" is used to indicate that more than one megaspore is involved; there is a possibility that a single megaspore may some time be found to develop into an embryo, in which case the designation "simple megasporic embryogeny" could be employed.

The above report has been disputed. It is claimed that degeneration of the ovules begins at the megasporocyte stage and that few if any functional megagametophytes are formed. The embryos are said always to originate from the nucellar epidermis; a single such embryo may also cleave, bud or proliferate. In short, it is stated that no evidence whatever of megasporic embryos can be found. All that can be done at the moment is to present both sides of the matter.

The following citations cover the apomictic topics thus far dealt with.
The observations on *Fuchsia* and *Godetia* are original.

AFZELIUS, K. 1936. Apomixis in der Gattung *Arnica*. Svensk Bot. Tidskr. 30:

BERGMAN, B. 1935. Zytologische Studien über die Fortpflanzung bei den Gattungen *Leontodon* und *Picris*. Diss., Stockholm.

BERGMAN, B. 1935. Zytologische Studien über sexuelles und asexuelles *Hieracium umbellatum*. Hereditas 20: 47–64.

BERGMAN, B. 1935. Zur Kenntnis der Zytologie der Skandinavischen *Antennaria*-Arten Hereditas 20: 214–226.

CARANO, E. 1921. Nuovo ricerche sulla embriologia delle *Asteraceae*. Ann. di Bot. 15: 97–196.

CHIARUGI, A., e E. FRANCINI. 1930. Apomissa in *Ochna serrulata* Walp. Nuovo Giorn. Bot. Ital. 37: 1–250.

EDMAN, G. 1931. Apomeiosis und Apomixis bei *Atraphaxis frutescens* C. Koch. Acta Horti Bergiani 11: 13–66.

ERNST, A. 1914. Embryobildung bei *Balanophora*. Flora 106: 129–159.

ESAU, K. 1946. Morphology of reproduction in Guayule and certain other species of *Parthenium*. Hilgardia 17: 61–120.

FAGERLIND, F. 1940. Die Terminologie der Apomixis-prozesse. Hereditas 26: 1–22.

FAGERLIND, F. 1940. Zytologie und Gametophytenbildung in der Gattung *Wikstroemia*. Hereditas 26: 23-50.

FAGERLIND, F. 1944. Die Samenbildung und die Zytologie bei Agamospermischen und sexuellen Arten von *Elatostema* und einigen nahestehenden Gattungen nebst Beleuchtung einiger damit zusammenhängender Probleme. Kungl. Svenska Vetenskap. Handl. III, Bd. 21, Nr. 4.

GELIN, O. V. E. 1934. Embryologische und cytologische Studien in *Helinatheae-Coreopsidinae*. Acta Horti Bergiani 11:

GENTSCHEFF, G. 1937. Zytologische und embryologische Studien über einige *Hieracium*-Arten. Planta 27: 165–195.

GENTSCHEFF, G., and A. GUSTAFSON. 1940. Parthenogenesis and pseudogamy in *Potentilla*. Bot. Notiser 1940: 109–132.
GUSTAFSON, A. 1935. Studies on the mechanism of parthenogenesis. Hereditas 21: 1–112.
GUSTAFSON, A. 1939. The interrelation of meiosis and mitosis. I. The mechanism of agamospermy. Hereditas 25: 289–322.
GUSTAFSSON, A. 1946. Apomixis in higher plants. Part I. The mechanism of apomixis. Lunds Univ. Arsskr. N.F. Avd. 2, Bd. 42, Nr. 3.
HAGERUP, O. 1944. On fertilisation, polyploidy and haploidy in *Orchis maculatus* L. sens. lat. Dansk Bot. Arkiv 11: 1–25.
HAGERUP, O. 1945. Facultative parthenogenesis and haploidy in *Epipactis latifolia*. Kgl. Danske Vidensk. Selskab., Biol. Meddel. 19: 3–14.
HOLMGREN, I. 1919. Zytologische Studien über die Fortflanzung bei den Gattungen *Erigeron* und *Eupatorium*. Kungl. Svenska Vet.-Akad. Handl. 59, Nr. 7.
JUEL, H. O. 1900. Vergleichende Untersuchungen über typische und parthenogenetische Fortpflanzung in der Gattung *Antennaria*. Kungl. Svenska Vet.-Akad. Handl. 33: 1–59.
KIELLANDER, C. L. 1935. Apomixis bei *Poa serotina*. Bot. Notiser 1935: 87–95.
KIELLANDER, C. L. 1937. On the embryological basis of apomixis in *Poa palustris* L. Svensk Bot. Tidskr. 31: 425–429.
KUYADA, Y. 1928. On occurrence of restitution nuclei in the formation of the embryo-sac of *Balanophora japonica*. Bot. Mag. Tokyo 42: 117–129.
LOTSY, J. P. 1899. *Balanophora globosa* Jungh. Ann. Jard. Bot. Buitenzorg II, 1: 174–186.
McDONALD, C. C. 1927. A study of seed development in three species of *Erigeron* with special reference to apogamy. Bull. Torrey Bot. Club 54: 479–497.
MUNTZING, A. 1940. Further studies on apomixis and sexuality in *Poa*. Hereditas 26: 115–190.
NOACK, K. L. 1939. Über *Hypericum*-Kreuzungen. VI. Fortpflanzungsverhältnisse und Bastarde von *Hypericum perforatum* L. Zeitschr. Ind. Abst. Vererb. 76: 569–601.
OEHLER, E. 1927. Entwicklungsgeschicktlich-zytologische Untersuchungen an einiger saprophytischen *Gentianaceen*. Planta 3: 641–733.
OKABE, S. 1932. Parthenogenesis bei *Ixeris dentata* Nakai. Bot. Mag. Tokyo 46: 518–523. (In Japanese.)
OSAWA, J. 1913. Studies on the cytology of some species of *Taraxacum*. Arch. Zellforsch. 10: 450–469.
PACE, L. 1913. Apogamy in *Atamosco*. Bot. Gaz. 56: 376–394.
PODDUBNAJA-ARNOLDI, W. A. 1933. Geschlechtliche und ungeschlechtliche Fortpflanzung bei einigen *Chondrilla*-Arten. Planta 19: 46–86.
ROSENBERG, O. 1912. Über die Apogamie bei *Chrondrilla juncea*. Svensk Bot. Tidskr. 6: 915–919.
ROSENBERG, O. 1917. Die Reduktionsteilung und ihre Degeneration in *Hieracium*. Svensk Bot. Tidskr. 11: 145–206.
ROSENBERG, O. 1927. Die semiheterotypische Teilung und ihre Bedeutung für die Entstehung verdoppelter Chromosomenzahlen. Hereditas 8: 305–338.
SESHAGIRIAH, K. N. 1941. Morphological studies in *Orchidaceae*. I. *Zeuxine sulcata* Lindley. Jour. Indian Bot. Soc. 20: 357–365.
STRASBURGER, E. 1909. Zeitpunkt der Bestimmung des Geschlechts, Apogamie, Parthenogenesis und Reduktionsteilung. Histol. Beitr. 7.
STRASBURGER, E. 1910. Sexuelle und apogame Fortpflanzung bei *Urticaceen*. Jahrb. Wiss. Bot. 47: 245–288.
SWAMY, B. G. L. 1946. The embryology of *Zeuxine sulcata* Lindl. New Phytol. 45: 132–136.
TAHARA, M. 1921. Cytologische Untersuchungen an einigen *Compositen*. Jour. Coll. Sci. Univ. Tokyo 43, No. 7.
TINNEY, F. W. 1940. Cytology of parthenogenesis in *Poa pratensis*. Jour. Agr. Res. 60: 351–360.
TREUB, M. 1906. L'apogamie de l'*Elatostemma acuminatum* Brongn. Ann. Jard. Bot. Buitenzorg II, 5: 141–152.
YORK, H. H. 1913. The origin and development of the embryo sac and embryo of *Dendrophthora opuntioides* and *D. gracile*. II. Bot. Gaz. 56: 200–216.

Apogamety:- Under the designation "apogamety" are included such phenomena as the development of one or both synergids, antipodal cells and polar or primary endosperm nuclei into embryos. All the nuclei involved are derived from the same original nucleus as is that of the egg, being once, twice or thrice removed according to the type of megagametogeny; and hence all, with the possible exception of the polar nuclei, may be considered as being potential egg nuclei. The question as to whether or not fertilization occurs seems not to have been taken into account in many investigations. Authentic descriptions of fertilization of all these cells or nuclei have been presented; such instances do not primarily come under apogamety for the reason that the cell or nucleus involved has assumed the function of the egg. Only those instances when the cell or nucleus develops into an embryo without actual fertilization involve apogamety. It is, of course, extremely difficult in most cases to establish whether fertilization occurred; the chromosomal constitution of the embryo is the most reliable criterion, but again difficulties are apparent. The chromosomes of the other megagametophytic cells and nuclei should be determined, both during megasporogenesis or megagametogenesis and subsequent to the development of the embryo, whatever the origin of the latter. If, on the other hand, the plant concerned is one in which fertilization does not normally occur, the demonstration of the apomictic nature of all embryos observed should be relatively easy.

Synergid Embryos:- The genetical literature abounds with descriptions of what are commonly called "haploid-diploid twins," but most of the explanations advanced for their origin are presumptive and are not based on embryological studies. In some instances, while one of the pair may be diploid, cytological study of the other member has shown it to be triploid rather than diploid. There are fortunately several accounts, however, which reveal the exact origin of the extra embryos. This is the concurrent growth of one synergid into an embryo.

One of the earliest descriptions, published in 1881, is that of the development of one or occasionally of both synergids in *Mimosa denhartii*. Although nothing is said of their cytological constitution, the embryos are presumably haploid and are essentially similar structurally to embryos derived from fertilized eggs. Some ten years later the occasional occurrence of two embryos in the same embryo sac of *Lilium martagon* was reported. It was suggested that the second embryo arose as the probable result of the fertilization of a synergid by a second microgametophyte, but it is now recognized that such an occurrence is highly improbable.

Many reports of the occurrence of two embryos within the same embryo sac require reinterpretation in the light of current comprehension of the real nature of embryos derived from other than the egg. The extra embryos of *Ulmus americana*, *Linum usitatissimum*, *Triticum spelta*, in several species of *Allium* and probably in some other recorded instances, all are of indubitably synergid origin, although most of the authors appear not to have recognized the possibility of such an origin.

One interesting end-result of certain interspecific or intergeneric crosses is the stimulation of a synergid into growth. In the cross *Avena strigosa* × *A. fatua*, for instance, the larger of the two embryos depicted undoubtedly arose from a fertilized egg while the smaller one developed from a synergid. That this can actually happen was demonstrated in the cross *Nicotiana glutinosa* (n-12) × *N. tabacum* (n-24). Here the larger of the twin embryos, presumably derived from the fertilized egg, is triploid; the smaller one, derived from a synergid, is haploid. When *Zygopetalum mackayi* is pollinated with pollen from *Odontoglossum crispum* the microgametophytes do not actually penetrate the megagametophytes but the stimulus which they exert is sufficient to activate both the egg, one or both synergids, and cells of the nucellus and inner integument into various manifestations of embryonal growth.

In the case of haploid-diploid twin seedlings of *Triticum vulgare* and *T. durum* it was speculated that the diploid member arose from a fertilized synergid and the haploid member from an unfertilized egg. This assumption is highly questionable; the actual origin of the respective members of the pair is probably the reverse of the assumption.

Careful cytological study of the development of synergid embryos began in

1926 and involved *Plantago lanceolata*. The synergid embryo was ascertained to be haploid, although structurally it closely simulated the diploid embryo derived from the fertilized egg. The most substantial and accurate account is the one involving the origin and development of synergid embryos in seven species of *Lilium*: *L. regale, superbum, elegans, henryi, hansoni, martagon* and *longiflorum*. Approximately 1 per cent of the ovules in each species possessed twin embryos. One is diploid and originated from the fertilized egg; the other is haploid and arose from an unfertilized synergid (FIG. 76). The division of the synergid begins at early post-fertilization stages; no evidence of such

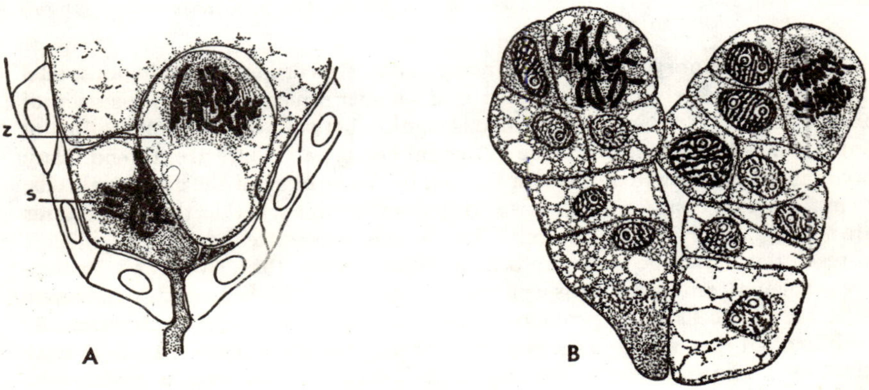

FIG. 76. — *Lilium martagon*. A. Micropylar end of ovule with dividing nuclei in zygote (at right) and synergid; with remnants of microgametophyte. B. Developing twin embryos, the one originating from a synergid at left. A and B × 400. (*Figures from* D. C. COOPER; *see also* COOPER 1943).

division was observed prior to fertilization. The stage of ovular development when a synergid began growth varied somewhat in the various species.

In *Hieracium vulgatum* and *H. ramosum* one or both of the synergids may develop into embryos. In *Sagittaria graminea* one or both synergids sometimes assume an egg-like appearance and develop into embryos. Since more than one microgametophyte may enter a megagametophyte, there is a possibility that the synergids may be fertilized.

Well developed synergid embryos are said to be common in *Alnus rugosa*.

It is evident, at least in *Lilium*, that fertilization and growth of the egg are necessary prerequisites to the development of a synergid embryo. On the other hand, in *Calamagrostis obtusata* and *C. purpurea* fertilization is not effected; one or all three of the cells comprising the egg apparatus may develop into presumably haploid embryos.

COOPER, D. C. 1943. Haploid-diploid twin embryos in *Lilium* and *Nicotiana*. Amer. Jour. Bot. 30: 408–413.

GENTSCHEFF, G. 1937. Zytologische und embryologische Studien über einige *Hieracium*-Arten. Planta 27: 165–195.

GUIGNARD, L. 1881. Recherches sur l'embryogénie des *Légumineuses*. Ann. Sci. Nat., Bot. VI, 12: 5–166.

HABERLANDT, G. 1922. Die Vorstufen und Ursachen der Adventivembryonie. Sitzungsber. Preuss. Acad. Wiss. 1922: 386–406.

JOHRI, B. M. 1937. Studies in the family *Alismaceae*. IV. *Alisma plantago* L.; *Alisma plantago-aquatica* L. and *Sagittaria graminea* Mich. Proc. Indian Acad. Sci., B. 4: 128–138.

KAPPERT, H. 1933. Erbliche Polyembryonie bei *Linum usitatissimum*. Biol. Zentralbl. 53: 276–307.

KIHARA, H., and I. NISHIYAMA. 1932. Different compatibility in reciprocal crosses of *Avena* with special reference to tetraploid hybrids between hexaploid and diploid species. Jap. Jour. Bot. 6: 245–305.

OVERTON, E. 1892. Beitrag zur Kenntnis der Entwicklung und Vereinigung der Geschlechtsprodukte bei *Lilium martagon*. Beih. Bot. Centralbl. 50: 336–337.
SHATTUCK, C. H. 1905. A morphological study of *Ulmus americana*. Bot. Gaz. 40: 209–223.
SOUEGES, R. 1926. Polyembryonie chez le *Plantago lanceolata* L. Bull. Soc. Bot. France 73: 955–957.
STENAR, H. 1934 (1933). Parthenogenesis in der Gattung *Calamagrostis*. Embryobildung bei *Calamagrostis obtusata* Trin. und *Calamgarostis purpurea* Trin. Arkiv f. Bot. 25A, Nr. 6.
SUESSENGUTH, K. 1923. Über die Pseudogamie bei *Zygopetalum mackayi* Hook. Ber. D. Bot. Ges. 41: 16–23.
WAKAKUWA, S. 1934. Embryological studies of the different seed-development in reciprocal interspecific crosses of wheat. Jap. Jour. Bot. 7: 151–186.
WEBER, E. 1929. Entwicklungsgeschichtliche Untersuchungen über die Gattung *Allium*. Bot. Archiv 25: 1–44.
WOODWORTH, R. H. 1929. Parthenogenesis and polyembryony in *Alnus rugosa* (Duroi) Spreng. Science 70: 192–193.

Antipodal Embryos:- The occurrence of embryos originating from antipodal cells has very rarely been described in a manner which leaves no question of authenticity. It is important to distinguish between mere proliferation of antipodal cells or nuclei and true antipodal embryos. There are a good many accounts of the growth and multiplication of one or more of the antipodal cells, as in *Alangium lamarckii*, but it is highly questionable whether these structures are actually embryonic in nature. The problem is admittedly one very difficult of resolution. It is necessary to bear in mind the fact that, as far as is known, the nuclei of antipodal cells contain the haploid number of chromosomes. Whether such cells which proliferate into embryolike structures are homologous with haploid eggs which develop into parthenogenetic embryos is again a moot question. The evidence so far appears to be contrary to such a conception. If the evidence is to be considered as positive, there should be far more reports of the occurrence of antipodal embryogeny than are actually found in the literature.

In *Ulmus americana* one of the antipodal cells may enlarge and assume the appearance of an egg. Microgametophytes were observed in the chalazal region of the ovule, which indicates that chalazogamy may occur, but there was no evidence that the antipodal "egg" was fertilized. In any event, this cell may develop into an embryo, which does not have the same structural regularity displayed by the normal embryo developed at the micropylar end of the same embryo sac. One of the antipodal cells in *Allium roseum* may develop into an embryo, as may one or two of those in *A. odorum* and in *Eragrostis cilianensis*.

There are a number of accounts of the development of embryos at the chalazal end of the embryo sac, but the majority of such embryos are derived from nucellar cells rather than from the antipodal cells.

CAPPELLETTI, C. 1932. Sull'azione dei prodotti del ricambio di miceli micorizogeni sulle piante capiti. Ann. di Bot. 29: 1–62.
GOPINATH, D. M. 1943. Cases of antipodal polyembryony in *Alangium lamarckii* Thw. Current Science: 329–330.
HABERLANDT, G. 1922. Die Vorstufen und Ursachen der Adventivembryonie. Sitzungsber. Preuss. Akad. Wiss. 1922: 386–406.
SHATTUCK, C. H. 1905. A morphological study of *Ulmus americana*. Bot. Gaz. 40: 209–223.
SWAMY, B. G. L. 1944. A reinvestigation of the embryo-sac of *Eragrostis cilianensis* (All.) Link. Current Science: 103–104.
TRETJAKOW, S. 1895. Die Betheiligung der Antipoden in Fällen der Polyembryonie bei *Allium odorum* L. Ber. D. Bot. Ges. 13: 13–17.

Endosperm Embryos:- Whether the polar nucleus (haploid) or the primary endosperm nucleus (diploid or tetraploid according to the type of megagametogeny; representing two or more fused polar nuclei) can give rise to an embryo without or subsequent to fertilization by the secondary male nucleus is a decidedly moot question. It would appear that these nuclei, alone among those constituting the megagametophyte, have lost the potentiality for the formation of an embryo. It should be kept in mind that these nuclei, unlike the others, are always naked and are never enclosed within individual cell walls, which may, at least for the most part, explain their inability to develop into embryos.

A somewhat questionable instance is that of *Isomeris arborea*, belonging to the *Capparidaceae*. The megagametophyte arises directly from a megaspore without the intervention of meiosis and consists of two synergid-like structures

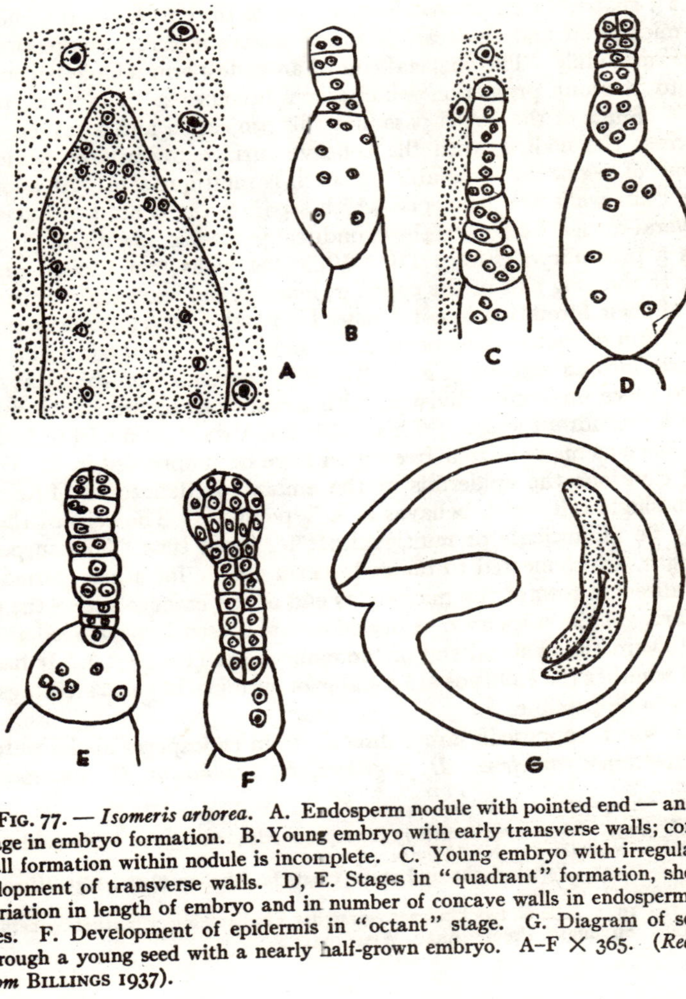

FIG. 77. — *Isomeris arborea*. A. Endosperm nodule with pointed end — an early stage in embryo formation. B. Young embryo with early transverse walls; concave wall formation within nodule is incomplete. C. Young embryo with irregular development of transverse walls. D, E. Stages in "quadrant" formation, showing variation in length of embryo and in number of concave walls in endosperm nodules. F. Development of epidermis in "octant" stage. G. Diagram of section through a young seed with a nearly half-grown embryo. A–F × 365. (*Redrawn from* BILLINGS 1937).

and a primary endosperm nucleus. There is no egg or egg nucleus. The chromosome number is 17 in all body regions; the plants are therefore haplodiplont. Although a microgametophyte may enter an ovule, fertilization of the primary endosperm nucleus is not effected. This nucleus presently divides, then the daughter nuclei repeatedly divide to form a coenocytic mass in the enlarging embryo sac. Later a second and different type of endosperm arises from this mass. This second type has been designated as nodular endosperm; it begins when the cytoplasm in the vicinity of certain nuclei becomes more dense and coarse granules appear (FIG. 77A). These dense areas tend to extend and fuse into cylindrical or even globular masses which become still more coarsely granular. At first no walls surround or divide the nodules, but the later formation of walls separates the nodules from the surrounding cytoplasm and the free nuclei (FIG. 77B). Distribution of endosperm nodule formation varies in different ovules, and the quantity of nuclei going into nodule formation is also variable. No walls are formed at first within these nodules. A third, cellular type of endosperm now arises, and as its amount increases the nodular masses become surrounded by it. The greatly enlarged embryo sac becomes curved, assuming a J-shape; the embryos originate close to the curve, about equidistant from the micropylar and chalazal ends, and always in direct connection with an endosperm nodule. The chalazal end of an endosperm nodule becomes extended into a blunt projection which soon becomes a cylindrical process (FIG. 77C). Some of the nuclei pass into this projection and a curved wall is formed across the nodule, with the concave surface toward the projection. The number of segments is variable but each is multinucleate. At about the same time cross walls are developed which separate the contained nuclei into one- to several-nucleate cells and the cylindrical process is now distinctly recognizable as a proembryo (FIG. 77D). While the walls first formed are perpendicular to the long axis of the cylinder, longitudinal walls early appear but the order of their formation is not constant. Further development results in the organization of that portion next to the nodule into a suspensor and of the outer portion into an embryo (FIG. 77E). The terminal hemispherical cell of the embryo twice undergoes division, with a wall formed after each mitosis, to produce a "quadrant stage." This cell is not always uninucleate from the beginning, since as many as four free nuclei have been observed in it. During succeeding divisions the epidermis of the embryo is delineated (FIG. 77F). There is no single cell which behaves as a hypophysis. The cells of the suspensor may be uninucleate or multinucleate for some time. The suspensors vary in length, are connected to the endosperm nodule for a long period, and are always directed toward the micropylar end of the embryo sac — the exact normal polarity of the suspensor. Cotyledons are presently developed and the tips point toward the chalazal end of the embryo sac (FIG. 77G). It has not been stated whether the embryos are capable of germinating into seedlings, but this appears to be possible.

Embryos which supposedly arose directly from endosperm nuclei were observed in *Hieracium canadense*, *H. flagellare*, *H. aurantium*, *H. ramosum* and *H. pilosella*.

BILLINGS, F. H. 1937. Some new features in the reproductive cytology of Angiosperms, illustrated by *Isomeris arborea*. New Phytol. 36: 301–326.

GENTSCHEFF, G. 1937. Zytologische und embryologische Studien über einige *Hieracium*-Arten. Planta 27: 165–195.

HABERLANDT, G. 1921. Die Entwicklungserregung der Eizellen einiger parthenogenetischen Kompositen. Sitzungsb. Preuss. Akad. Wiss. 1921: 861–881.

Chapter XX

ADVENTITIOUS EMBRYOGENY

Adventitious embryogeny embraces all instances in which an embryo is developed directly from a nucellar or integumental cell or from a group, isolated or not, of such cells. It was formerly described as nucellar embryogeny but a more descriptive term is required to include embryos originating from cells of the inner integument adjacent to the nucellus.

In many species in which adventitious embryony occurs it is necessary for the egg to be fertilized before the nucellar embryos are able to develop. This feature is not so apparant in the case of integumentary embryos. In other instances, stimulation exerted by the entrance of a microgametophyte into the megagametophyte is sufficient (or necessary) to cause the development of nucellar embryos.

The two types of adventitious embryos may be discussed separately.

Integumentary Embryos:- All but two of the following instances concern dicotyledonous families; and in all cases, moreover, only the inner integument is involved, usually in the near neighborhood of the micropyle. There are no reports of the development of embryos from cells of the outer integument, when one is present.

Rosaceae.— A single integumentary embryo was observed in *Potentilla reptans.* It was oriented obliquely lateral to the normal embryo; its dimensions were greatly reduced in comparison to those of the latter. Its internal differentiation was less clearly indicated, but in the region of the embryo proper the same fundamental cellular organization was recognizable.

Onagraceae.— A callus-like tissue may develop at one side of the embryo sac in *Oenothera lamarckiana*; it is derived from the inner integument and embryolike outgrowths may proceed therefrom.

Myrtaceae. — In *Eugenia malaccensis* adventitious embryos develop from the epithelial-like cells of the inner integument.

Solanaceae. — Integumentary embryos similar to those just described for *Oenothera lamarckiana* occur in *Scopolia carniolica,* but here they are developed from the inner tapetal layer of the inner integument adjacent to the micropyle.

Myrsinaceae. — In *Ardisia crispa* cells of the inner integument occasionally develop into embryos. Other types of apomixis appear to be present in this and in other species of *Ardisia.*

Celastraceae. — "Polyembryony" was reported for *Evonymus europaeus* in 1814, but was not confirmed when this species was reinvestigated 130 years later. It was also reported for other species, but was not confirmed for any of them. However, it was rediscovered for still other species, notably *E. latifolius,* and also for *Celastrus scandens.* Two to several apomictic embryos occur in most ovules of *E. latifolius* and all are derived from cells of the inner integument. The embryo originally derived from a fertilized egg frequently becomes aborted. There is also a possibility that embryos arise from the endosperm, but the evidence is inconclusive. The apomictic embryos in *C. scandens* are all integumental in origin.

Apiaceae. — Integumentary embryos are reported for *Ammi majus.*

Amaryllidaceae. — An embryo may develop from the inner integument in the micropylar region in *Allium roseum* and *A. odorum.*

Liliaceae. — Cells of the inner integument adjacent to the micropyle in *Hosta (Funkia) ovata* may proliferate to produce embryolike structures.

ANDERSON, A. 1931. Studien über die Embryologie der Familien *Celastraceae, Oleaceae* und *Apocynaceae.* Lunds Univ. Årsskr., N. F. Avd. 2, 27: 1–110.

CAPPELLETI, C. 1932. Sull'azione dei prodotti del ricambio di miceli micorizogeni sulle piante ospiti. Ann. di Bot. 29: 1–62.

DAHLGREN, K. V. O. 1916. Zytologische und embryologische Studien über die Reihen *Primulales* und *Plumbaginales.* Kungl. Svenska Vetensk. Handl. 56, Nr. 4.

HABERLANDT, G. 1921. Über experimentelle Erzeugung von Adventivembryonen bei *Oenothera lamarckiana.* Sitzungsb. Preuss. Akad. Wiss. 1921: 695–725.

HABERLANDT, G. 1922. Die Vorstufen und Ursachen der Adventivembryonie. Sitzungsb. Preuss. Akad. Wiss. 1921: 386–406.

HAKANSSON, A. 1923. Studien über die Entwicklungsgeschichte der *Umbelliferen.* Lunds Univ. Årsskr., N. F. Avd. 2, 18, Nr. 7.

PIJL, L. VAN DER. 1934. Über die Polyembryonie bei *Eugenia.* Rec. trav. bot. Néerl. 31: 113–187.

SOUÈGES, R. 1935. Polyembryonie chez le *Potentilla reptans* L. Bull. Soc. Bot. France 82: 381–384.

Nucellar Embryos:- An attempt has been made to separate the various manifestations of nucellar embryogeny into three main divisions. The scheme is workable in the main, but only so far as all the circumstances surrounding the origin of the nucellar embryos are known. There are numerous reports which do not present sufficient detail to permit the classification of the species concerned. This scheme will be presented first, following which all the species will be listed according to their respective families.

A. No actual apomictic propagation exists; the formation of nucellar embryos is merely a sporadic tendency. *E.g.: Clarkia elegans, Potentilla geoides.*

B. Induced nucellar embryony (equivalent to pseudogamy).

- I. The formation of nucellar embryos is dependent upon fertilization and subsequent division of the egg. *E.g.: Citrus aurantium, C. trifoliata,* some varieties of *Mangifera indica, Eugenia hookeri* (?).
- II. The formation of nucellar embryos is dependent upon fertilization of the egg or upon stimulation by the microgametophyte. This type is almost similar to the preceding one. *E.g.: Opuntia vulgaris, Funkia ovata, Euphorbia dulcis, Spathiphyllum patinii, Tulipa gesneriana, Smilacina racemosa.*
- III. The formation of nucellar embryos is caused by the entrance of the microgametophyte of some other species. The egg, when such is formed, is fertilized but develops only into a short-lived monstrosity. *E.g.: Epilobium hirsutum* pollinated by *E. dodonaei*; *E. montanum* pollinated by *E. angustifolium*; *Oenothera muricata* pollinated by *Oenothera biennis*; *Potentilla nepalensis* × *P. splendens* pollinated by *P. nepalensis*; *Nicotiana rustica* pollinated by *Petunia* sp.
- IV. Formation of nucellar embryos is induced by injection of growth-promoting substances. *E.g.: Petunia violacea.*
- V. The formation of nucellar embryos is autonomous but the entrance of a microgametophyte is necessary to stimulate the development of seeds and fruits. *E.g.: Nothoscordum fragrans.*
- VI. The formation of a cellular endosperm is necessary to initiate the development of nucellar embryos. *E.g.: Calycanthus* spp., *Chimonanthus* spp.

C. Autonomous nucellar embryony.

- VII. The entrance of a microgametophyte is not necessary but when this occurs it apparently does accelerate the tendency to the formation of nucellar embryos. A mature megagametophyte is not produced. *E.g.: Nigritella nigra, Zeuxine sulcata.*
- VIII. Neither the entrance of a microgametophyte nor fertilization is required, but the formation of nucellar embryos is apparently stimulated by the degeneration of aposporous megagametophytes. *E.g. Ochna serrulata.*
- IX. Neither the entrance of a microgametophyte nor fertilization is required. Nucellar embryo formation, as well as seed and fruit development, is

autonomous. *E.g.: Xanthophyllum bungei, Caelobogyne ilicifolia, Sarcococca pruniformis, Allium roseum,* species of *Citrus* other than those listed under BI above, *Mangifera indica* (at least some varieties), *Mallotus japonica, Colchicum autumnale* (?), *Euphorbia dulcis* (?).

Onagraceae. — Nucellar embryos occasionally occur at the micropylar end of the embryo sac of *Clarkia elegans.* A normal embryo is developed.

Hypertrophied proliferations of nucellar cells, which have the general appearance of embryos, are formed in *Epilobium hirsutum* and *E. montanum* when these species are pollinated with pollen from *E. dodonaei* and *E. angustifolium* respectively. Single embryolike bodies are similarly produced when *Oenothera muricata* is fertilized by pollen from *O. biennis.*

Nucellar embryos originate at the chalazal end in *Oenothera lamarckiana* or may occur at the micropylar end or at both ends. A few bear a slight resemblance to normal embryos, but most are in the form of monstrosities.

Solanaceae. — A growth-promoting substance (called "*Belbitan*" but otherwise undescribed) injected into ovules of *Petunia violacea* induced the formation of small nucellar embryos. Other ovular tissues were stimulated into enlargement but none developed apomictic embryos.

Calycanthaceae. — Nucellar embryos occur in *Calycanthus florida, C. occidentalis* and *Chimonanthus praecox,* while only a single massive, very irregularly constructed structure occurs in several other species of *Chimonanthus.* The nuclei of the megagametophytes in all species degenerate.

Myrtaceae. — From two to twenty-two embryos occur in the seeds of *Eugenia hookeri*; most of the extra embryos are presumably nucellar in origin.

In *Eugenia jambos* the formation of nucellar embryos is dependent upon fertilization of the egg.

Buxaceae. — *Sarcococca pruniformis* develops a megagametophyte as usual but this degenerates as a result of the failure of fertilization. The polar nuclei fuse and develop a nuclear endosperm. Several cells of the nucellar cap above the megagametophyte in the micropylar region become richer in cytoplasm and presently divide to produce from two to as many as eleven embryos (FIG. 79A, B). One or two of these develop to maturity and the embryos have an essentially normal appearance.

Somewhat similar phenomena occur in *Sarcococca ruscifolia* and probably in other species in this genus.

Betulaceae. — Nucellar budding is claimed to be present in *Alnus rugosa.*

Rosaceae. — In the hybrid *Potentilla nepalensis* × *P. splendens* the megasporocyte fails to develop beyond the diaphase stage and degenerates. If flowers of this hybrid are pollinated with pollen from *P. nepalensis,* nucellar embryos are produced in from 10 to 20 per cent of the ovules. Two days after pollination three large nucellar cells have appeared in the space formerly occupied by the megasporocyte. Pollination with pollen from one parent appears to be necessary for the stimulation of the nucellar cells into embryonal development. By the further division of these cells an embryonal mass is formed, but the development of the structure to maturity was not followed out.

Aposporic megagametophytes which may total as many as three in a given ovule, as in *P. argyrophylla,* are also known in the above mentioned hybrid. The egg in such a megagametophyte may be stimulated into growth (but is not fertilized) by pollen from *P. nepalensis* since young embryos have been observed in such aposporic megagametophytes. It is apparent, therefore, that both adventitious embryogeny and somatic apospory occur here.

Capparidaceae. — Typical nucellar polyembryony is extremely common in *Capparis frondosa,* as many as twenty having been observed in a single ovule. A normal embryo is usually present.

Anacardiaceae. — Behavior in the mango, *Mangifera indica*, depends upon the variety. In the Strawberry mango fertilization does not occur; numerous adventitious embryos may arise from nucellar cells, both those directly abutting on the micropyle and those on the sides of the megagametophyte opposite the funiculus.

In the Carabao and Pico varieties the egg is fertilized and gives rise to a

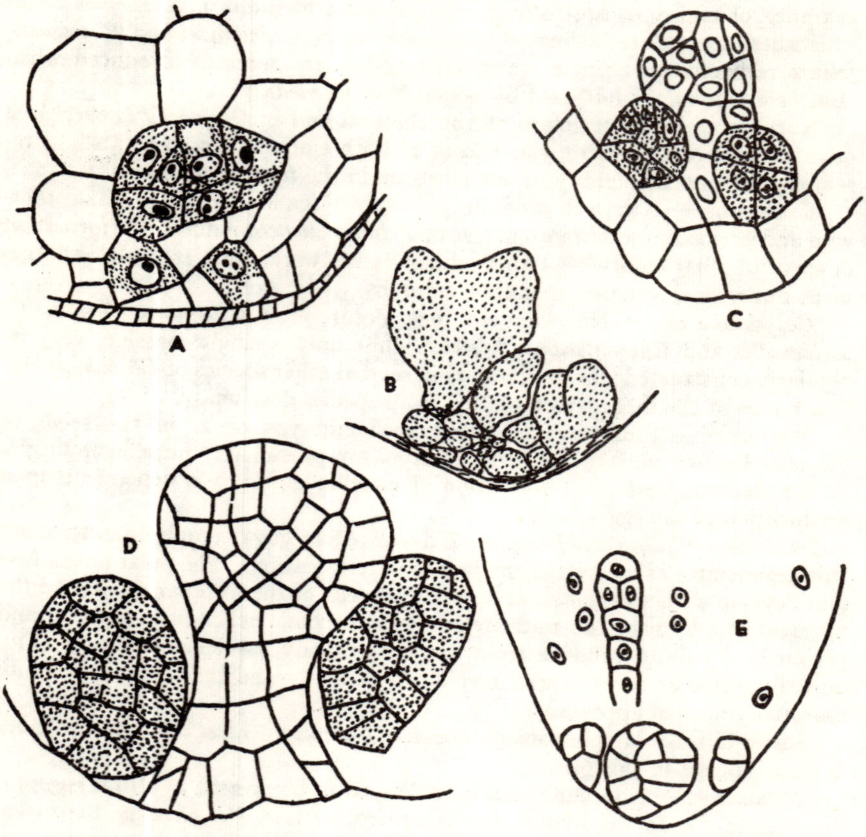

FIG. 78. — *Sarcococca pruniformis.* A. Nucellar cap with one developing and two potential adventitious (nucellar) embryos. B. Eleven potential adventitious embryos. *Spathiphyllum patinii.* C. Normal and nucellar (stippled) young embryos. D. Older normal and nucellar (stippled) embryos. *Euphorbia dulcis.* E. Three nucellar embryos without suspensors and one with a suspensor. A × 335; B × 60. (A, B *redrawn from* WIGER 1930; C. D *redrawn from* SCHÜRHOFF and JÜSSEN 1925; E *redrawn from* HEGELMAIER 1903).

normal embryo, but it may degenerate in the former. In both varieties nucellar embryos arise from the epidermal nucellar cells around the micropyle.

Cactaceae. — The development of embryos from nucellar cells was first described with sufficient accuracy in *Opuntia vulgaris.* Isolated groups of nucellar cells, one to three cell layers deep and located at or close to the micropylar end of the ovule, become meristematic and grow outward as irregularly shaped protuberances. From one to several embryos begin development on each protuberance. One of these gains an advantage over the others and becomes more fully developed although one to three of the others may also attain a more or less mature condition and all are capable of germination.

Euphorbiaceae. — In *Euphorbia dulcis* from two to six embryos may be produced, all arising at or near the micropylar end of the ovule and originating independently of the others. A nucellar cell becomes enlarged and divides approximately transversely; the second division is usually perpendicular in the outer cell but it may be transverse (FIG. 79E). Later divisions are decidedly

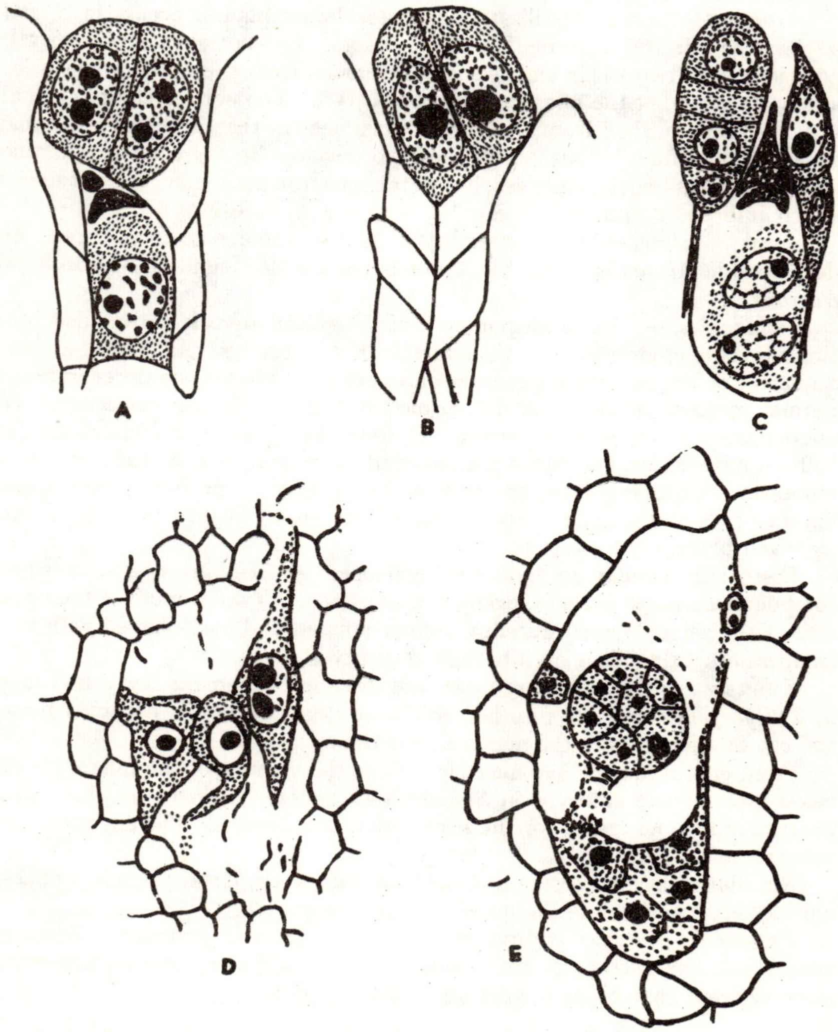

FIG. 79. — *Nigritella nigra.* A. Nucellus with quartet and apical proembryo. B. Nucellus with apical proembryo but without megaspores. C. Two-nucleate megagametophyte, two degenerated megaspores and two overlying proembryos. *Potentilla nepalensis* × *P. splendens.* D. Early nucellar embryonal cells. E. Adventitious embryo five days after pollination. A–C × 1000; D, E × ca. 500. (A–C *redrawn from* AFZELIUS 1928; D, E *redrawn from* GENTSCHEFF 1938).

irregular; either a globular mass of cells or a cylindrical "proembryo" not more than two cells in width may be formed. Although all embryos may develop to maturity, it appears that the one with the longest suspensor gains the ascendancy over the others.

In *Mallotus japonicus* fertilization does not occur. Nucellar (or possibly integumental?) embryos, usually two in number, arise at the micropylar end of the embryo sac.

Rutaceae. — Nucellar embryony is very prevalent in cultivated species and varieties of *Citrus*. In some it is the sole means of reproduction, but in *C. trifoliata* and *C. aurantium* normal embryos are also produced.

Araceae. — Normal fertilization (in greenhouse plants) occurs in *Spathiphyllum patinii* and a normal embryo is usually but not always present. In addition, from two to six embryos are developed from the nucellar cap below and at each side of the micropyle (Fig. 79C, D). The normal embryo can always be distinguished from the adventitious ones in that it possesses a broad suspensor or "foot." The normal embryo develops to a greater extent than do any of the apomictic embryos, but as the seeds are not viable, it is not known which embryo or embryos are capable of producing seedlings.

Amaryllidaceae. — Nucellar embryos, in the absence of fertilization, are developed in *Allium roseum*. Similar embryos are also found in *Nothoscordum fragrans*.

Orchidaceae. — The megasporocyte of *Nigritella nigra* is surrounded by a single layer of epidermal cells, the equivalent of a nucellus. As in most orchids, a quartet of megaspores is produced; all save the chalazal one degenerate. A normal megagametophyte is not formed in the functioning megaspore. No microgametophytes were observed. Despite the absence of fertilization, an embryo may be formed within the enlarged megaspore cell; its exact nature is unknown. Ordinarily one or more embryos arise from the epidermal-like nucellar cells at the apex of the ovule. These grow outward from the micropyle and beyond the ovule.

Essentially similar embryos are developed in *Zeuxine sulcata*, in which compound megasporic embryogeny is also present. Two or more of the upper cells of the single-layered nucellus become inflated and flask-shaped with very narrow necks, then develop into diploid embryos.

Liliaceae. — The proembryo does not develop beyond the few-celled stage in *Tulipa gesneriana*; its position, following degeneration, is occupied by an embryo developed from the nucellus near the micropyle.

From one to four or five nucellar cells in the vicinity of the micropyle develop into mature embryos in *Smilacina racemosa*. Entrance of the microgametophyte is necessary for the stimulation into growth. The egg rarely develops into an embryo.

Nucellar embryony occurs in *Colchicum autumnale*; fertilization is probably unnecessary to stimulate the nucellar cells into growth.

Poaceae. — Nucellar embryony is extensive in *Poa pratensis*. Embryos arising from the micropylar and chalazal ends of the embryo sac are larger and more vigorous than those originating in other regions.

Afzelius, K. 1928. Die Embryobildung bei *Nigritella nigra*. Svensk Bot. Tidskr. 22: 82–91.

Afzelius, K. 1932. Zur Kenntnis der Fortpflanzungsverhältnisse und Chromosomenzahlen bei *Nigritella nigra*. Svensk Bot. Tidskr. 26: 365–369.

Bambacioni-Mazetti, V. 1931. Sullo sviluppo dell'embrione in *Tulipa gesneriana* L. Ann. di Bot. 19: 145–155.

Biraghi, A. 1930. Impollinazioni tra *Nicotiana rustica* ♀ var. *basilia* e *Petunia* sp. ♂ e loro effeti. Ann. di Bot. 18: 216–222.

Brofferio, I. 1930. Osservazioni sullo sviluppo delle *Calycanthaceae*. Ann. di Bot. 18: 387–394.

Cappelletti, C. 1930. Sull'azione dei prodotti del ricambio di miceli micorizogeni sulle piante ospiti. Ann. di Bot. 19: 1–62.

Carano, E. 1926. Ulteriori osservazioni su *Euphorbia dulcis* L., in rapporto col suo comportamento apomittico. Ann. di Bot. 17: 50–79.

Chiarugi, A., e E. Francini. 1930. Apomissa in "*Ochna serrulata*" Walp. Nuovo Giorn.

Bot. Ital. 37: 1–250.
FURLANI, J. 1904. Zur Embryologie von *Colchicum autumnale*. Oesterr. Bot. Zeit. 54: 373–379.
GENTSCHEFF, G. 1938. Über die pseudogame Fortpflanzung bei *Potentilla*. Genetica 20: 398–408.
GANONG, W. F. 1898. Upon polyembryony and its morphology in *Opuntia vulgaris*. Bot. Gaz. 25: 221–228.
HABERLANDT, G. 1921. Über experimentelle Erzeugung von Adventivembryonen bei *Oenothera lamarckiana*. Sitzungsb. Preuss. Akad. Wiss. 1921: 695–725.
HEGELMAIER, F. 1903. Zur Kenntnis der Polyembryonie von *Euphorbia dulcis* Jacq. (*purpurata* Thuill.). Ber. D. Bot. Ges. 21: 6–19.
JOHANSEN, D. A. 1930. Studies on the morphology of the *Onagraceae*. II. Embryonal manifestations of fasciation in *Clarkia elegans*. Bot. Gaz. 90: 75–91.
JOHNSON, A. M. 1936. Polyembryony in *Eugenia hookeri*. Amer. Jour. Bot. 23: 83–88.
JULIANO, J. B. 1934. Origin of embryos in the Strawberry Mango. Philip. Jour. Sci. 54: 553–561.
JULIANO, J. B. 1937. Embryos of Carabao Mango (*Mangifera indica* L.). Philip. Agr. 25: 749–760.
JULIANO, J. B., and N. L. CUEVAS. 1932. Floral morphology of the Mango (*Mangifera indica* Linn.) with special reference to the Pico variety from the Philippines. Philip. Agr. 21: 449–472.
MCALLISTER, F. 1913. On the cytology and embryology of *Smilacina racemosa*. Trans. Wisc. Acad. Sci., Arts and Letters 17: 599–660.
MAURITZON, J. 1935. Die Embryologie einiger *Capparidaceen* sowie von *Tovaria pendula*. Arkiv f. Bot. 26A, Nr. 15.
MICHAELIS, P. 1925. Zur Cytologie und Embryoentwicklung von *Epilobium*. Ber. D. Bot. Ges. 43: 61–67.
NISHIMURA, M. 1922. Comparative morphology and development of *Poa pratensis*, *Phleum pratense* and *Setaria italica*. Jap. Jour. Bot. 1: 55–85.
NISHIMURA, M. 1922. On the germination and the polyembryony of *Poa pratensis* L. Bot. Mag. Tokyo 36: 47–54.
ORR, M. Y. 1923. Polyembryony in *Sarcococca ruscifolia* Stapf. Notes Roy. Bot. Gard. Edinburgh 14: 21–23.
OSWAA, J. 1912. Cytological and experimental studies in *Citrus*. Jour. Coll. Agr. Imp. Univ. Tokyo 4: 83–116.
PETER, J. 1920–26. Zur Entwicklungsgeschichte einiger *Calycanthaceen*. Beih. Biol. Pflanzen 14: 59–86.
PIJL, L. VAN DER. 1934. Über die Polyembryonie bei *Eugenia*. Rec. trav. bot. Neerl. 31: 113–187.
POPOFF, A. 1935. Über die Fortpflanzungsverhältnisse der Gattung *Potentilla*. Planta 24: 510–522.
RENNER, O. 1915. Befruchtung und Embryobildung bei *Oenothera lamarckiana* und einigen verwandten Arten. Flora 107: 115–150.
SCHURHOFF, P. N., und J. JÜSSEN. 1925. Nuzellarpolyembryonie bei *Spathiphyllum patinii* (Hogg.) N. E. Br. Ber. D. Bot. Ges. 43: 454–456.
SESHAGIRIAH, K. N. 1941. Morphological studies in *Orchidaceae*. I. *Zeuxine sulcata* Lindley. Jour. Indian Bot. Soc. 20: 357–365.
STENAR, H. 1932. Studien über die Entwicklungsgeschichte von *Nothoscordum fragrans* Kunth und *N. striatum* Kunth. Svensk Bot. Tidskr. 26: 25–44.
STRASBURGER, E. 1878. Ueber Polyembryonie. Jenaische Zeitschr. Naturw. 12: 647–670.
VENTURA, M. 1934. Sulla poliembrionia di *Malloius japonicus* Muell. Arg. Ann. di Bot. 20: 568–578.
WIGER, J. 1930. Ein neuer Fall von autonomer Nucellarpolyembryonie. Bot. Notiser 1930: 368–370.
WIGER, J. 1935. Embryological studies on the families *Buxaceae*, *Meliaceae*, *Simarubaceae* and *Burseraceae* Thesis, Lund.
WOODWORTH, R. H. 1929. Parthenogenesis and polyembryony in *Alnus rugosa* (Duroi) Spreng. Science 70: 192–193.
YASUDA, S. 1940. A preliminary report on the artificial parthenogenesis induced by application of growth promoting substances. Bot. Mag. Tokyo 54: 506–510. (In Japanese.)

Chapter XXI

POLYEMBRYONY

Definition:- The term "polyembryony" requires redefinition. It is too loosely employed, usually to designate occurrences of one or more types of apomictic phenomena. Such instances should properly be described by the nature of the apomictic process or processes and not as polyembryony, since the use of the latter designation in this sense fails to signify the nature of the embryos. For example, if one of twin seedlings is haploid and the other diploid, it can ordinarily be taken for granted that the diploid seedling arose from a fertilized egg, while the haploid seedling developed from an unfertilized synergid. Such an example should be classified as normal embryogeny accompanied by apogamety, not as polyembryony. Again, nucellar embryos often arise as groups; this may be polyembryony in one sense, but the proper designation is adventitious embryogeny.

Other phenomena which should be ruled out of classification as polyembryonic include:

A. Spurious or false polyembryony:

1. Two or more incompletely developed ovules may have a single megagametophyte, from which usually only a single embryo arises.

2. The nucellus becomes divided early in the ontogeny of the ovule and each nucellus develops independently within the integuments. One megagametophyte may be fertilized, rarely both may give rise to embryos; usually the ovule perishes. This phenomenon is very common in the *Loranthaceae* (*Viscum, Razoumofskya*), *Onagraceae* (*Clarkia, Fuchsia*), *Orchidaceae* (*Orchis, Gymnadenia, Platanthera*), and others.

3. There may be two to several megagametophytes in the nucellus of the same ovule; they may have arisen from different mother cells of a multicellular archesporium, or from the same megasporocyte.

4. A normal and an apomictic megagametophyte may occur in the same ovule. Such are common, for instance, in apomictic species of *Hieracium.*

B. So-called "true polyembryony."

1. In a three-celled egg apparatus there may be no typical differentiation between egg and synergids. (It may be recalled that in many genera the egg and synergid nuclei are sisters, either immediate or once removed.) In some species of *Burmannia* more than one of these cells may develop into an embryo; this matter requires a careful cytological investigation to determine the exact nature of the embryos.

2. In the older literature there are accounts of the development of embryos from "two or more potential eggs." These indubitably require confirmation.

3. Embryos derived from antipodal cells, from the endosperm, whether coenocytic or cellular, and from synergids.

With all these eliminations, there remain, in the strict sense, only three causes of true polyembryony in the Anthophyta:

1. A normal egg gives rise to a more or less multicellular body from which several potential embryos arise. *Tulipa* and *Erythronium* (FIG. 80E-H) are classical examples. Or, again, a normal embryo is first developed and an extra embryo may arise by epidermal budding, as in *Nicotiana rustica* (FIG. 80D).

2. A filamentous proembryo first develops from the egg, then its terminal cell splits longitudinally into two embryonal initials. The evidence in all recorded cases is circumstantial. *Nymphaea advena* was the first instance to be described.

3. The zygote may divide longitudinally and the two daughter cells become more or less separated and each develops independently into an embryo. The two embryos are equivalent to identical twins. *Cymbidium bicolor* furnishes one authentic instance (FIG. 80A–C).

Polyembryony in the Anthophyta may therefore be described as the formation of more than one embryonal initial by a fertilized egg in either of three ways: (*1*) the zygote divides vertically and each derivative develops independently; (*2*) the terminal cell of a filamentous proembryo splits into two embryonal initials, or (*3*) an embryogenic mass of cells is produced from which several potential embryos arise. In most instances only one embryo attains maturity.

Discussion:- *Cymbidium bicolor* affords perhaps the best authenticated instance of cleavage polyembryony among the Anthophyta. The first division in the zygote is customarily transverse, but may occasionally be either vertical or pronouncedly oblique (FIG. 80A). When such walls are produced, the two resulting cells may become more or less separated and each develops independently into an embryo complete with the characteristic embryonal tubes (FIG. 80B, C). Out of some 2,000 ovules which were examined, about 2 per cent. contained two embryos. One of the embryos is usually slightly smaller than the other.

One case of polyembryony was observed in *Nymphaea advena*. The exact origin is somewhat dubious, but the presumptive evidence points to the splitting of a very young embryo. The normal embryo has from two to four basal cells (equivalent to a suspensor), whereas each of the two embryos has only one basal cell.

The question of the origin of polyembryony in *Crotalaria sagittalis* is equally dubious. Two instances were noted: in one there were two and in the other four embryos. It is improbable that the extra embryos were derived from the synergids since these cells normally disintegrate very early. The primary embryo most likely split during an early stage of growth. Similar embryos have been described for *Mimosa denhartii*; the extra embryos were presumed to have arisen from synergids but their origin could equally well have been similar to those of *Crotalaria*.

It will be noted that the preceding cases of polyembryony occur among dicotyledonous genera; the other type of polyembryony, consisting of the formation of an embryogenic mass from which several potential embryos arise, occurs only among monocotyledonous genera. This type of polyembryony was first discovered in *Erythronium americanum*, and subsequently in several other species of this genus. The zygote divides first transversely, then both daughter cells are partitioned longitudinally (FIG. 80E, F). It is not clear as to whether the basal or the terminal cell of the two celled proembryo contributes the greater part of the embryogenic mass of cells which is presently produced. It would appear, however, that the entire basal portion of the embryogenic mass is produced by derivatives of the basal cell while the outer potentially embryonic cells are derived from the terminal cell (FIG. 80G). This conclusion is further supported by the fact that the basal portions of the embryogenic mass disintegrates after the several embryos have been formed, exactly as a suspensor withers when a normal embryo approaches maturity.

The succeeding divisions in the quadrant embryo exhibit no fixed sequence or planes (FIG. 80F). A broad and somewhat flattened mass of embryogenic cells is presently formed (the "corps embryogené" of GUIGNARD) and on the free outer surface two or more protuberances make their appearance. In *Erythronium hartwegi* the embryogenic mass gives rise regularly to only one embryo; in the other species from two to four potential embryos is the customary number. The embryogenic mass is much more extensive in *Tulipa gesneriana* than in *Erythronium*, but the number of potential embryos is about the same. One of the young embryos soon gains the ascendancy over the others and develops into a viable embryo which has all the appearances of being normal.

Precisely how the multiple embryos in *Erythronium dens-canis* arise is not very clear. They are stated to originate from the terminal cell of the two-celled proembryo (which divides transversely) although it would appear from the original illustrations that they could equally well originate from the derivatives of the basal cell (which divides vertically). The extra embryos, in any event, arise very early in the ontogeny of the proembryo; a massive cellular tissue is not first formed.

The cells in the embryogenic mass of *Tulipa gesneriana* have been described and figured as being bi- to multinucleate; there are also bi- and trinucleate cells figured in the oldest stages depicted for *Erythronium americanum*. Later investigations on other species of *Erythronium* have not confirmed this cytological phenomenon.

It is highly probable that splitting of the proembryo or young embryo occurs in *Zea mays* and *Asparagus officinalis*.

In *Limnorchis emarginata* the basal cell of the two-celled stage never divides again when the embryo develops normally. A few instances are known in which this cell divides repeatedly to engender a globular embryogenic mass from which several potential embryos arise. Later development was not followed out.

In *Linum usitatissimum* the normally fertilized zygote may presumably split in various ways, usually more or less transversely or both transversely and longitudinally, after the first two or three divisions, to form from two to three "identical twin" embryos. However, since the exact origin of the extra embryos is not entirely clear, confirmation is required.

An authentic instance of budding from the epidermal layers of the upper part of a well developed embryo is afforded by *Nicotiana rustica* (FIG. 80D). If the embryos both developed to maturity they would afford an excellent example of true polyembryony among the Anthophyta. Polyembryony

FIG. 80. — *Cymbidium bicolor*. A. Longitudinal section of zygote. B. First transverse division in one daughter cell. C. Developing embryos with embryonal tubes. Further description in text. *Nicotiana rustica*. D. Secondary embryo budding from primary embryo. *Erythronium americanum*. E. Two-celled proembryo, with mitosis in basal cell. F. Four-celled embryo. G. Developing embryonic mass. *Erythronium revolutum*. H. Embryonal mass with early differentiation of several potential embryos. A–C × 560; D × 800; E–H × 250. (A–C *redrawn from* SWAMY 1942; D *original from* D. C. COOPER — *see also* COOPER 1943; E–G *redrawn from* JEFFREY 1895; H *redrawn from* GUÉRIN 1937).

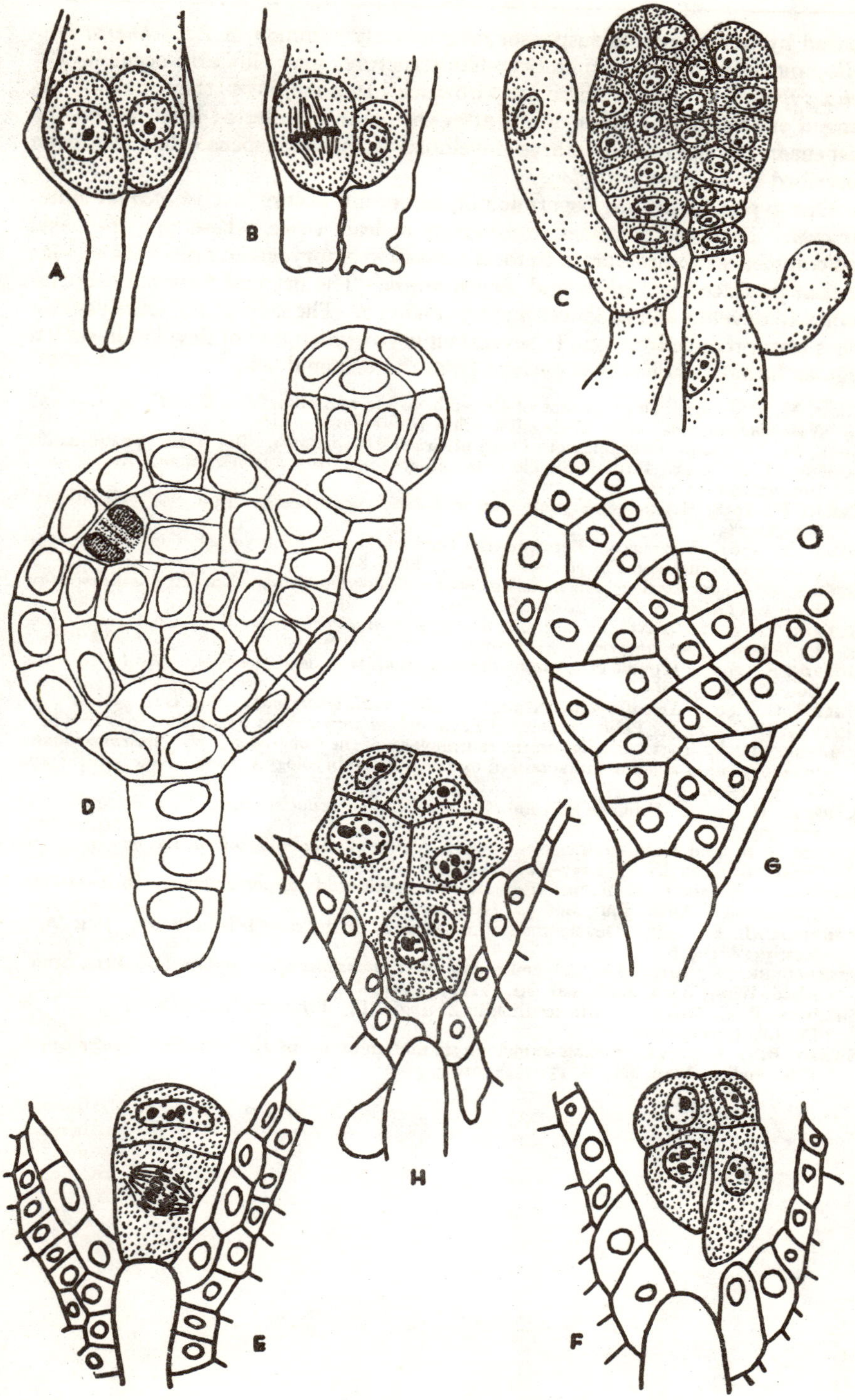
A
B
C
D
G
H
E
F

caused by budding from suspensor cells is fairly common in *Zauschneria latifolia*, and from one to two embryo-like structures may similarly occur in *Lobelia syphilitica*. When there are two such extra embryos they may be arranged one beside the other or one after the other at different locations on the suspensor. A single bud-embryo developed from the suspensor has also been described in *Isotoma longiflora*.

One type of true Angiospermic polyembryony occurs in *Cynanchum vincetoxicum*. The zygote divides transversely at least twice, whereupon the basal cell commences to proliferate to form a massive embryogenic mass very similar to that produced in *Tulipa* and *Erythronium*. The original terminal cell continues to develop and produces a single embryo. The cells of the embryogenic mass may produce from one to several embryos which do not develop in such a regular fashion as does that derived from the terminal cell.

COOK, M. T. 1902. Development of the embryo sac and embryo of *Castalia odorata* and *Nymphaea advena*. Bull. Torrey Bot. Club 29: 211–220.

COOK, M. T. 1924. Development of seed of *Crotalaria sagittalis*. Bot. Gaz. 77: 440–445.

COOPER, D. C. 1943. Haploid-diploid twin embryos in *Lilium* and *Nicotiana*. Amer. Jour. Bot. 30: 408–413.

CRÉTÉ, P. 1938. La polyembryonie chez le *Lobelia syphilitica* L. Bull. Soc. Bot. France 85: 580–583.

ERNST, A. 1901. Beiträge zur Kenntniss der Entwicklung des Embryosaches und der Embryo (Polyembryonie) von *Tulipa gesneriana* L. Flora 88: 37–77.

GUÉRIN, P. 1930. Le développement de l'oeuf et la polyembryonie chez l'*Erythronium dens-canis* L. (*Liliacées*). C. R. Acad. Sci. Paris 191: 1369.

GUÉRIN, P. 1931. L'ovule et la graine de l'*Erythronium dens-canis* L. La polyembryonie. Le Botaniste, sér. 23, fasc. 3–4: 185–195.

GUÉRIN, P. 1937. L'ovule et la graine des *Erythronium* et des *Calochortus*. Ann. Sci. Nat., Bot. X, 19: 255–265.

HALL, J. G. 1902. An embryological study of *Limnorchis emarginata*. Bot. Gaz. 33: 214–219.

JEFFREY, E. C. 1895. Polyembryony in *Erythronium americanum*. Ann. Bot. 9: 537–541.

JOHANSEN, D. A. 1931. Studies on the morphology of the *Onagraceae*. V. *Zauschneria latifolia*, typical of a genus characterized by irregular embryology. Annals New York Acad. Sci. 33: 1–20.

KAPPERT, H. 1933. Erbliche Polyembryonie bei *Linum usitatissimum*. Biol. Zentralbl. 53: 276–307.

KAUSIK, S. B., and K. SUBRAMANYAM. 1946. A case of polyembryony in *Isotoma longiflora* Presl. Current Sci. 15: 257–258.

RANDALL, T. E., and C. M. RICH. 1945. A cytogenetic study of polyembryony in *Asparagus officinalis* L. Amer. Jour. Bot. 32: 560–569.

RANDOLPH, L. F. 1936. Developmental morphology of the caryopsis in maize. Jour. Agr. Res. 53: 881–916.

SEEFELDNER, G. 1912. Die Polyembryonie bei *Cynanchum vincetoxicum* (L.) Pers. Sitzb. Akad. Wiss., Wien, math.-nat. Kl. 121, I: 273–296.

SHARMAN, B. C. 1942. A twin seedling in *Zea mays* L. Twinning in the *Gramineae*. New Phytol. 41: 125–129.

SWAMY, B. G. L. 1942. Female gametophyte and embryogeny in *Cymbidium bicolor* Lindl. Proc. Indian Acad. Sci., B, 15: 194–201.

GLOSSARY
and
INDICES

GLOSSARY

The following list of terms and their definitions is intended to include all those peculiar to plant embryology.

Explanation of abbreviations. — A, Anthophyta; G, all gymnospermous phyla; Co, Coniferophyta; Cy, Cycadophyta; E, Ephedrophyta; Gi, Ginkgophyta. If no symbol is included, the term is common to all groups.

Acotyledonous. A: without cotyledons.

Agamogony. A: in apomictic reproduction, those instances in which an alternation of generations is involved.

Agamospermy. A: reproduction by means of seeds, whose embryos are generally of apomictic origin.

Anticlinal. Referring to walls, those which are perpendicular to the external surface or at right angles to periclinal walls.

Antipodal (*Nuclei* or *Cells*). A: those megagametophytic structures occupying the chalazal end; not always present.

Apical Cell. G: a cell initial with from one to three cutting faces.

Apomeiosis. A: absence of chromosome reduction in the development of a megagametophyte.

Apomixis. A: a reproductive process in which sexual fusion of gametes does not occur.

Apospory. A: in apomictic plants, those cases in which the first division in the megasporocyte is strongly mitotic; otherwise, the megagametophyte develops from a somatic (nucellar) cell.

Archegonium. Co, Cy, Gi: one of from two to numerous large uninucleate (rarely binucleate) cells produced at or near the apex of the megagametophyte, in which the embryo arises subsequent to fertilization; absent in certain Ephedrophyta.

Calyptrogen. A: the tier of cells or the histogen from which the cells of the root cap originate.

Calyptroperiblem. Co: the gymnospermous equivalent of the angiospermous root cap.

Cap, Proembryonic. Co: the cap-like cells of the proembryo of the Araucariaceae.

Central Cylinder. The stele or vascular system of an axis such as the hypocotyl.

Chalaza. A: the basal region of an ovule or seed, where the nucellus and integument(s) converge; opposite from the micropyle.

Coleoptile. A: in the Poaceae, that portion of the embryo which encloses the stem apex.

Coleorhiza. A: in the Poaceae, the sheath surrounding the embryonal root tip.

Corpus. The internal portion (below the tunica or epidermis) of the stem apex.

Cortex. That portion of an axis between the epidermis and the stele.

Cotyledon. G: lateral or terminal protuberances, two to many in number, arising on the upper periphery of the embryo. A: a single terminal protuberance located precisely on the embryonic axis (monocotyledony), or two terminal protuberances located laterally with reference to the embryonic axis (dicotyledony).

Cotyledonary Node. The point of attachment of the cotyledons (in other than monocotyledonous plants) to the embryonal axis.

Dermatogen. A: the tier of cells or the histogen from which the secondary epidermis of the root is derived; located above the calyptrogen.

Dermatocalyptrogen. A: the calyptrogen and dermatogen combined into a single histogen, with the respective functions also united.

Dicotyledonous. With two cotyledons; these always lateral in origin.

Diplospory. A: in apomictic plants, those cases in which the first division in the megasporocyte is strongly meiotic in character.

Embryo, Adventitious. A: one arising from cells of the nucellus or inner integument.

Embryogenergy. The destination and functions of the embryo.

Embryogenesis. The origin of the embryo.

Embryogeny. The origin, development, structure and functions of the embryo.

Embryo, Hibernal. G: the overwintering late proembryo, or, in certain Taxaceae, the very young embryo.

Embryology. That division of Plant Morphology concerned with all phases of embryogeny.

Embryonal Tubes. G: tube-like cells produced near the base of multicellular embryos, originating or not in successive waves, and which grow backward to constitute a generally massive secondary suspensor. Called "supernumerary suspensors" by the earlier investigators. A: similar cells, produced directly by the suspensor initial or indirectly from the suspensor, in certain Orchidaceae.

Embryonic, Embryonal. Of or pertaining to the embryo.

Embryonomy. Exposition of the laws of embryonic development.

Embryo, Nucellar. A: same as adventitious embryo, q.v.

Embryo, Rosette. G: one derived from a cell of the rosette tier.

Embryo Sac. A: the ovular cavity, formerly occupied by the megagametophyte, in which an embryo develops (often erroneously applied to the megagametophyte).

Embryotectonics. The architecture of the embryo.

Endosperm. A: a coenocytic or cellular, evanescent or permanent, tissue arising in the embryo sac; derived from the fertilized or unfertilized polar nucleus (haploid) or primary endosperm nucleus (which may be diploid, triploid or tetraploid) as a result of fusion with the secondary male nucleus.

Epiphyseal. A: of or pertaining to the epiphysis and its derivative cells.

Epiphysis. A: the cell (equivalent to epiphysis initial) or group of cells in the late proembryo or early embryo from which the stem tip originates. Not always present. Essentially the angiospermous equivalent of the gymnospermous apical cell.

Epicotyl. That portion of the embryonic axis (usually not recognizable until after germination) above the cotyledonary node.

Filamentous. A: referring to proembryos which consist of more than four cells arranged in a linear row; or to suspensors with cells similarly arranged.

Free-nucleate. When the nuclei lie free and mitosis is not followed by wall formation.

Histogen. Layers or regions of meristematic cells from which the primary tissues are derived. In the young embryo the histogens are the epidermal initials, periblem and plerome; later the calyptrogen, dermatogen, dermatocalyptrogen, pericycle, corpus, tunica, etc., may be developed.

Hypocotyl. That portion of the embryonal axis between the root and the cotyledonary node.

Hypophyseal. A: of or pertaining to the hypophysis and its derivative cells.

Hypophysis. A: the cell (equivalent to hypophysis initial) or group of cells in the proembryo or young embryo from which part or all of the root tip and associated structures originate. It usually is, or is a descendant of, cell *m* of the proembryonic tetrad. It is not always present.

Initial(s). A cell or group of cells from which a structure is developed.

Integument. A: one (single) or two (inner and outer) protective coverings of the ovule, usually only a few cells in thickness.

Intercalary. A: in embryogeny, referring to an embryo, such as that of *Hydnora africana*, which develops in the approximate center of a filamentous proembryo.

Intranuclear. Occurring within the nuclear membrane.

Megagametophyte. G: the tissue in which the archegonia appear, originally developed from a megaspore. A: the female gametophyte, with from four to sixteen (or rarely more) nuclei or cells, originally developed according to various types from a megaspore.

Megaspore. G: one of the four (or often only three) cells arising from the megasporocyte as a result of meiosis. A: one of four cells arising similarly; one of the four develops into a megagametophyte.

Megasporocyte. The mother cell from which the megaspores arise subsequent to meiosis.

Microgamete. G: a nucleated male cell in the microgametophyte; there are usually two, often unequal in size; the sperm.

Microgametophyte. The pollen tube or male gametophyte, an outgrowth from the microspore or pollen grain.

Micropylar. Pertaining to the micropyle; more commonly employed to designate the apical end of an ovule.

Micropyle. The orifice between the integument or integuments at the apical end of an ovule.

Monocotyledonous. A: with a single cotyledon; this always terminal in origin.

Nucellus. A: all that portion of an ovule exclusive of the integument(s), archesporium, megasporocyte, megaspores or megagametophyte.

Nuclei, Relict. G: free proembryonic nuclei in the archegonial cytoplasm above the prosuspensor cells.

Pattern. A: in the Orchidaceae, a subdivision based on other than true embryonomic characters; a designation serving as a temporary expedient.

Periblem. The histogen from which the cortex is derived.

Periclinal. Referring to walls which are parallel to the external surface or whose configuration parallels that of the curved periphery of the embryo.

Pericycle. A: the outermost layer of cells of the stele, lying adjacent to the endodermis.

Plerome. The histogen from which the stele is derived.

Plumule. A: sometimes applied to the growing point or stem apex of the embryo; not in common use in embryology.

Polarity Units. Co: initials capable of developing into embryos.

Polyembryony. A: the origin of more than one embryo from a single fertilized egg; commonly misused to indicate a plurality of embryos regardless of their origin.

Polyembryony, Cleavage. Co: derivation of more than one embryo from each fertilized archegonium in an ovule.

Polyembryony, Simple. Co: derivation of a single embryo from each fertilized archegonium in an ovule.

Proembryo. G: the earliest phase of embryogenesis. It may be free-nucleate at first, later cellular, or cellular from the beginning. This stage ends when the suspensor or prosuspensor cells commence to elongate and the embryonal initials to divide. A: a more or less filamentous row of cells constituting the earliest phase of embryogenesis. It ends with the fourth cell generation when it consists of approximately sixteen cells. In all instances the proembryo originates from the zygote.

Prosuspensor. G: a suspensor derived from a tier of cells in the proembryo which elongate as a group, not independently, and generally becomes coiled and twisted.

Pseudomonocotyledonous. A: a condition arising when one cotyledon is aborted or fails to develop fully.

Radicle (or *Radicular Region*). A: the entire root region of an embryo.

Root Cap. A: the external covering of the root tip; derived from the calyptrogen or dermatocalyptrogen on older and germinating embryos, but originating from the lower cells of the proembryo in various ways.

Semiaposporv. A: in apomictic plants, those cases in which the first division in the megasporocyte is pseudohomotypic.

Stalace. A: in certain species, a column of cells in the root cap derived from an initial.

Stele. The central, or vascular, cylinder of an axis.

Stem Apex. The growing point of the stem.

Supernumerary Suspensors. Same as embryonal tubes, q.v.

Suspensor. A: the basal cell (*cb*) of the two- or later-celled proembryo or embryo, if it does not divide further; or the derivatives of the basal cell, plus, under certain circumstances, cells derived from the terminal cell (*ca*) of the two-celled proembryo. G: for the corresponding structures, see Prosuspensor, Primary Suspensor, Secondary Suspensor, Embryonal Tubes.

Suspensor, Primary. G: a single suspensor cell elongated by itself from one of the individual lower proembryonal units.

Suspensor, Secondary. G: produced on developing embryos by embryonal tubes.

Synergid. A: one of usually two cells situated above or adjacent to the egg at the micropylar end of a megagametophyte.

Syngamy. Co, Cy, Gi: union of a microgamete nucleus and the archegonial nucleus. E:

union of a generative cell nucleus and the egg tube nucleus. A: union of the primary male nucleus and the egg nucleus.

Tetrad, Proembryonic. A: the four primary cells of the proembryo at the second cell generation; they may be disposed in a linear arrangement, or in a "T-shape" with twó upper juxtaposed cells and two lower superposed cells.

Tier, Open. G: the upper tier of incomplete cells in the proembryo which have no walls toward the apex of the archegonium.

Tier, Relict. Same as Open Tier, q.v.

Tier, Rosette. G: that tier of cells in the proembryo from which rosette embryos may later arise, or which perishes without producing such embryos.

Tissue. A group of cells of common origin possessing the same structure and functions.

Tricotyledonous. A: when there are three in place of the usual two cotyledons; often due to splitting (schizocotyly) of one of the latter.

Tunica. The epidermal layer or layers of the stem apex.

Type. A: a fundamental embryonomic unit.

Variation. A: a subdivision of a Type, differing from the latter in minor characteristics.

Ventral Canal Nucleus. G: a sister of the archegonial nucleus, formed when the nucleus of the central cell divides; it usually perishes, but in *Ginkgo* it may be fertilized in place of the archegonial nucleus and in *Pinus* both nuclei may fuse with microgamete nuclei.

Zygote. The cell, originally the egg in A or a portion of an archegonium in G, but with certain exceptions in *Gnetum*, containing the fused male and female nuclei. In Co, Cy, Gi and E, the parental chromosomes, following nuclear fusion, usually become arranged on separate spindles at first; they are indistinguishable in all other groups.

INDEX TO PHYLA, ORDERS AND FAMILIES

INDEX TO GENERA AND SPECIES

Recapitulatory tables (first, second, third and fourth cell generation) for:—